Go to the table of c[...]

The World's Greatest Physical Science Textbook for Middle School Students in the Known Universe and Beyond! Volume Two

Motion
Forces
Physics

By Michael A. Ritts

January 2017

Note:
Volume One – Scientific method, Matter, and Energy is available here.
E-book version
Paperback version

Volume Three will be complete in the future and will include the Periodic Table and chemistry.

Ethics and honesty

This book is meant to be a textbook, treat it as one. If you buy a personal copy you can lend it just like any book, but do not copy it. If you are a School and you have 500 students per year you should pay for 500 copies, but you should not have to buy it again the next year, or the next ten. Just like any hardback text, you buy it once, and are good for quite some time. You get your textbook and I get my money. Honesty is a great thing.

The Internet and computers are wonderful things, but full of copyright infringement. Do not lower yourself because you can get "free stuff". Someone is getting ripped off, whether it is a writer, singer, or movie company. A lot of time, money and talent goes into these products, and those people deserve to be compensated. As the cost of products, in the form of profits, goes down, so do the rewards to these hard working people. Bad for everyone. Everyone wants "something for free", but when you think about it, it does not pay off in the long run. The day music is free, no one will make **new** music, the day medicine is "free" no one will make **new** medicine, the day textbooks are free, no one will write **new** textbooks.

So please, be honest, it is an honorable thing to be.

Why does this book exist?

In a nutshell, efficiency and cost. This is meant to be a "new age" middle school textbook, organized in such a way that the teacher can follow it chronologically, in a sensible pattern. Written immaturely, so that middle school students will actually read it. It is not watered down, the vocabulary is there, but admittedly it looks like it was written by a 13 year old. That was my goal. With the Internet and this on-line book I have the ability to include links to videos, simulations, demonstrations, and anything I believe will make the book more interesting. These are things a physical book cannot do.

With school budgets in crisis, and current textbooks being very expensive, this book is made to be as affordable as practical. My competition are the major textbook companies who charge in the neighborhood of $100 per text, granted they come with a lot of junk, worksheets, test, videos, etc. But to tell you the truth, these are not really very useful. For a school of 500 students, that comes to quite a chunk of change. How about if we could cut that cost by a factor of 20? The biggest selling point of the old textbook my district spent a ton of money on was a free stick drive for every teacher. Really that is why we chose it. A bribe. On top of that, all the different middle school textbooks are nearly the same. Not much choice out there.

Current textbooks are huge; they contain *way* too much stuff. They are *not fun* to read. They were written by intelligent people, who really do not understand the middle school student's mind. I am a teacher, I know how they think, and I am not sitting in an office somewhere imagining it. I may not be as smart as those writers, but I do not have to be, I only need to accurate and interesting. This is what I have tried to do. Granted there are some "free" textbooks available online but they are rather out dated and not "aimed" at the middle school student. This one is. There may be some mistakes and if you find something you cannot tolerate, e-mail me at junglecat3388@gmail.com and I will fix it (if I agree).

So what do we have here? An inexpensive e-book textbook that is easy to use and fun to read. Time will tell……

So what am I visualizing here?

I hope to make the price affordable for any student or school district. It took two years to write, I should get something. I am also saving up for a new ping-pong ball, so I need the money.

You will find many links to videos of some of the demonstrations I do in my class, along with other on-line resources (simulations and things).

I wrote this book exactly like I teach the lessons. The stories and examples are the same ones I use in class, so I know they work.

Textbook Organization
Each chapter is organized into four parts.

1. **A link** to a video of the actual power point lesson I use in class. It is not the same as an actual lesson but it is a good review or introduction. It will also give the teacher some good ideas.
2. **An anecdote about science** – it may not have anything to do with the chapter's topic, it is not meant to, it is to try and spark interest and curiosity.

3. **The Body** – This is the information part – accurate but full of silly examples to help explain the topic. **There is a link to flash cards at the end of each chapter.**
4. **A day in the life of Earwig Hickson III** – the creative part. Earwig is an imaginary child full of wonder, curiosity, and imagination, like we all should be. Since he is imaginary (think a Saturday morning cartoon character) he can do things that really are not possible. His imagination can run wild without any dangerous consequence.

How to use this book
To the teacher
It is a textbook, an outline, and a teaching strategy. You as the teacher still have to present the material, come up with lesson plans, write your own tests and review exercises. A good teacher is not lazy; they just need a foundation to work with. This book will give you a good start though, including a video of how my lesson is organized, a flashcard review on quizlet and many links to demonstrations and such.

To the student
Read it, check out the web links and watch the videos. Find out that science is fun. You will be smarter after reading this book that is something to be proud of. I hope to teach you, if nothing else, that science is fun, I mean awesome. Science rules!!!!!

Table of contents

Volume One – Not this book

Unit 1 – What the Heck is science?

Chapter 1 – How to think like a scientist
Chapter 2 – The scientific Method
Chapter 3 – Physical Science
Chapter 4 – Lab safety
Chapter 5 – The controlled experiment

Unit 2 – What is Matter

Chapter 6 – Measuring Matter
Chapter 7 – Atoms
Chapter 8 – Combining matter into new stuff
Chapter 9 – The common states of matter

Unit 3 – The Properties of matter

Chapter 10 – Properties of matter
Chapter 11 – Changing states of Matter
Chapter 12 – Using properties

Unit 4 – Energy

Chapter 13 – Forms of energy
Chapter 14 – Energy transitions
Chapter 15 – Energy technology

Unit 5 – Heat

Chapter 16 – Temperature
Chapter 17 – Heat
Chapter 18 – The movement of heat

Volume 2 – this book!

Unit 6 – Motion	8
Chapter 19 – Relative motion	9
Chapter 20 – Speed	24
Chapter 21 – Velocity	38
Chapter 22 – Acceleration	50
Unit 7 – Forces	60
Chapter 23 - Force	61
Chapter 24 - Simple machines	76
Chapter 25 - Newton's first law of motion	98
Chapter 26 - Newton's second law of motion	106
Chapter 27 - Newton's third law of motion	119
Chapter 28 - Law of conservation of momentum	126
Unit 8 – Types of force	136
Chapter 29 – The law of universal Gravitation	137
Chapter 30 – Falling objects	152
Chapter 31 - Gravity in the solar system	167
Chapter 32 – Space	178
Chapter 33 – Friction	187
Chapter 34 – Pressure	196
Chapter 35 – Fluids	207
Chapter 36 – Pressure in fluids	221
Chapter 37 – Pressure in gases	235
Chapter 38 – Buoyancy and Archimedes principle	253
Chapter 39 – Bernoulli's principle	270
Chapter 40 – Hydraulics Pascal's principle	292

Link to all my science videos:
https://www.youtube.com/channel/UCNERGz70pAoK8A5oe2Wkw_Q

Unit 6
Motion

Using Newton's first law of motion to do the old tablecloth trick.
https://www.youtube.com/watch?v=nRcBQrX3FrU

Chapter 19
Relative motion

Video of the actual power point lesson:
https://www.youtube.com/watch?v=gEKFyiMaZKg

The wonders of Science
   It was so obvious. For thousands of years it was so obvious that no one even questioned it. It was known by everyone that the earth was the center of the universe and everything went around it. It seems obvious, even today. We see the sun rise in the east and set in the west, the stars rise and set each night, so does the moon. Some people noticed problems with this theory. The idea did not explain some things such as; the Seasons, the Tides, the changing positions of the stars from day to day, and the planets wandering all over the night sky (the word planet actually means wanderer).
   Now we know that the Earth isn't the center of anything. The Sun is the center of the Solar System, but not the center of the Milky Way Galaxy. The Earth is a normal planet in a random part the universe. It was not easy discovering this. It was not until a man named Nicolaus Copernicus came up with a theory that the Earth (everything else) rotated around the Sun back in 1543. In 1610 Galileo Galilei actually saw the moons of Jupiter orbiting around the planet with the newly invented telescope. He was placed under house arrest for mentioning this (laws were strict back then) because his ideas were considered "dangerous". Go figure.
   How could so many people have not noticed this for so long? The problem was their point of view. They were standing on the Earth, which did not feel like it was moving so they assumed it was stationary. They saw the sun and stars moving, so they assumed they actually *were* moving. If these people were out in deep space instead, they would have seen things completely different. For a short movie explaining how this was proven:
http://www.pbslearningmedia.org/resource/ess05.sci.ess.eiu.galileosys/galileo-sun-centered-system/

Frames of Reference
   Einstein explains frames of reference to Queen Elizabeth about using a 600 miles/hour toilet: https://www.youtube.com/watch?v=zdi7x3IfsmA

   How do you know you are moving? You can easily tell can't you? I am not so sure you can because actually everything in the universe is moving, even you. In fact how can you tell the difference between moving and stationary? As far as science goes you cannot. How can this be?
   We can start with you right now. You are probably sitting someplace reading these words and both of us would agree that you are stationary, but you really are not. You are sitting on a planet that spins around once per day (causing day and night). At my latitude in Pennsylvania that comes out to about 1000 miles per hour, so you are moving at 1000 miles per hour and you never noticed! The reason you never noticed is because everything around you is moving at 1000 miles per hour also. If the Earth stopped

spinning you would still be moving. You would notice then (and paint your walls red). So if you say you are stationary, you have to say *compared* to what. It is true that you are stationary compared to your chair or anything else in the room you are in, but not to the sun. You are moving at 1000 miles/hour compared to the sun. On top of that, the Earth is circling the sun at about 67,000 miles per hour. I almost forgot our Sun is circling the Milky Way Galaxy at another 43,000 miles per hour. In other words you are depending on your surroundings to tell you if you are moving (compared to them). These background objects are called *Frames of Reference*, and are the objects you use to detect motion. Your brain automatically assumes that whatever you are using as a frame of reference is stationary (even if you know it isn't). *All frames of reference are assumed to be stationary.*

Check out this video of a very good magician (illusionist) using frames of reference to make you think a playing card is floating.
https://www.youtube.com/watch?v=NObQp8r4Wzg
A Video of the Earths actual path around the sun:
https://www.youtube.com/watch?v=82p-DYgGFjI
The helical model - our solar system is a vortex:
https://www.youtube.com/watch?v=0jHsq36_NTU
Frames of Reference (1960) Educational Film:
https://www.youtube.com/watch?v=aRDOqiqBUQY

- YOU ARE <u>NOT</u> MOVING RIGHT NOW RELATIVE TO YOUR CHAIR

- YOU <u>ARE</u> MOVING RIGHT NOW RELATIVE TO THE SUN

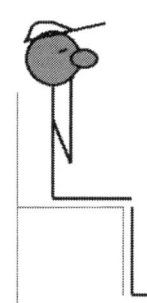
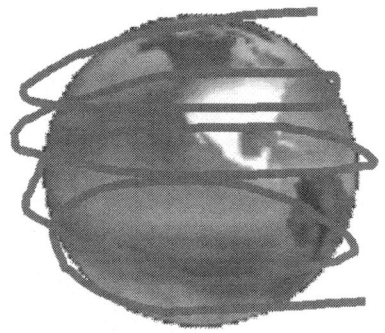

Some frames of reference are more useful than others. It is very useful to use your chair as a frame of reference right now instead of the sun. If you used the sun you would get very dizzy. Wherever you are, *that is where your frame of reference needs to be.*

Now let's imagine you are in a moving car (relative to the road) and I am standing alongside the road. I see you and the car go by and I say you are moving, and you are compared to the trees and signs I am using as my *frame of reference*. But you see me fly by because a better frame of reference to you is the inside of the car. So who is moving, you or me? We both are, it depends on which frame of reference you are using. I am moving compared to the inside of the car, and you are moving compared to the trees.

Now let's put you in a big jet plane flying 6 miles high in very smooth air at 600 miles per hour. If the windows were painted black you would have no clue you were moving. You could do anything in that 600 mile/hour plane that you could do in a parked plane. You could drink a glass of water without spilling any; you could get up and walk to the back of the plane to the bathroom with no problem. Since all your frames of reference are moving with you, you feel stationary.

Sometimes your brain can get confused and pick the wrong frame of reference, this can cause you to get dizzy, car sick, or temporarily confused into thinking you are moving when you are not. A nasty thing that someone could do when stopped at a red light beside another car is to slowly drift backwards and watch the other driver. The driver will be momentarily confused into thinking they are drifting forward into traffic and slam on the brakes. The driver's brain gets confused because it just saw a frame of reference start to move (and they do not do that) and thinks it is moving instead.

An amazing video of how Fred Astaire was filmed dancing on the ceiling. It is all about tricking your brain with the wrong frame of reference:
https://www.youtube.com/watch?v=i0g3g6AvLtM

Relative motion

The word *motion* really does not mean anything in science unless you compare the motion to *something*. This is called *Relative Motion*, which is the change in position of one object *compared* to another object (the frame of reference).

This is a cool video taken from a moving train of another moving train, see if you can tell when the first train stops: https://www.youtube.com/watch?v=AKhvqO5UBsA

Now for something really weird, an object can be *stationary and moving at the same time*, and we would actually disagree about what that object is actually doing. This is because in science *stationary* and *moving at a constant velocity* is actually *the same thing*! Let me explain.

Imagine you are in a school bus going down the road at a constant 60 miles/hour, in a straight line (no turning, speeding up, or slowing down). You are sitting in your seat and you throw a coin straight up into the air. What does the coin do? Well it comes right back down into your hand, no big deal. The coin went straight up and straight down, compared to the seats on the bus. You do not get into any trouble, you did nothing wrong. Go ahead and try it. You can even take a movie of it and it will just go up then down (as long as the camera was on the bus).

THIS IS WHAT THE PERSON ON THE BUS SEES

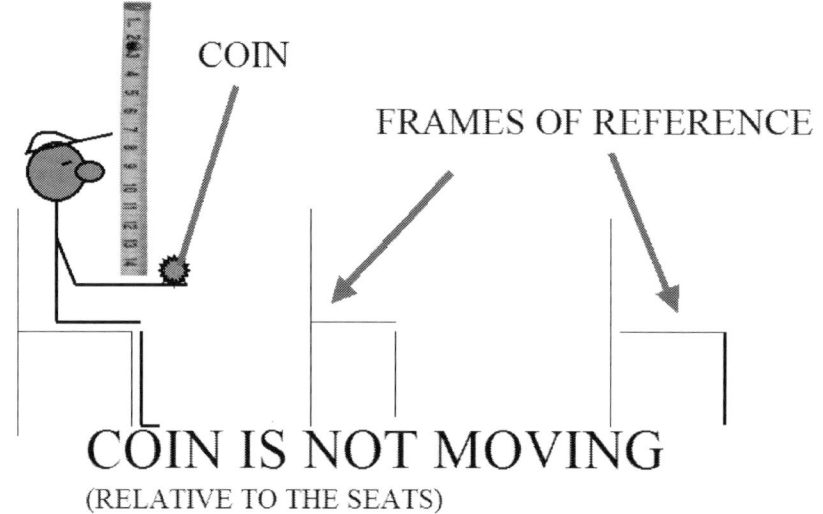

COIN IS NOT MOVING
(RELATIVE TO THE SEATS)

COIN IS THROWN UP

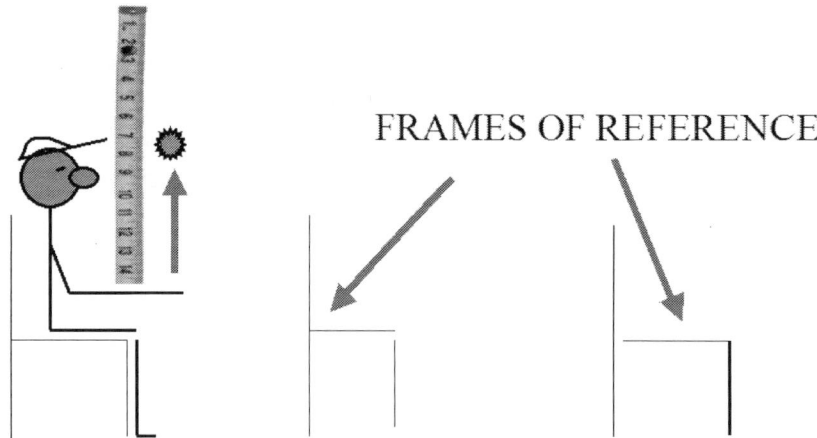

THE PERSON SEES THE COIN GO STRAIGHT UP

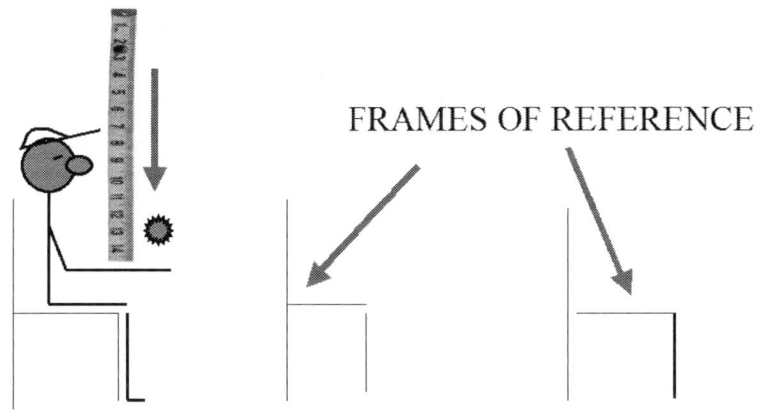

COIN FALLS BACK DOWN AND PERSON CATCHES IT

FRAMES OF REFERENCE

THE PERSON SEES IT GO STRAIGHT DOWN

As the bus drove by and you were throwing the coin upward, I was standing alongside the road and saw you through the window. What I saw was not what you saw. I saw you throw the coin but it did not go straight up then straight down. I saw it fly forward at 60 miles/hour in a big curve. And that is what it did, *compared* to the trees along the road. I call your mom and accuse you of throwing the coin at the bus driver. You are in big trouble now! I could even take a video and it would show the coin flying forward.

BUT TO THE PERSON OUTSIDE THE BUS THINGS ARE DIFFERENT

FRAME OF REFERENCE

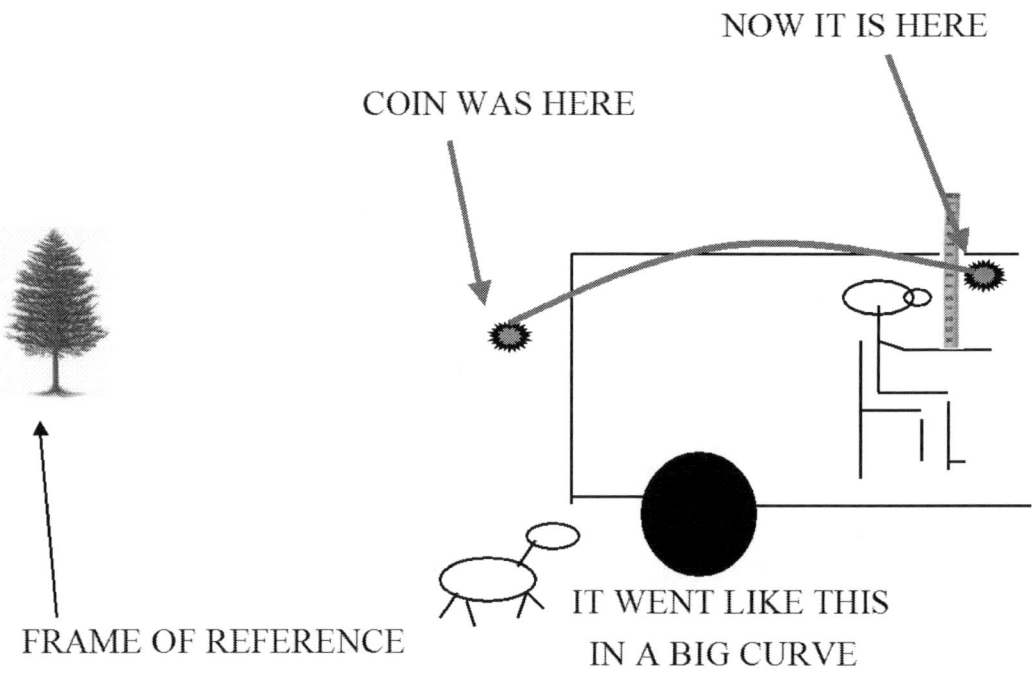

This is a great video showing how a ball shot straight up from a moving cart lands at exactly the same place in the cart: https://www.youtube.com/watch?v=v0gg1F0sz0E

Merry go round

Sometimes in science we use a thought experiment, which means we do not have the proper equipment to make it work in real life. This is such an experiment. Imagine you are on a giant merry go round, spinning very fast, and you throw a snowball at someone on the other side. What will the snowball do? To someone *off* the merry go round watching, the ball will go in a straight line. To the people *on* the merry go round, the ball will go in a circle. Let's look from the point of view of someone *off* the merry go round. He sees the person throw the snowball in a straight line and hit themselves with the snowball, since the merry go round went half way around the circle.

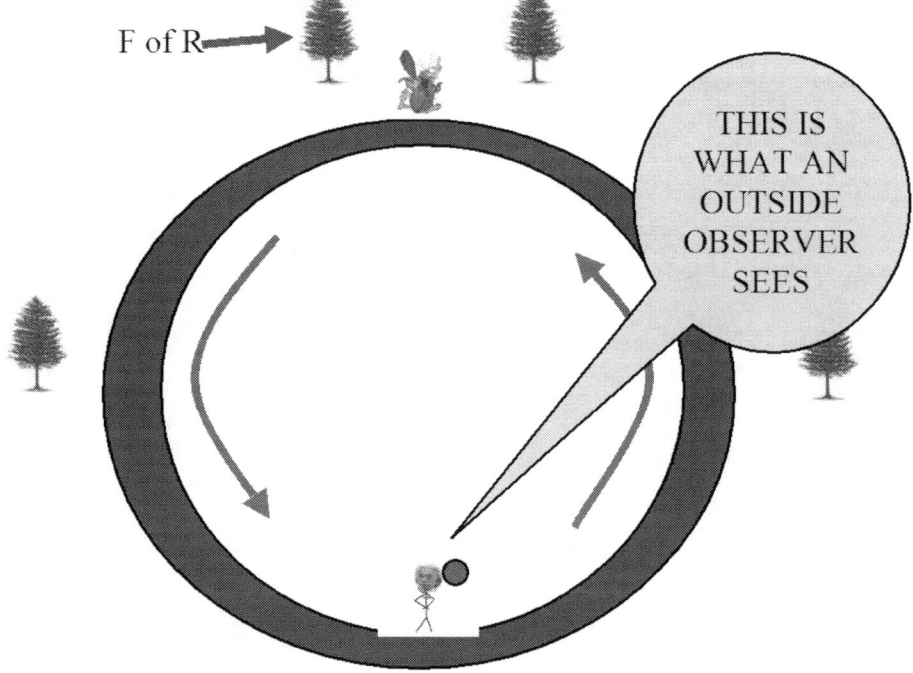

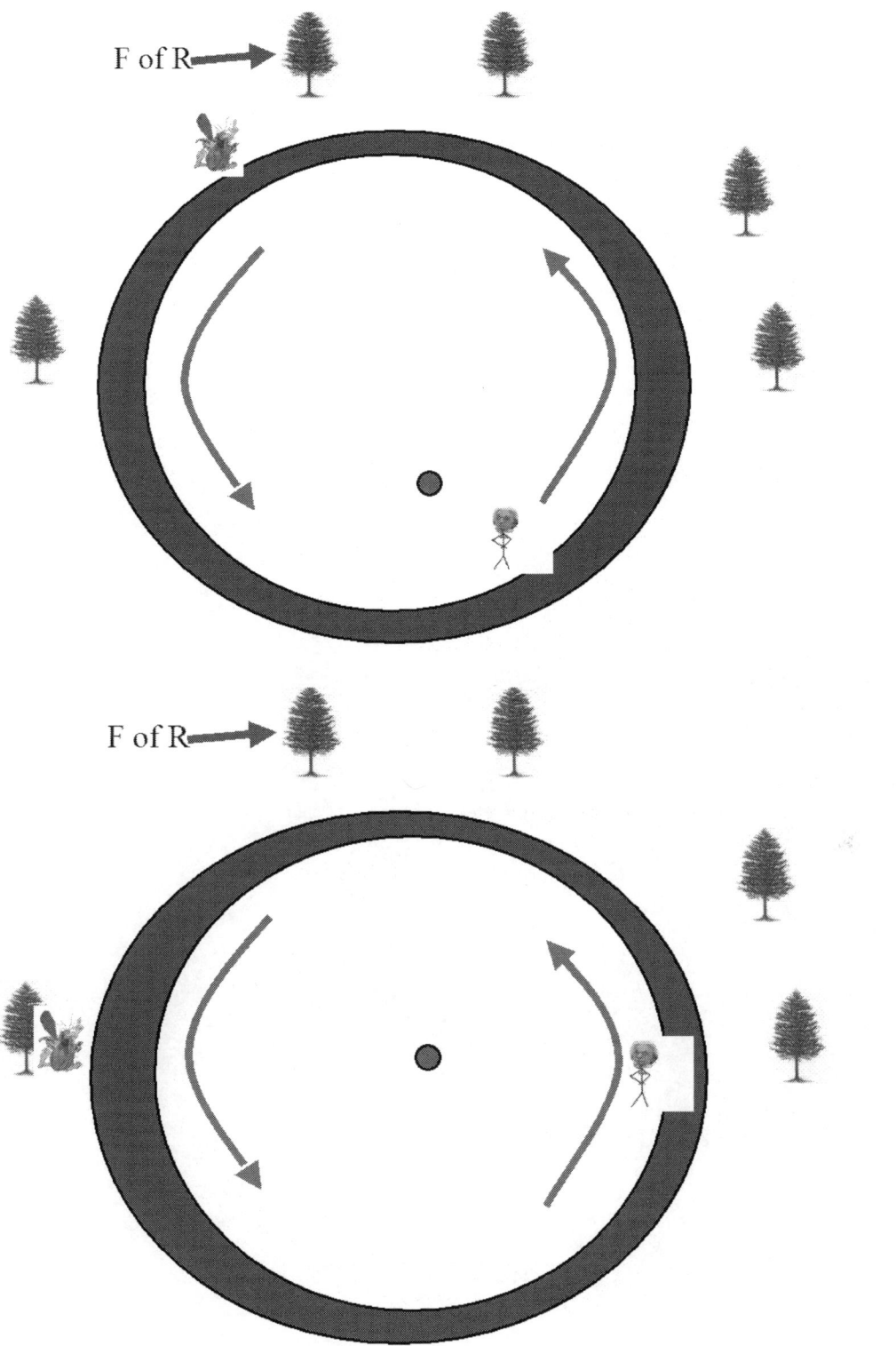

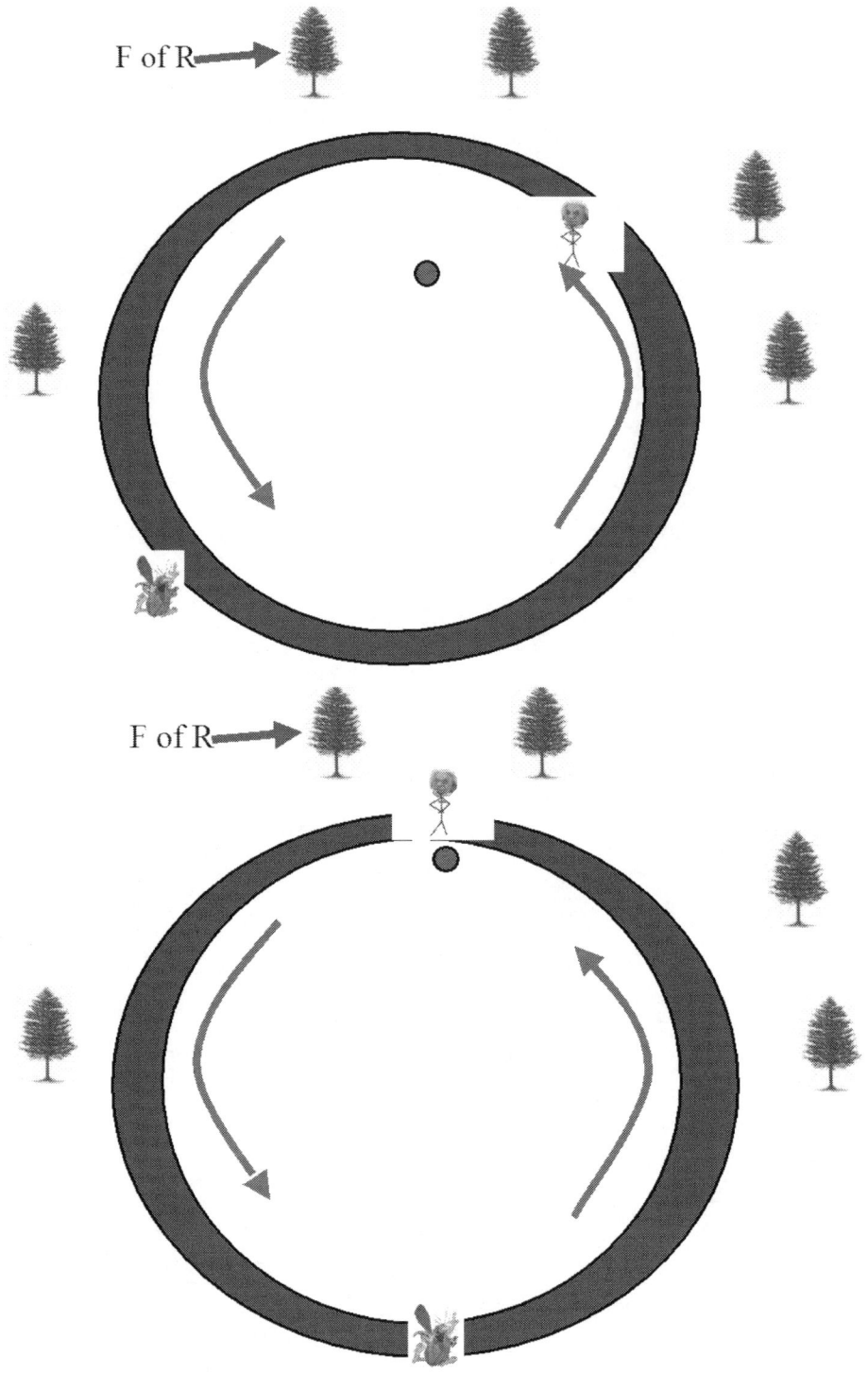

The person *throwing the snowball* sees it curve and come back like a boomerang.
https://www.youtube.com/watch?v=Fp8QlkNjHqM

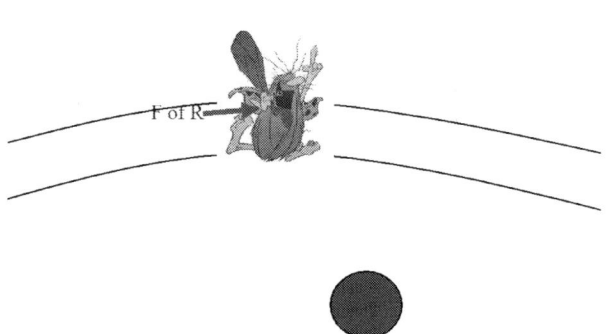

The person who the snowball *was thrown at* sees it turn around and hit the other person.

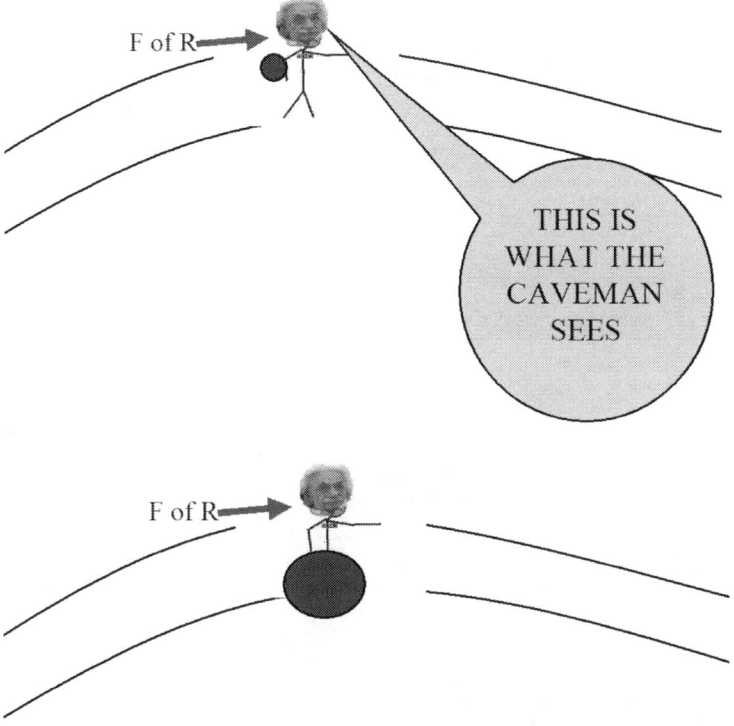

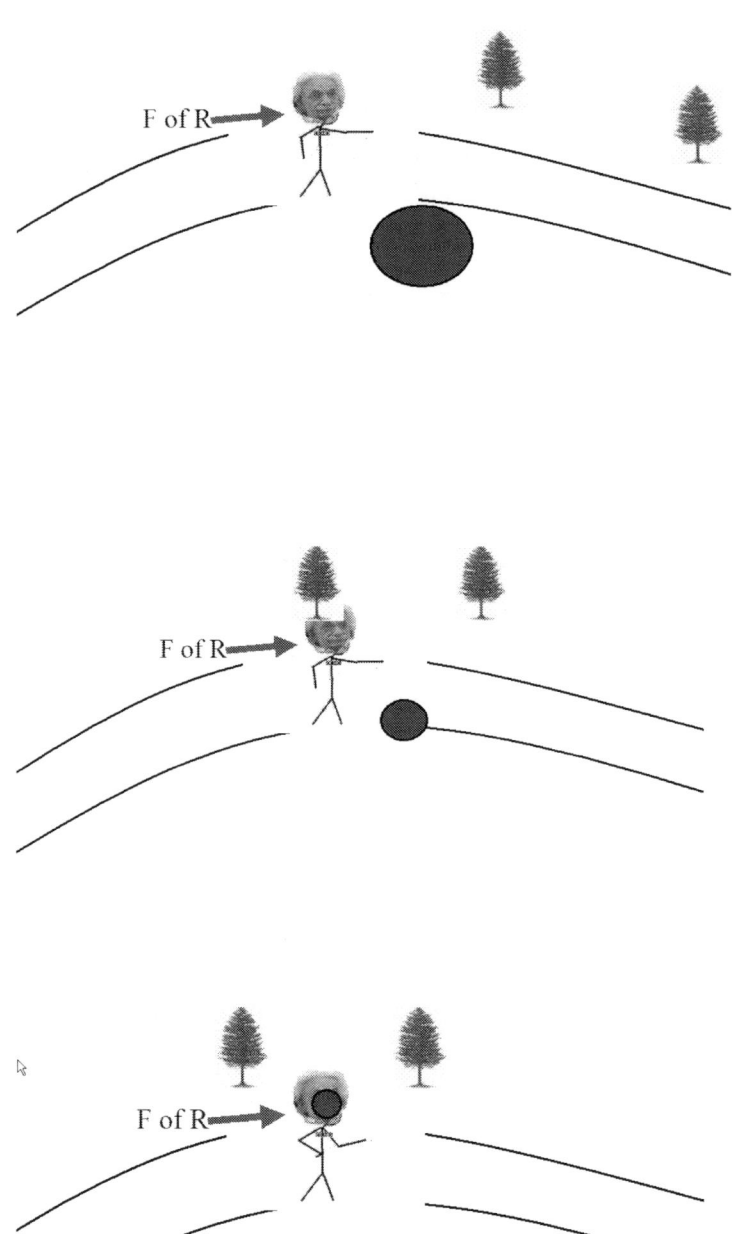

An experiment where some people are rolling a ball on a small merry go round with a camera view both on and off the spinning platform. The video part is cool, the math part is a little nasty, but you can ignore that:
https://www.youtube.com/watch?v=KDOLQ6P0vog

A video of two kids trying to play catch on a merry go round:
https://www.youtube.com/watch?v=PTjVQDR9jTg

Review

All motion is relative to something, and that something is called a frame of reference. Wherever you are, *that is where your frame of reference must be*. If you are on the bus, that is where your frame of reference is. If you are outside the bus, your frame of reference is also.

Sir Isaac Newton explaining relative motion to Napoleon, and how to use a bathroom: https://www.youtube.com/watch?v=I3D9OZAW3to

A day in the life of Earwig Hickson III

My science teacher is a monster. He built a machine for torturing us students called the "Drum of Nausea". It consisted of a giant upside down trashcan hanging from the ceiling on a string. There were pictures inside. He made me get inside the contraption and told me to look at the pictures. I started spinning around and getting dizzy. Suddenly I reversed direction and almost fell down. I had spun around so many times I lost track of what direction I was facing. I could barely stand up. I was starting to get sick from all the spinning. How could he do this to me? What did I do to deserve this?

When he finally let me out, the classroom was spinning, but that was not possible. I was spinning. I was confused. Then I found out that I had never spun at all. I never even moved. I was stationary the whole time. It was the drum with the pictures that was spinning. My brain thought the pictures were stationary (when they were actually moving) and it made me think I was spinning instead. I could not believe it. I wanted to go back in and try again. While inside the second time I knew for a fact that I was stationary but it felt like I was spinning just the same. I could not believe I could be fooled so easily just by using a moving frame of reference.

This is what it looked like: Drum of nausea: https://www.youtube.com/watch?v=AOT6fO5F9V0

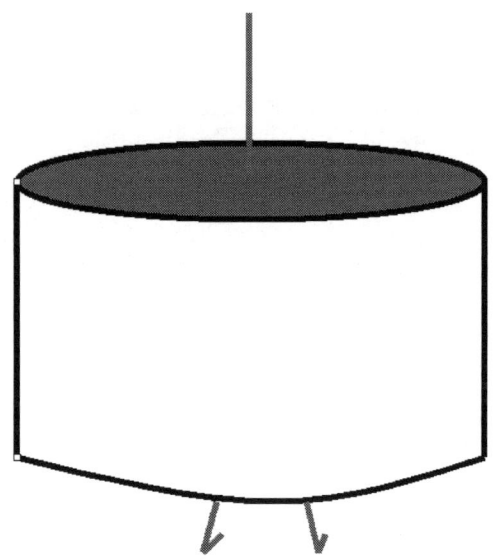

Review of terms - quizlet
https://quizlet.com/124209191/ch-1-volume-2-flash-cards/?new

Fun things go Google
Coriolis Effect
What direction does water swirl down a toilet in the northern hemisphere?

Links
Mythbusters shoots soccer ball out the back of a moving truck at exactly 60 miles/hour. The truck is also going at 60 miles/hour. What does the ball do?
https://www.youtube.com/watch?v=DXkmc2p_Zio

A motor cycle rider hits a moving car, and since he and the car are going the same speed he lands on top of the car. https://www.youtube.com/watch?v=17ig4WziQzM

Wing walking on an airplane. The frame of reference is the plane wing.
https://www.youtube.com/watch?v=t_vTivvXrC0

Drum of nausea: https://www.youtube.com/watch?v=AOT6fO5F9V0

Sir Isaac Newton explaining relative motion to Napoleon, and how to use a bathroom:
https://www.youtube.com/watch?v=I3D9OZAW3to

An experiment where some people are rolling a ball on a small merry go round with a camera view both on and off the spinning platform. The video part is cool, the math part is a little nasty, you can ignore that part:
https://www.youtube.com/watch?v=KDOLQ6P0vog

A video of two kids trying to play catch on a merry go round:
https://www.youtube.com/watch?v=PTjVQDR9jTg

History Channel – Galileo's Battle for the Heaven – PBS NOVA Documentary 2016
https://www.youtube.com/watch?v=JUAsLcFPeNw

Galileo Galilei Documentary https://www.youtube.com/watch?v=PPMc7Q5zSv8

This is a great video showing how a ball shot straight up from a moving cart lands at exactly the same place in the cart: https://www.youtube.com/watch?v=v0gg1F0sz0E

Einstein explains frames of reference to Queen Elizabeth about using a 600 miles/hour toilet: https://www.youtube.com/watch?v=zdi7x3IfsmA

An amazing video of how Fred Astaire was filmed dancing on the ceiling. It is all about tricking your brain with the wrong frame of reference:
https://www.youtube.com/watch?v=i0g3g6AvLtM

This is a cool video taken from a moving train of another moving train, see if you can tell when the first train stops: https://www.youtube.com/watch?v=AKhvqO5UBsA

Some silliness but it does explain frames of reference:
https://www.youtube.com/watch?v=5oSrDrDLylw

Check out this video of a very good magician (illusionist) using frames of reference to make you think a playing card is floating.
https://www.youtube.com/watch?v=NObQp8r4Wzg

Copernicus and the Scientific Revolution – Past is Present (2011)
https://www.youtube.com/watch?v=zHUWP9zu4W8

Frames of Reference (1960) Educational Film:
https://www.youtube.com/watch?v=aRDOqiqBUQY

Chapter 20
Speed

The power point video of this lesson:
https://www.youtube.com/watch?v=uU36WK_Ok4E

The wonders of science

Speed is what the Air Force needed. The jet engine had been developed a few years earlier, and airplanes were flying faster and faster. It was shown in WWI that the faster aircraft had an advantage in combat, and the cold war with the Soviet Union was at its beginning, but a serious worry. The problem was that as planes flew faster and faster, they seem to reach a limit where the aircraft became unstable and dangerous. Some people thought of this limit as a wall, or a speed that could not be passed. It was known as the "sound barrier". Any airplane that reached the speed of sound was feared to become unstable, and could break up into tiny pieces of airplane and pilot. Other people saw the sound barrier as something that could be overcome with new designs and experimentation, so the X-program was started (List of X-planes). The first airplane in the X-program was called the X-1 , a rocket-powered plane that was designed to do one thing and one thing only, fly faster than sound (called Mach) in controlled manned flight. All the X planes were experimental, and the knowledge learned would be incorporated into new military aircraft. It was known at that time that a .50 caliber bullet could fly faster than sound, so the X-1 was shaped like a bullet. The wings were razor sharp thin to cut air resistance. To save fuel and weight it was dropped from a larger airplane (mother ship) like a bomb. It was a lot like a flying bomb.

On October 14, 1947, test pilot Charles "Chuck" Yeager waited to be dropped from the mother ship He had broken some ribs he day before and was feeling the pain. The X-1 was dropped. Yeager lit the engines and history was made. The first aircraft to pass through the sound barrier, and the first human made sonic boom, Yeager was the fastest man alive. The X-program continued, hitting its climax in the 1960s when the North American X-15 rocket plane reached speeds near Mach 7 (4520 miles/hour) and an altitude of over 50 miles, high enough to earn the pilots astronaut wings (https://en.wikipedia.org/wiki/North_American_X-15). The research from the X-15 resulted in the greatest airplane of them all, the SR-71 Blackbird.

Here is a cool video of the X-1: https://www.youtube.com/watch?v=2-mXNPhTdtk
The X-15: https://www.youtube.com/watch?v=akZ4BHMrCsk
The amazing SR-71 in flight: https://www.youtube.com/watch?v=60zniMMifFA

Speed

Speed is the *rate* at which an object *changes position*. How fast it moves. It compares the distance something travels to how long it took. *Distance per unit of time*. It is useful to express speed as how far something moved in **one** second, or **one** hour, so we can compare different things easily. If my car can go 70 miles in *one* hour and yours moves at

240 miles in *five* hours, it is harder to tell which car goes faster. This is why we have to calculate speed.

Leonardo explains speed: https://www.youtube.com/watch?v=uvW7vRwJo8U

Speed Formula

Ah, math! Don't you love math? OK, so not everyone loves math, but we scientists do because it is useful to us; it is a great tool. It has been my experience that as soon as a student sees a math formula, they have a tendency to get turned off from science quite quickly. This does not have to be the case Follow along and you will see that math is actually pretty easy when it has a goal. By the way, this is why you have to study math in school, so you can learn science! You need to study reading and writing so you can read and write about science! History tells us all the ways science has screwed up the world or helped it.

The formula for calculating speed is:

Speed = Distance/Time

Or in shortened form:

$$S=d/t$$

SPEED FORMULA

$$SPEED = \frac{DISTANCE}{TIME}$$

$$S = \frac{d}{t}$$

UNITS - M/S   KM/H   M/H   KM/S

To solve an equation such as this all you have to do is substitute the proper number for the proper variable. It is actually quite easy. It is important to show your work by writing three steps.
1. The **formula** (so people know what you are calculating)
2. The **set up** (plug the numbers in with the correct units)
3. The **answer** (with the proper units)

Units are very important. Without a unit after the number (called a naked number) no one knows what you are talking about. What if I asked you if you wanted 20? Would you say yes or no? Or would you ask, 20 what? Well, I might be talking about 20 *dollars* or 20 *whacks with a baseball bat*. They are not the same. What you are asking for are *units*. A number by itself means nothing.

Let's try a typical speed calculation:

EXAMPLE SPEED PROBLEM

- A COW RUNS 8 METERS EVERY 2 SECONDS. WHAT IS THE COW's SPEED?

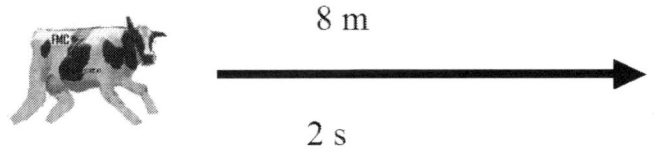

Notice I underlined the important information and drew a simple sketch to help me think because I hate word problems. Pictures are easier.

The first thing I do for this problem is write the formula:

$$S = \frac{d}{t}$$

Now I look at my sketch (or underlined information) and decide which number goes where. Now you should know that *seconds are a unit for time* so that goes in place of **t**. You should also know that *meters are a unit for distance*, so that goes in place of **d**. If you do not know the units for distance and time, you really need to learn them, or this will be really hard. Time to write my set up.

$$S = \frac{d}{t}$$

$$S = \frac{8 \text{ m}}{2 \text{ s}}$$

Now I can calculate my number answer.

$$S = \frac{d}{t}$$

$$S = \frac{8 \text{ m}}{2 \text{ s}}$$

$$S = 4$$

So, I got an answer of 4, but remember a number by itself is not really an answer, it needs units. This is actually the easy part because the units are always in the set up; just write down what you see.

$$S = \frac{d}{t}$$

$$S = \frac{8 \, m}{2 \, s}$$

$$S = 4 \, M/S$$

That's it, we now know the cow was running at 4 meters every one second https://www.youtube.com/watch?v=h46ry-hpM3o and if we want to, we could compare its speed to the speed of other cows, cars, snails, horses, or even space ships. If you want or need to see some more examples watch the lesson at the beginning of this chapter.

Reading a distance vs. time graph, a speed graph

Speed can be expressed on a graph too. Reading graphs is easy because the answer is always on the graph; the trick is looking for it.

In this graph the zebra has walked 40 m in 20 seconds. You can see the line is a gentle slope.

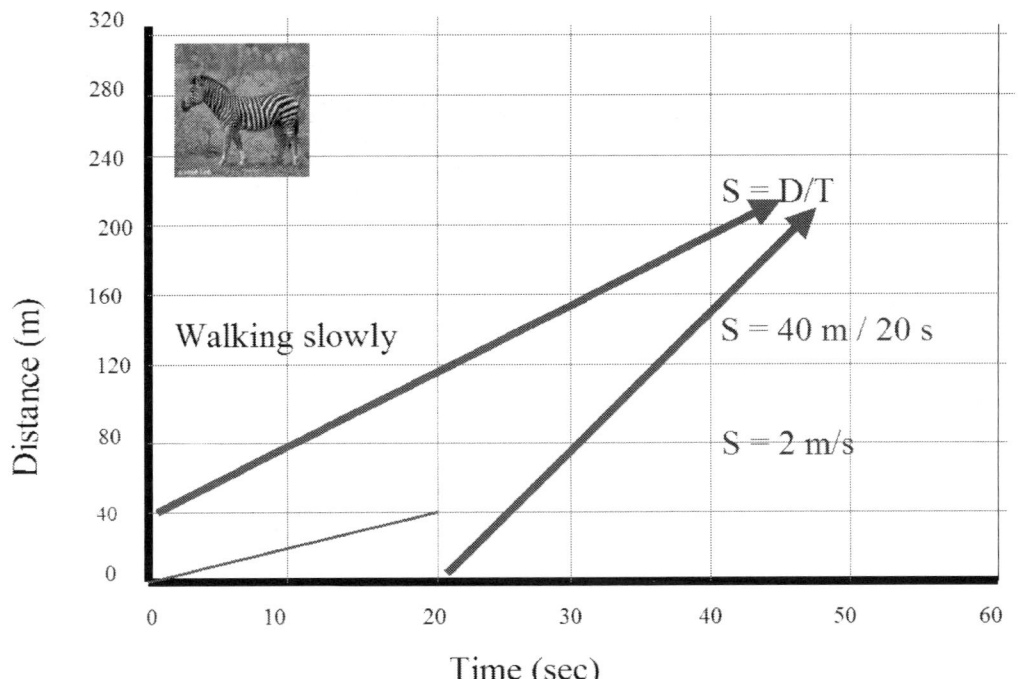

In the next graph the zebra is not moving, therefore his distance is zero, since it did not change. The time he spent stopped was 20 seconds (40 s – 20 s).

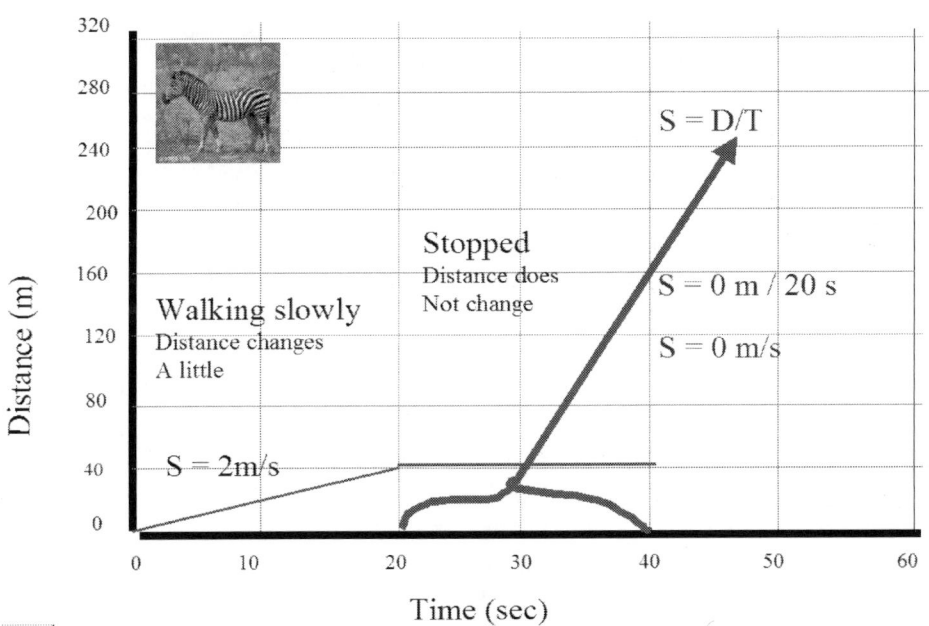

Now the zebra picks up speed and runs from a lion. The distance he ran was 240 m (280 m – 40 m) and his time was 20 s (60 s – 40 s).

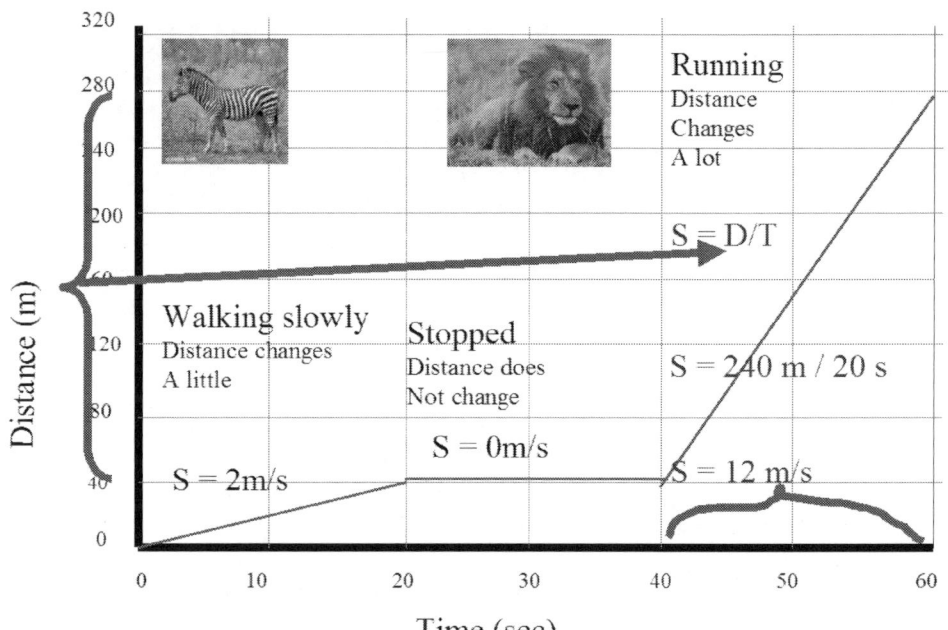

Now let's look at the whole graph without the math. Notice that the gentle slope is a slow speed and a steep slope is a fast speed.

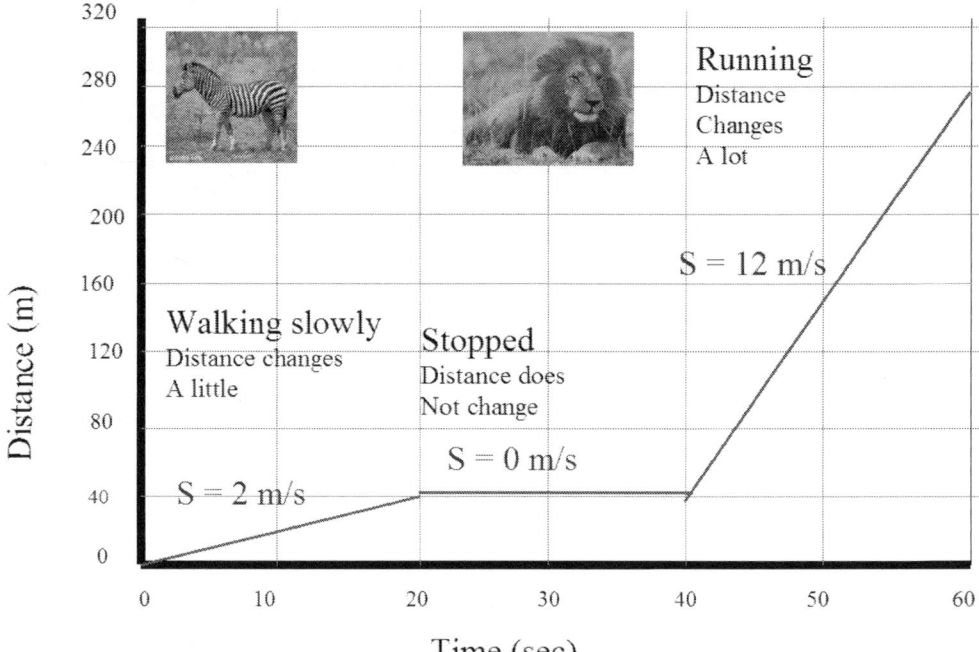

Now we can take it one step farther. What follows is an acceleration graph, which shows if the zebra is speeding up or slowing down as he runs. This graph shows the zebra gaining speed because it is shallow at the beginning and gets steeper at the end.

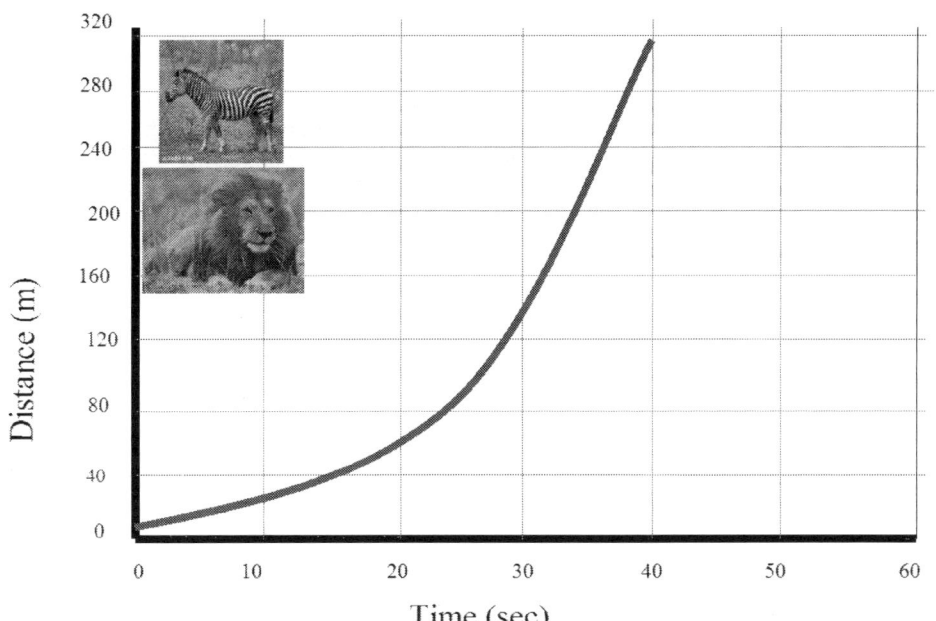

The next graph shows the zebra slowing down since it was steeper at the beginning and gets shallower at the end. I am not sure why.

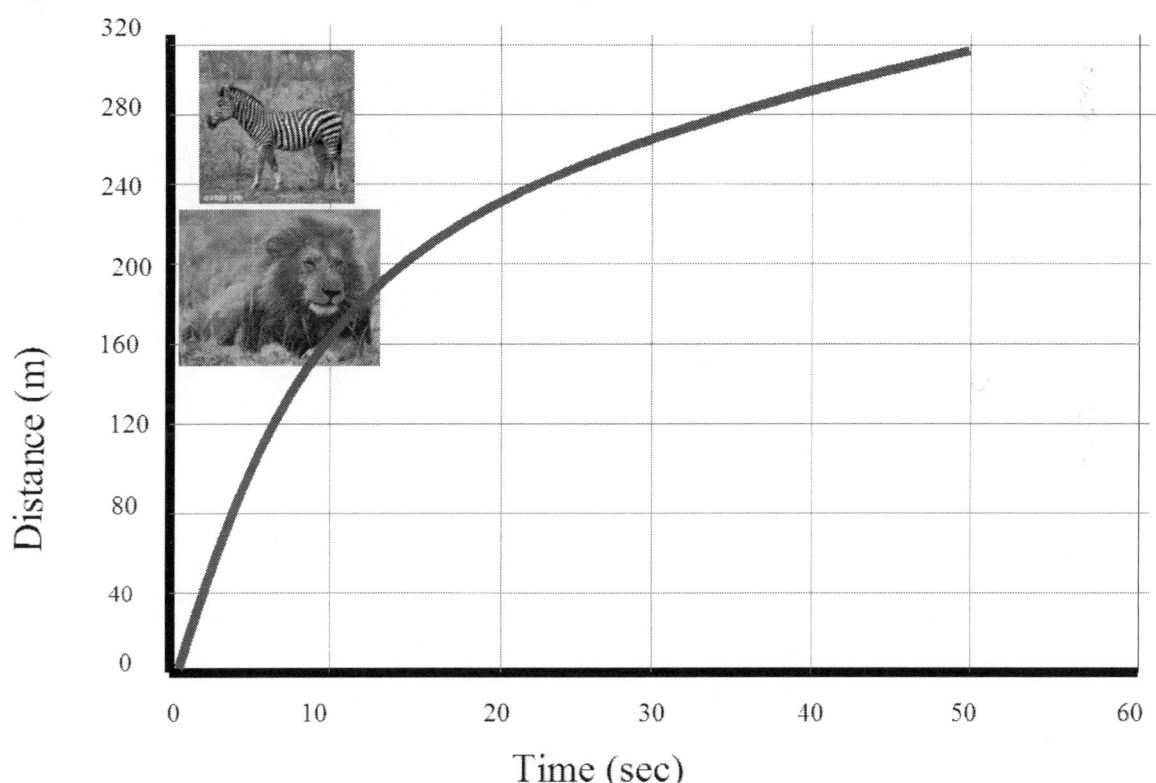

The next graph shows a constant speed since the slope does not change.

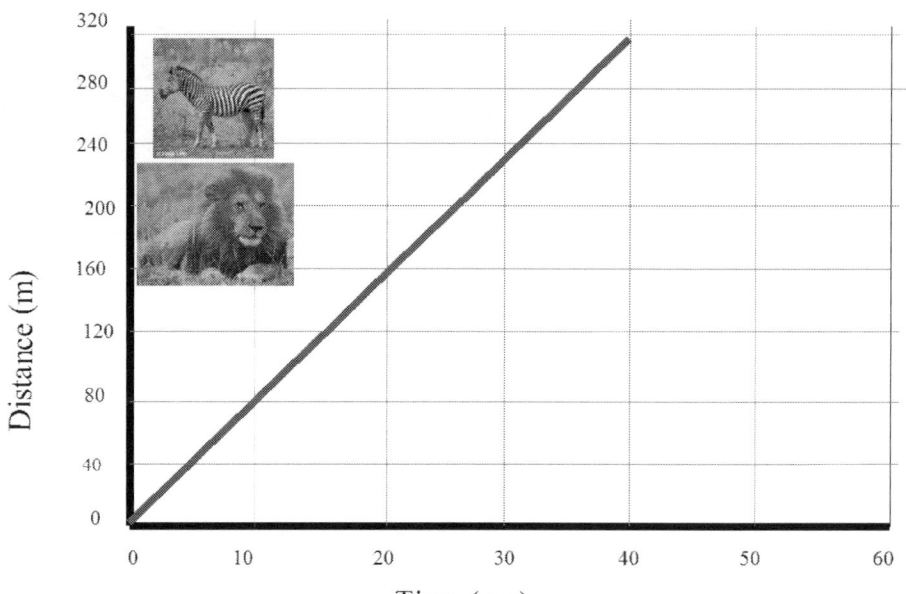

The last graph shows a story. Can you figure it out before I tell you?

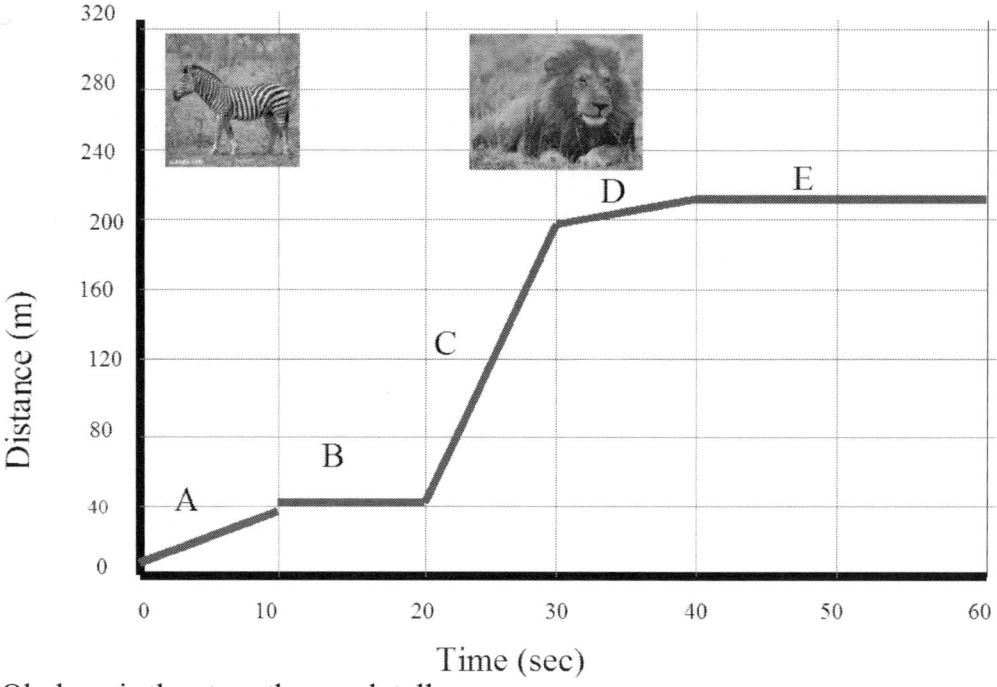

Ok, here is the story the graph tells:
- A. The zebra is walking along slowly having a fine old time.
- B. The zebra sees the lion and hides.
- C. The lion sees him and the zebra starts running very fast.
- D. The zebra slows down when he gets tired.
- E. The zebra stops and has dinner with the lion, or is dinner.

They lived happily ever after (at least the lion did).

A day in the life of Earwig Hickson III

There is a big hill by my house, and I mean a BIG HILL! We call it *the terminator* because at the bottom is a forest of trees, sticker bushes, a rock wall, a creek and a farm pasture full of really mean cows. It was a name left over from long ago when kids who are now grandparents told stories about the hill. I was sure they were just that, stories, to scare us kids into staying away from the steep hill. In the winter it was our sledding hill. No one ever makes it to the forest, rock wall or creek, no one wants to. So at the bottom we piled a giant barrier of hay bales. There is a big sign at the top that says, "DANGER! NO SLEDDING, HAZARDOUS OBSTICALS". Too bad we were all too young to read, that would have been useful information.

Whenever we had a big snowstorm, we kids would flock to the hill with our big plastic sleds, tubes and saucers. The snow would pile up in the front as we slid down the hill and finally stop us at the bottom. It was fun but I wanted more speed. The wide flat bottoms of our sleds just had too much friction, too much surface area; they were kid safe. I figured if I had to climb all the way back up the hill, I wanted the ride down to be awesome! I wanted value for my exercise. Dad says this is a character flaw, and I should appreciate what I have.

I suffered all winter with my friction-filled, slow sled, climbing the hill, sliding down at a gradual pace and climbing back up again. It was fun but where was the excitement? Where was the speed? Why did I never make it to the hay bales? One day that all changed.

I was playing in the garage and noticed way up in the rafters, the point of something interesting. I climbed up and found an old wooden sled that must have been 50 years old. The wood was gray and the paint faded, but I could just make out the words, "flexible flyer, the world's fastest sled". This was interesting, why had my father never mentioned this? It looked wonderful. It had thin metal runners, a steering bar and a rope handle, much better looking than the plastic speed-sucking saucer he gave me. Even at my young age I could see this thing was made to reduce friction, it was made for speed. With less friction to slow down the sled, I knew I would find the speed I wanted! Into the shop area I went. It was an old sled so I had to polish the metal runners with sand paper and steel wool. When it looked shiny and smooth I added a layer of wax to reduce friction even more. The thin metal runners with hardly any surface area to touch the snow and cause the evil friction, reflected light like a mirror. The oiled steering mechanism, flexed the runners so the sled could steer. All I needed was a snowstorm.

Winter was giving its last gasp when a small snowstorm was predicted. I waited patiently all night for the snow to fall, but the next morning, no snow. Instead the weather had deposited a thick layer of ice, a glaze storm, freezing rain. Everything was covered in ice, school was cancelled, power lines were heavy with the sticking ice, the roads were un-drivable, and the trees looked like crystals as the ice-covered branches refracted the sunlight. It was beautiful but no snow. I was disappointed.

Since I was bored, I decided to check out terminator hill anyway. To my delight, there were others there, sliding down on their plastic tubs and actually reaching the hay bale barrier. I was in luck, that ice-covered hill was just made for speed!

I fastened my boots, checked my hat and gloves, gave one last check on the steering bar and gave myself a starting push. I was doing what my grandfather called a belly whopper. The world at that moment seemed to end. On the hard ice, my metal-waxed runners accelerated on the steep hill. I passed the saucers, tubes and plastic sleds as if they were stationary. Branches from small shrubs slapped my face as they blurred by. My face was stretched to the side of my head. The sled went faster. There came a point where I had achieved more speed than I really wanted. My sled had no brakes! My feet, hanging off the sides tried to pierce the hard ice to slow down, but to no avail. My speed increased. I saw the hay bales up ahead getting bigger as they got closer. The blurry scenery on my sides going by like a blink, my short life starting to flash before my eyes. As the bottom of the hill came nearer my eyes started watering from the wind, my boots flew off, and my hat was far behind. The hay bales looked big, they would save me. By the time I reached the bottom of the hill my cheeks were stretched way behind where they used to be, my lips were flapping in the wind like a flag in a hurricane. They were making a thumping sound like a helicopter.

I held on for my life when I hit the hay bales, but they did not stop me. I blasted right through like a train hitting a paper wall, the hay bales went flying. The forest was next. A bunny ran, birds took for the sky, my flexible flyer a flash before their eyes. I pulled on the steering bars with all of my little muscles, weaving between trees, snapping the sticker bush vines grabbing me, snapping smaller saplings at the base. That is when I became air born. The rock wall below zipped by; my eyebrows and eyelashes snapped off, the creek was just a memory. The mean cows on the other hand did not know what to think. They panicked and ran, as only cows can run, in every direction, smashing though fences and running onto the local highway.

I eventually came to a stop far from the hill, far from the cows, and far from the sign that said, "DANGER! NO SLEDDING, HAZARDOUS OBSTICALS". I lived. I had found my speed. I was the fastest kid alive.

My father was actually kind of proud of me after the police dropped me off at home. He bought me an ice cream cone. He said no one had ever dared to try *the terminator* after an ice storm in a flexible flyer, and that I would be a legend! He said not to pay attention to the farmers, who complained that the cows refused to give milk anymore, or my mother who insisted I be chained to my bed each winter, or even the people who claimed my sonic boom broke their windows. He just said to invite him next time.

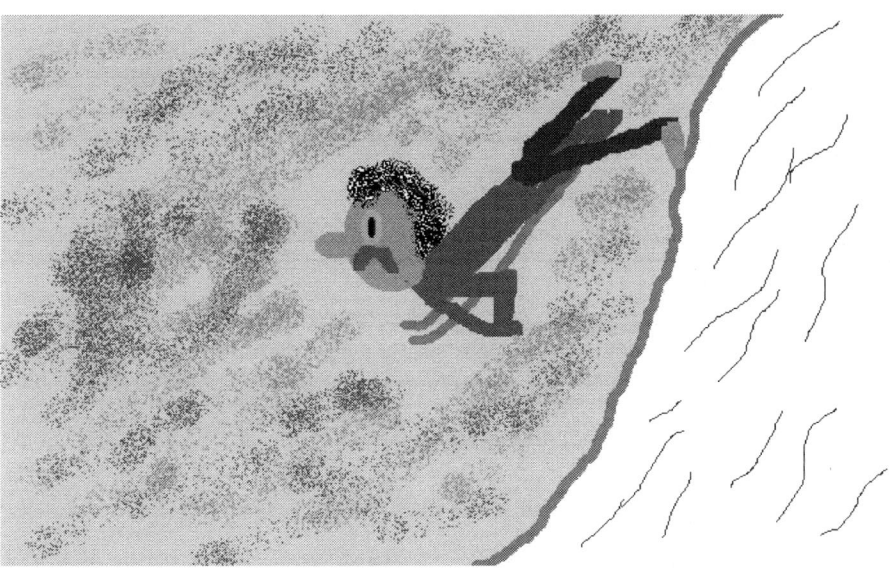

Review of terms – quizlet
https://quizlet.com/124209676/chapter-2-volume-2-flash-cards/?new

Fun things to Google
- SR-71
- X planes
- Sound barrier
- Land speed record
- Chuck Yeager
- X-1
- X-15
- Leonardo DeVinchi – artist and scientist
- V-2 Rocket

Links
SR-71 documentary: https://www.youtube.com/watch?v=s3tvBRs8_qU

X-15 documentary: https://www.youtube.com/watch?v=QCnmuAkrf9M

Building Aircraft Faster Than Sound - NOVA: https://www.youtube.com/watch?v=0Bv4mv50lXo

One of the fastest ping pong balls in the world, flying faster than sound. How I made the cannon and how I measured the speed of the ping pong ball.
https://www.youtube.com/watch?v=zIwz6XcndSk

The fastest animals on land, in air, and in water.
https://www.youtube.com/watch?v=sexgCGfgfCM

Supersonic jets and sonic booms.
https://www.youtube.com/watch?v=gWGLAAYdbbc

Chapter 21
Velocity

Video of this power point lesson: https://www.youtube.com/watch?v=Gjk9CW5Icsc

The wonders of science

It was called The Orteig Prize, a $25,000 award to the first airplane pilot to fly between New York City and Paris nonstop, across the Atlantic Ocean. Others had made shorter crossings of the Atlantic, or with many landings along the way but nothing this long. It was May 20, 1927, and a man named Charles Lindbergh was ready to try his skills, and make the first crossing. Although modern airplanes do this every day, back then it was a big deal, a very big deal. Lindberg made the 3,600 mile flight in 33.5 hours, piloting what amounted to a flying gas tank, called the *Spirit of St. Louis*. He was welcomed in Paris by a crowd of 150,000 people. He was given a ticker tape parade back home in New York. He was famous, and he still is.

Think about the problems of navigation he faced, flying across the ocean, with no landmarks to guide him, relying only on a compass, the sun, and stars when he could see them. Traveling over 3000 miles then having to find a small airport in France without GPS or other modern navigation equipment. A small miscalculation and he would end up lost or worse, run out of fuel and crash. Direction when flying a plane is important. You have to be perfect.

After Lindberg's historic flight many adventurers tried the dangerous trip, it was the thing to do. The United States was beginning to regulate air travel and for a pilot to take a long trip, they had to get permission, especially one from New York to Europe. They inspected the airplane for safety, and a mess of other things. Permission was hard to get.

In 1938 a young pilot named Douglas Corrigan wanted to try a flight from New York to Ireland. He had actually helped build the Spirit of St. Louis and wanted to imitate his hero Charles Lindberg. His plane was considered such a piece of junk by the Bureau of Air Commerce, that not only was he denied permission for the flight, his plane was considered too unstable to fly. They refused to license it to fly over water, but he could fly over land. It was easier and safer to fly from New York to California over land. This he was allowed to do and on July 17, 1938, that is exactly what he decided to try. His flight plan called for him to take off to the east, make a U-turn and fly west for 3000 miles. He *forgot* to make the U-turn. As he flew to the east he *accidentally* flew over the Atlantic Ocean and *accidentally* landed in Ireland instead. People accused him of doing this on purpose, but he never admitted it. He claimed it was a small navigation error, 3000 miles to the east, when he meant to go west. He is forever known as "Wrong Way" Corrigan, a legend.

What you do vs. what you get accomplished.

Imagine a football player at the 50 yard line. When he gets the ball, he runs straight ahead but then curves to the right and then back to the left again across the field. He switches direction once more and finally runs toward the end zone and scores a touchdown. What did he do? What did he get accomplished? Well, he may have run a total of 100 yards to escape the defenders; he certainly did not run in a straight line. On the other hand he started 50 yards from the end zone which is where he ended up. So did he run 100 yards (his actual path) or 50 yards (what he actually accomplished)? It comes down to what he got accomplished. This is the value that goes into the record books. It is how many yards he gets credit for on your fantasy football team. He only earned 50 yards, it does not matter how far he ran to earn those 50 yards, but that is all he really got accomplished.

A straight line from where he began to where he ended, even if he did not run in a straight line, is what is important in this case, and many other cases too.

Velocity

Velocity is kind of like speed, except it gives *direction* too. It is also a straight line, it is what you get accomplished. Direction is important sometimes. Speed only tells you how fast you are going, or your distance per unit time. Velocity tells you your speed and in what direction you are traveling. Imagine if our football player ran the wrong way to the wrong end zone, would that matter. You bet it would. *So velocity is defined as speed in a given direction.*

39

VELOCITY

- **SPEED IN A GIVEN DIRECTION**
    - WHAT YOU GET ACCOMPLISHED
- <u>**IS ALWAYS A STRAIGHT LINE**</u>
    - <u>EVEN IF IT DID NOT GO STRAIGHT!!</u>
- **DIRECTION IS GIVEN RELATIVE TO SOMETHING**
- **WEST, EAST, UP, DOWN, WITH THE WIND, AGAINST THE WIND**

If I meet a Grizzly bear in the woods and run at a speed of 20 miles/hour, did I escape? Do I need to tell you more information? What would you want to know? Is it better when I say, I ran *away* from the bear at 20 miles/hour. Now you know I did not run toward the bear. This is my velocity; **20 miles/hour away from the bear**, my speed was just **20 miles/hour**. Too bad bears can actually run at about 30 miles/hour, hopefully in the wrong direction.

The great cow race

Two cows are going to have a race. The first cow is very fast but her legs are all a different length. She cannot run in a straight line. The slower cow is very slow, but at least she runs in a straight line, and she is headed right to the finish line. The race begins and off they go, one cow running in a zigzag pattern and other one perfectly straight. They both reach the finish line at the same time.

THE GREAT RACE

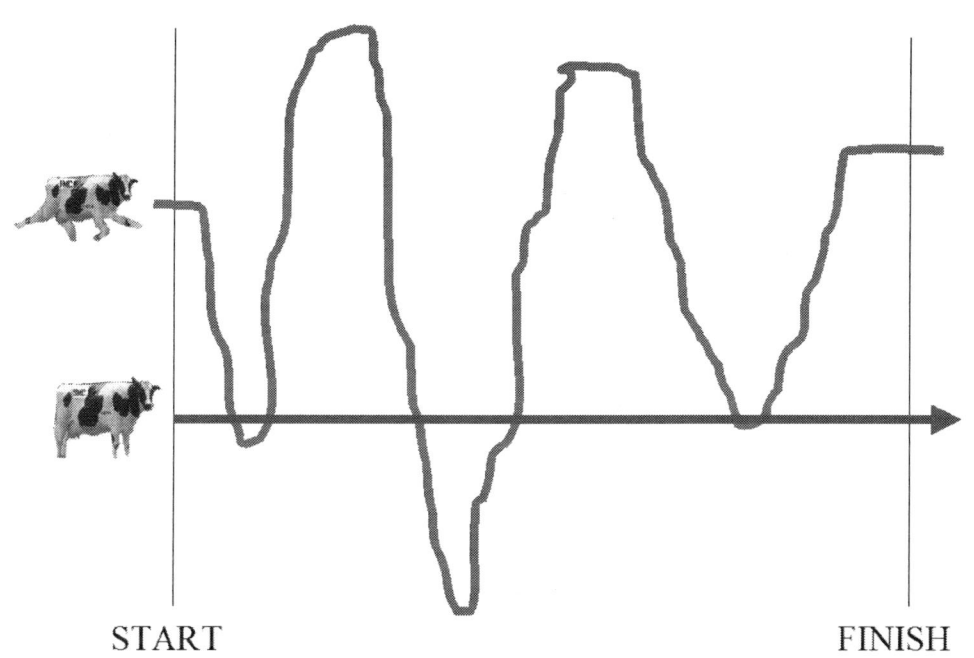

START   FINISH

Which cow ran the farthest? Which got the most accomplished? The first cow ran much farther than the second cow, so his distance was much greater. But they both got the same thing accomplished since they both started at the same place and ended at the same place. The path the cows actually ran is the *distance*, but what they accomplished is called *displacement*.

THE GREAT RACE

DISTANCE IS GREATER FOR THE FMC

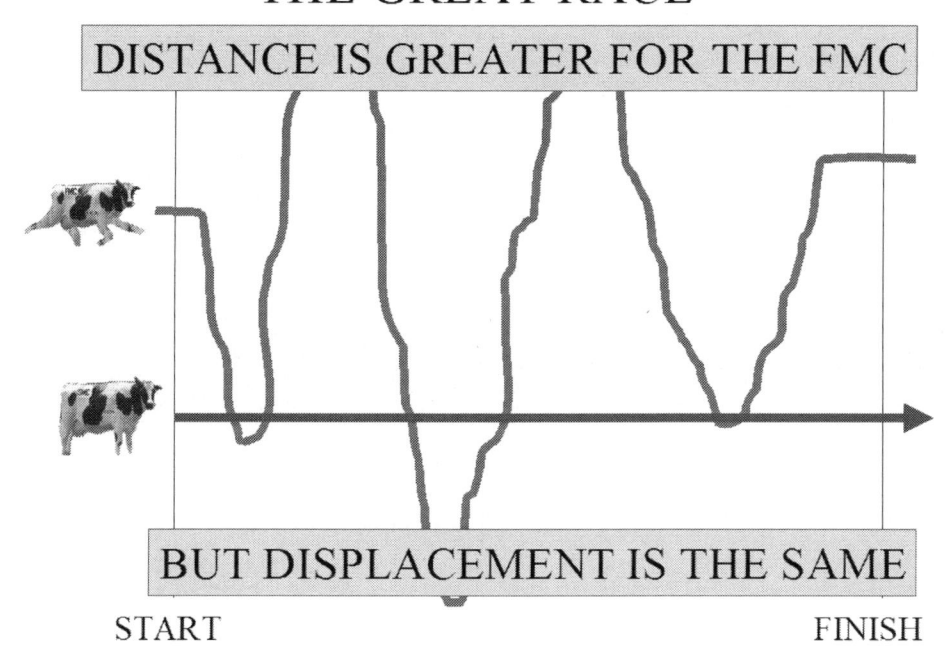

BUT DISPLACEMENT IS THE SAME

START   FINISH

Scalar quantity vs. Vector quantity

A vector quantity shows both *direction* and *magnitude* (magnitude is how big it is) and is shown with an arrow. A scalar quantity shows *only magnitude*, so no arrow. Now the name of the bad guy in the movie *Despicable me* should make sense, when he introduced himself by saying, "I am Vector, committing crime with both *direction* and *magnitude*."
https://www.youtube.com/watch?v=bOIe0DIMbI8

VECTOR QUANTITY

- HAS DIRECTION AND MAGNITUDE (SIZE)
- VELOCITY IS A VECTOR

⟶

SCALAR QUANTITY

- MAGNITUDE ONLY (SIZE)
- NO DIRECTION
- SPEED

Now let's watch the great cow race and see who wins.

THE GREAT RACE

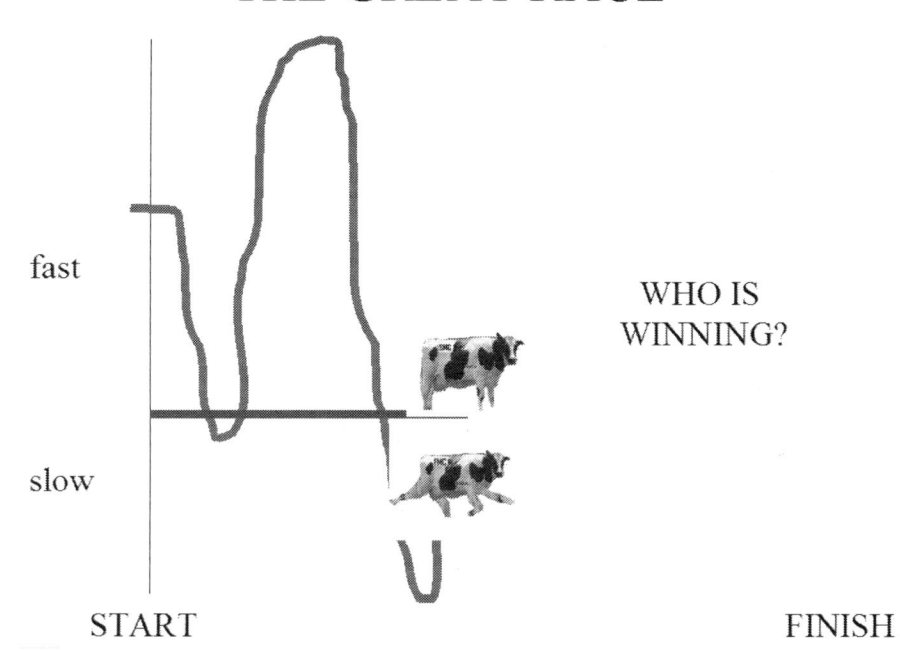

fast

slow

WHO IS WINNING?

START FINISH

THE GREAT RACE

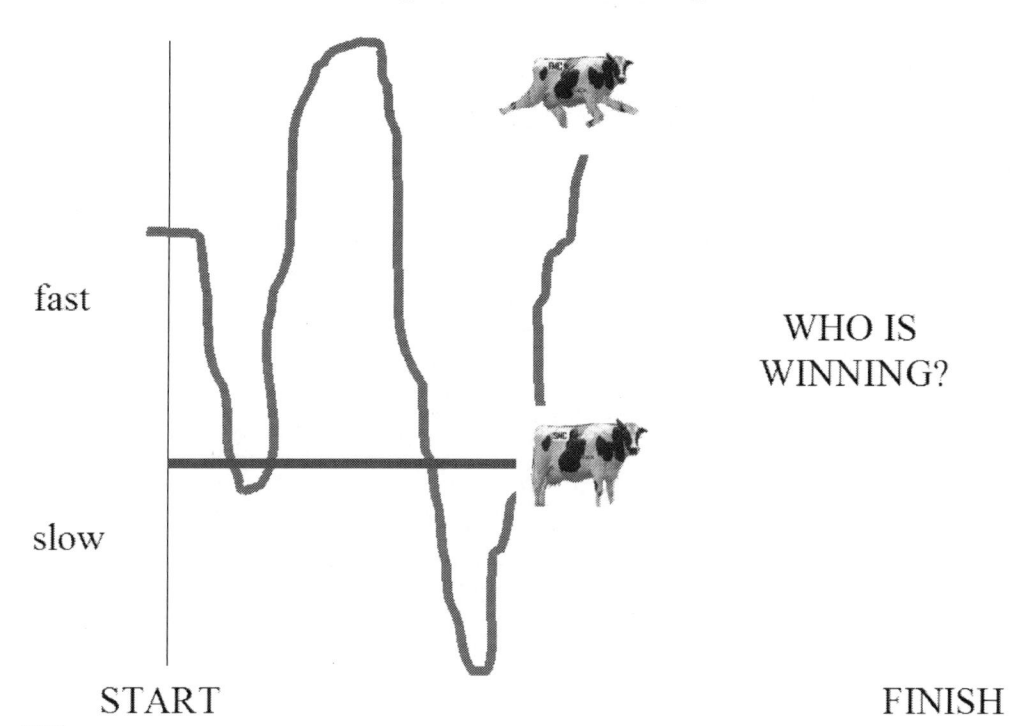

fast

slow

WHO IS WINNING?

START FINISH

THE GREAT RACE

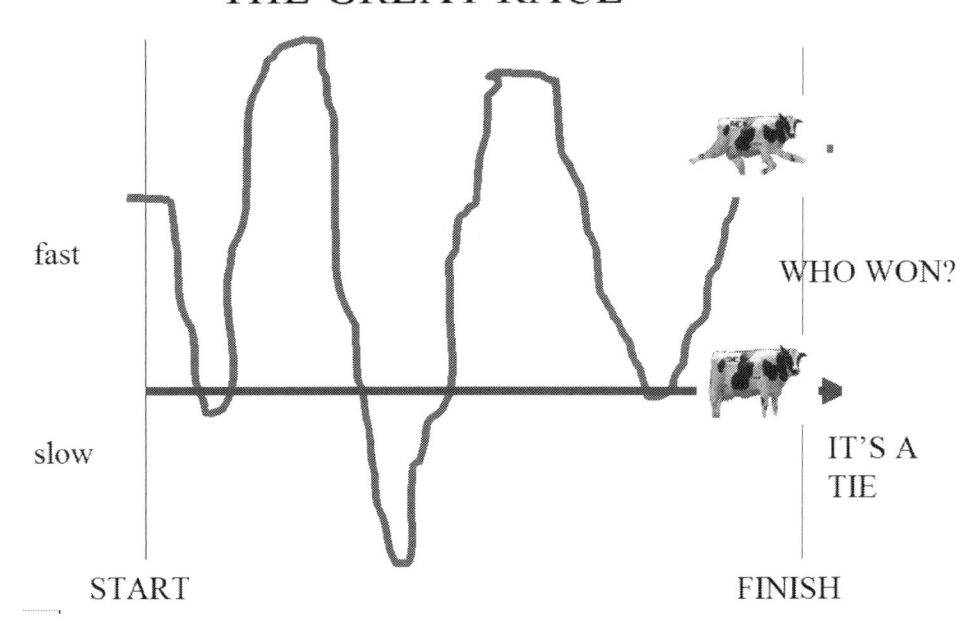

fast

slow

START　　　　　　　　　　　　FINISH

WHO WON?

IT'S A TIE

The race ended up as a tie, so the new question is who had the greater speed? Which had the greater velocity? Well, the first cow ran the greater distance in the same amount of time so she had the greater speed. They both got the same thing accomplished, so they had the same displacement, thus the same velocity.

> **THE FAST COW WENT A GREATER DISTANCE IN THE SAME AMOUNT OF TIME, SO HE SHE HAD THE GREATER SPEED**
>
> **THEY BOTH GOT THE SAME THING ACCOMPLISHED IN THE SAME AMOUNT OF TIME SO THEY HAD THE SAME VELOCITY**

A short video of a race between a tractor and a weasel.
https://www.youtube.com/watch?v=K-_GiE8VSh4

Calculating velocity

Velocity is calculated almost the same as speed was, and we will be using the same three step system. The only real difference is that displacement is used in place of distance, so direction is important. The formula for finding velocity is:

44

VELOCITY FORMULA

$$\text{VELOCITY} = \frac{\text{DISPLACEMENT}}{\text{TIME}}$$

$$V = \frac{D}{t}$$

UNITS M/S **WEST**   KM/H **UP**

The final answer for velocity must have a direction in the units.
Let's try a sample velocity problem.

EXAMPLE VELOCITY PROBLEM

- A COW RUNS 10 METERS WEST IN 2 SECONDS. WHAT IS THE VELOCITY OF THE FMC?

The first thing I choose to do since I do not like word problems is to simplify it with a sketch and by underlining the important information.

EXAMPLE VELOCITY PROBLEM

- A COW RUNS <u>10 METERS WEST</u> IN <u>2 SECONDS</u>. WHAT IS THE <u>VELOCITY</u> OF THE FMC?

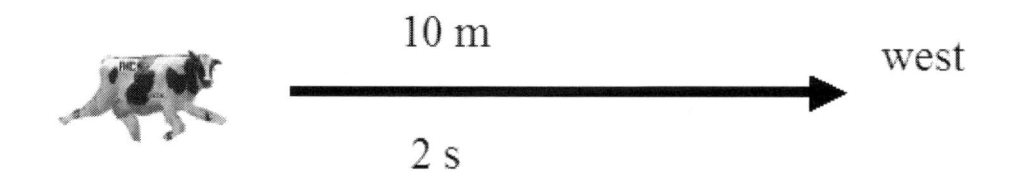

10 m → west

2 s

To solve the problem I just substitute the number values (with the units) into the formula. Remember to write the proper units by looking at the set up.

$$V = d/t$$

$$V = \frac{10 \text{ m WEST}}{2 \text{ s}}$$

$$V = 5 \text{ M/S WEST}$$

It is very good advice to NEVER SKIP STEPS, things may seem easy right now but as formulas become more complex you will mess up. Also remember a number by itself is not an answer. It is nothing. If you want to see more examples take a look at the lesson at the top of this chapter. https://www.youtube.com/watch?v=h46ry-hpM3o

Review

COMPARE AND CONTRAST SPEED AND VELOCITY

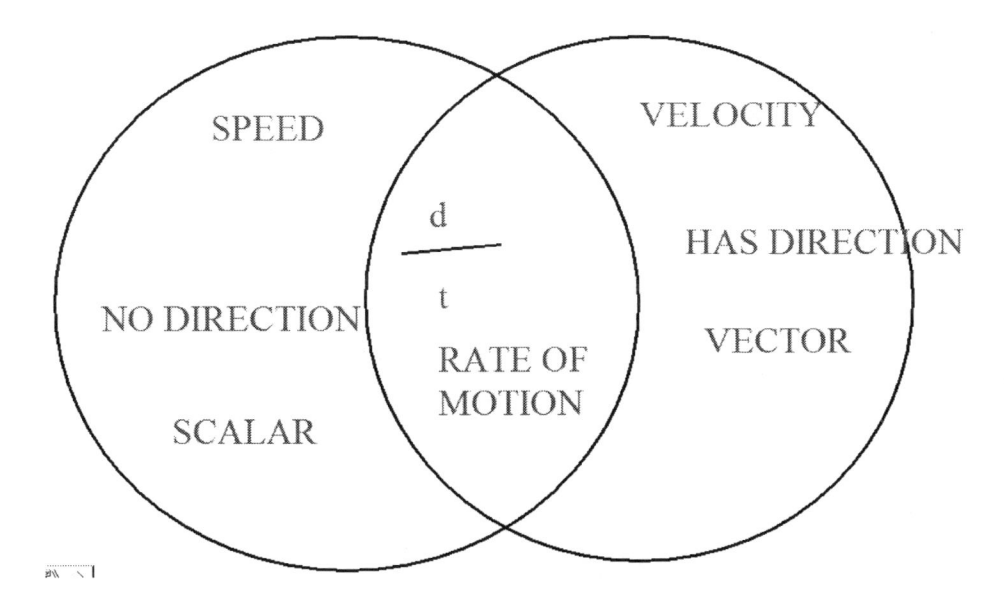

SPEED
NO DIRECTION
SCALAR

$\dfrac{d}{t}$
RATE OF MOTION

VELOCITY
HAS DIRECTION
VECTOR

A day in the life of Earwig Hickson III

My father gave me an old treasure map he found when he was a kid. It was old and faded. But it had directions on it to buried treasure, probably from pirates, or thieves, or maybe even ancient Indians. I had to find it, I had to make a great archeological discovery and maybe become rich so I could buy myself a new sled.

The directions on the map were easy to follow; all I had to do was use my compass to go a certain direction for a certain distance, which I realized was really a displacement. Pirates were so dumb back then, not knowing what displacement was. My first set of directions had me standing in my driveway and walking exactly one mile to the east. I found it odd that pirates would know where my driveway was so long ago, but I did not complain, I was going to be rich! I knew that one of my steps was exactly three feet, so I needed to walk exactly 1940 steps to go one mile. I had to go around a scary cat, a telephone pole, and I stepped in some dog dirt along the way, but I arrived at the exact spot I needed to be. From there I needed to turn due north and walk for two miles, or 3880 steps. This brought me into the local forest where I had to go through thick sticker

47

bushes, a swamp, a lot of mosquitoes, and a very old scary cemetery. The map said I would find a small pile of rock, and right there it was. I was surprised that such a pile of rocks had survived for 100s of years, but what did I care, I would be rich! The map said to walk for exactly half a mile to the northeast. When I arrived at the spot I was very surprised to see a big red X painted on the ground. I started digging. It took me all day and some of the night to dig that hole. Finally I hit something! It was a chest! I hurried as fast as my little arms could dig. I removed the chest and broke the lock with my shovel and then I opened the lid.

It was not filled with gold and jewels. No treasure at all. Instead there was a note and a small package wrapped in colorful paper. The note said, "Happy Birthday, Earwig". Inside the box I found a new pair of pants, a shirt, a new pair of socks, and 3 pairs of new underwear. It was my birthday, and this was my present. I wish He had included the directions to get home because I had no idea which direction to go. I was lost. That would have been a better present, or maybe a GPS.

Review of terms – quizlet
https://quizlet.com/124210309/chapter-3-volume-2-velocity-flash-cards/?new

Fun things to Google
X-1
X-15
Sr-71
Land speed record
Sound barrier
Sonic boom

Links
　Charles Lindbergh: https://en.wikipedia.org/wiki/Charles_Lindbergh

　Spirit of St. Louis: https://en.wikipedia.org/wiki/Spirit_of_St._Louis

The difference between speed and velocity. https://www.youtube.com/watch?v=-6lrr6-ADY0

Scalar vs. vector quantity. https://www.youtube.com/watch?v=0iz7nGEkf4Y

A short and exciting video of a race between a tractor and a weasel. https://www.youtube.com/watch?v=K-_GiE8VSh4

A criminal defines the word vector. https://www.youtube.com/watch?v=bOIe0DIMbI8

A fast mean cow chasing a man in a vector path. https://www.youtube.com/watch?v=h46ry-hpM3o

Chapter 22
Acceleration

Power point lesson about acceleration:
https://www.youtube.com/watch?v=Smjvm1vEfJw

The wonders of science
The fastest man an earth. That is what he was, not for fame or glory, but for science. It was 1946, the jet plane had only been in existence for a few years and the air force was experimenting with ejection seats. The question was; could a pilot safely eject from an aircraft going at extreme speeds? At what speed could a pilot safely eject?

A fellow named Dr. John Stapp had an answer. Well not an answer, what he had was an experiment to try. He would build a series of rocket sleds, and make them go as fast as possible. Cool, you might say but he was not interested in how fast he could go. What he wanted to know was how fast he could stop! How many gs could he survive? What would happen to a pilot at extreme g-force? What he needed was a volunteer to get into one of these rocket sleds and once up to speed, stop instantly to see what would happen. He was the volunteer. Beginning in 1947 he conducted 26 tests himself, 74 tests were made total, to see just what the human body could withstand. His last test on a rocket sled propelled him to 632 miles per hour! A record that made him the fastest man on earth, but the interesting part is that he stopped in 1.4 seconds! Decelerating from 632 mile/hour, resulted in poor Stapp experiencing 43 g's of force. A g would be his body weight and at 43 g's he had a weight of 6,800 pounds! He broke a wrist, some ribs and every blood vessel in his eyes ruptured, but he survived. Talk about guts!

Remember Indiana Jones on a rocket sled?
https://www.youtube.com/watch?v=Tke92CmoLLU

His research led to a lot of useful things. Not only was it used to keep air force pilots alive, it was the beginning of car safety. Seat belts were installed into cars from his research. The stuff he learned from being a human crash test dummy has undoubtedly saved quite a few human lives.

A short video of some of Dr. Stapp's experiments:
https://www.youtube.com/watch?v=s4tuvOer_GI

What you need to remember
Always remember and never forget, velocity is speed in a direction (straight line). So, to change velocity, you can *speed up, slow down, or turn in a new direction*. This is what this chapter is all about, changing velocity, because a change in velocity is acceleration.

Acceleration
Neil Armstrong talks about acceleration:
https://www.youtube.com/watch?v=7AJkC6H-fSg

Acceleration is the *rate* of change in velocity, how fast velocity is changing. If I have a car that accelerates from 0 to 60 miles per hour in 2.4 seconds, that is a rather good car. My other car may accelerate from 0 to 20 miles per hour in 2 hours, that would be kind of bad; the velocity change takes way too long. Of course, acceleration does not have to be how long it takes to go *faster*, since acceleration is a change in velocity. The car could be *slowing down* also, or even *turning*, since a change in direction is a change in velocity too. You can always recognize acceleration when you experience it because you can feel it. At least your stomach feels it. The reason you like Amusement Park rides is because they accelerate you. The faster the better, you call acceleration fun. The local amusement park near me (Hershey park) tries to bring me as close to death as possible, without killing me, since dead people do not buy more tickets. Acceleration is fun, acceleration makes you throw up, or as we say in science: *acceleration makes me blow chunks!*

ACCELERATION

- THE RATE OF CHANGE IN VELOCITY
- HOW **FAST** THE VELOCITY IS CHANGING
- THIS CAN BE A CHANGE IN SPEED
- (FASTER OR SLOWER)
- **OR** A CHANGE IN DIRECTION
- (TURNING)

Calculating acceleration

Another formula, uuggh, but no worries, if you follow the steps I showed you it is easy. It is simply how much the velocity changes in how much time.

ACCELERATION FORMULA

$$ACCELERATION = \frac{FINAL\ VELOCITY - ORIGINAL\ VELOCITY}{TIME}$$

Or more simply:

ACCELERATION FORMULA

$$\text{ACCELERATION} = \frac{\text{FINAL VELOCITY} - \text{ORIGINAL VELOCITY}}{\text{TIME}}$$

$$A = \frac{V_2 - V_1}{T}$$

UNITS - M/S/S   KM/H/H

The units are a little odd, but you can find them in the set up, just like before. The hard part to this formula is realizing that the second thing that happened is put first in the formula, V_2 (velocity second). The first thing that happened is V_1 (velocity first).

There are other versions of this formula but I think this is the easiest version. So the formula you need to know is:

$$A = \frac{V_2 - V_1}{T}$$

Where **V_1** is the first thing that happened
V_2 is the second thing that happened and
T is the time it took for the change in velocity.

Notice the funny units, don't worry about them yet, just write down what is in the set up.

Before we go on, I had better mention that sometimes when you calculate acceleration, the answer can be negative. This is not a problem, it just means that the object is slowing down or decelerating. *A negative acceleration means deceleration* (which is also acceleration – just slowing down). Confusing I know, but just remember that a positive answer is speeding up, and a negative answer is slowing down.

DECELERATION

- DECELERATION IS SLOWING DOWN
- IF YOUR ANSWER IS NEGATIVE IT IS DECELERATING
- IT IS A NEGATIVE ACCELERATION

Let's look at an example that is as hard as it gets.
The first thing I might suggest is to underline the important information and label it.

EXAMPLE

- A FMC IS RUNNING AT 20 M/S $[V_1]$, 5 S LATER IT IS WALKING AT 10 M/S $[V_2]$. WHAT IS THE ACCELERATION?

Or draw a simple sketch.

EXAMPLE

- A FMC IS RUNNING AT 20 M/S, 5 S LATER IT IS WALKING AT 10 M/S. WHAT IS THE ACCELERATION?

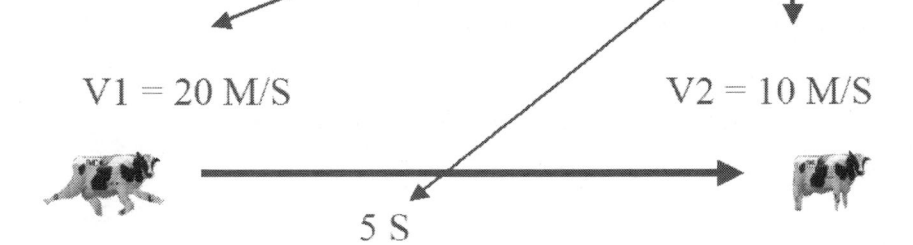

V1 = 20 M/S   V2 = 10 M/S

5 S

In the sketch I just drew, you may notice that the cow is slowing down. This means it is *decelerating*, so I know my answer is going to be *negative*. Remember this, if you make a mistake, you will know it.
The first thing we do in this example is to write the correct formula.

$$A = \frac{V_2 - V_1}{T}$$

Now we substitute the correct values for the variables (letters).

$$A = \frac{V_2 - V_1}{T}$$

$$A = \frac{10 \text{ M/S} - 20 \text{ M/S}}{5 \text{ S}}$$

Now we calculate and simplify the problem. Let's do the numerator (top number).

$$A = \frac{V_2 - V_1}{T}$$

$$A = \frac{10 \text{ M/S} - 20 \text{ M/S}}{5 \text{ S}}$$

$$A = \frac{-10 \text{ M/S}}{5 \text{ S}}$$

$$A = -2$$

I have calculated the number answer, and I see the units for the answer. Now for the final answer with the proper units.

$$A = \frac{V_2 - V_1}{T}$$

$$A = \frac{10 \text{ M/S} - 20 \text{ M/S}}{5 \text{ S}}$$

$$A = \frac{-10 \text{ M/S}}{5 \text{ S}}$$

$$A = -2 \text{ M/S/S}$$

We now know the cow was slowing down at −2 m/s/s. In other words, for *each second* the cow slows down it goes 2 m/s slower. Since it originally was running at 20 m/s, after one second, it is running at 18 m/s (-2 m/s), then after the next second slows down to 16 m/s (-2 m/s again). A sketch would look like this.

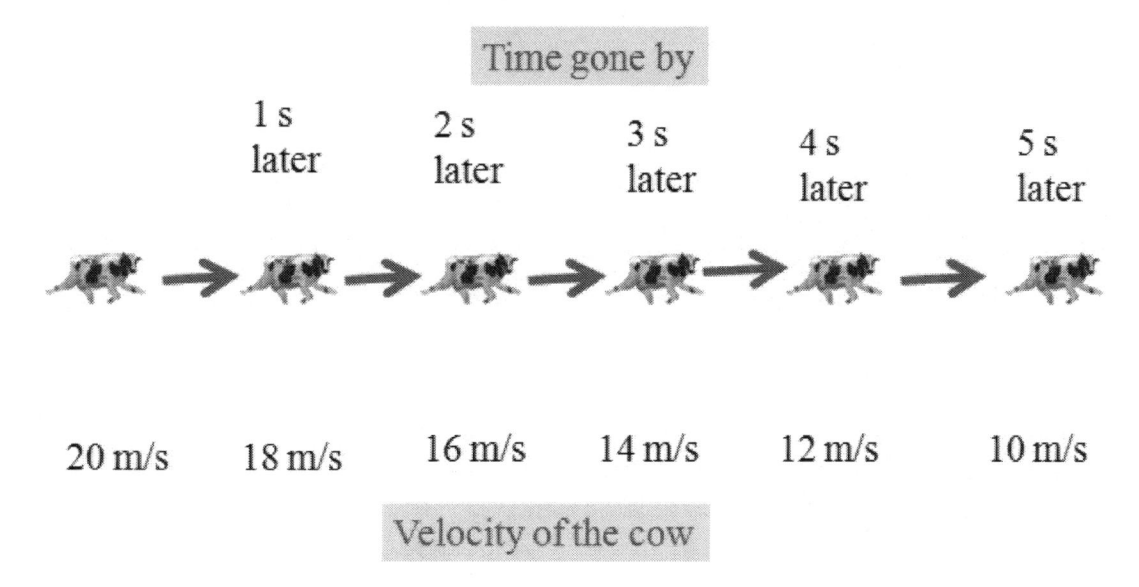

Time gone by

| 1 s later | 2 s later | 3 s later | 4 s later | 5 s later |

20 m/s   18 m/s   16 m/s   14 m/s   12 m/s   10 m/s

Velocity of the cow

What does m/s/s mean anyway?

55

If I say I am accelerating at **5 m/s/s**, it means I go faster at a rate of 5 m/s, *each second* I accelerate. So if I start at rest (0 m/s), one second later I am going at 5 m/s, and after the next second I am going at 10 m/s, then 15, then 20 and so on.

It would look like this (if I was a cow):

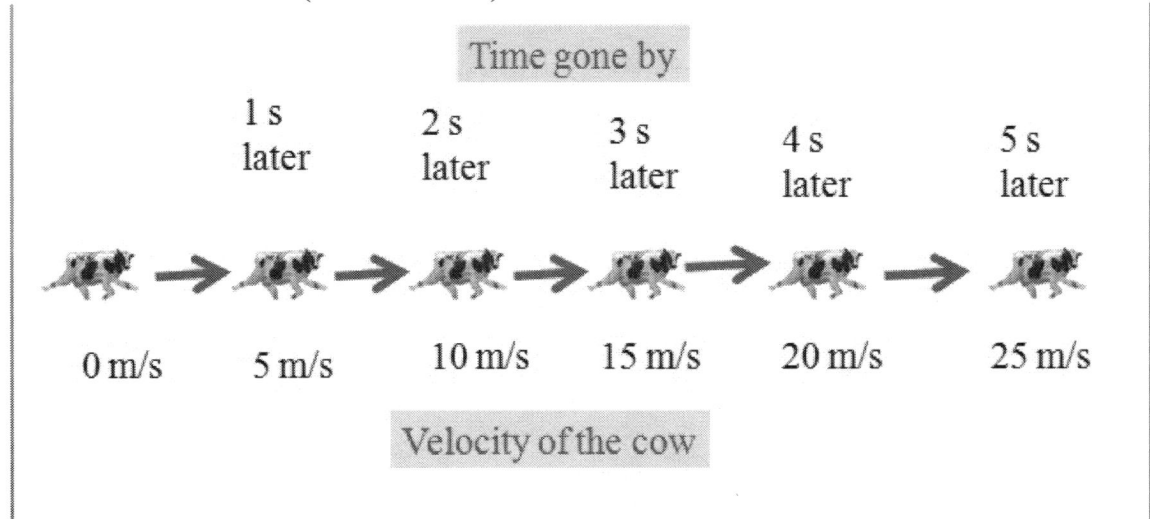

Time gone by

| 1 s later | 2 s later | 3 s later | 4 s later | 5 s later |

0 m/s   5 m/s   10 m/s   15 m/s   20 m/s   25 m/s

Velocity of the cow

Lincoln explains m/s/s to Washington:
https://www.youtube.com/watch?v=fJ8_zy0hFig

G-force

G's are a unit of acceleration, usually measured when something is turning, but it can be when something is speeding up or slowing down, all of which are a change in velocity, thus acceleration. Right now you are at 1 g, which is normal Earth gravity, because that is what g stands for, gravity. You may hear that astronauts are at 0 g's, that means no gravity and they float around. Sometimes when you are on a roller coaster you feel yourself squished into your seat because you are at 2 or 3 g's, and feel heavy.

G FORCES

- g MEANS GRAVITY
- GRAVITY IS ACCELERATION
- 1 g = 10 m/s/s   100 LBS
- 1g = NORMAL GRAVITY
    - YOUR WEIGHT   200 LBS
- 2 g = 2 TIMES GRAVITY
    - 2 TIMES YOUR WEIGHT   1000 LBS
- 10 g = 10 TIMES GRAVITY
    - 10 TIMES YOUR WEIGHT

Some useful g forces

G'S

- 1 g IS NORMAL
- 3 g's IS WHAT A PERSON CAN HANDLE WITHOUT DAMAGE
  - HERSHEY PARK
- 9 G'S IS THE MAX WITH A SPECIAL SUIT
  - PILOTS
- 10 G'S PLUS CAN BE SURVIVED FOR A SHORT AMOUNT OF TIME
  - CAR ACCIDENTS

Review
Acceleration is a rate of change in velocity, either speeding up, slowing down, or turning. The g-force is how many times gravity the force on something is.

Review of terms - quizlet
https://quizlet.com/124208067/accleration-ch-4-volume-2-flash-cards/?new

A day in the life of Earwig Hickson III

I finally fixed my flexible flyer sled, but no more sledding on ice. I was still the fastest sled on the hill but much slower than last time. I did have a problem though, and problems need solved. My problem was that going down the hill was fun, coming back up was not fun; it was WORK, and I am not a fan of work. What I needed was an easy way to get me and my sled back up the hill with less effort. I did some research on the internet and learned that at ski resorts they use a ski lift to carry people back up the hill, but these cost money. My first experiment consisted of a long rope tied to the top of the hill hanging down to the bottom. An improvement but I still had to pull myself up the hill. What I needed was for the rope to pull me, like they do at a ski resorts. So I needed a machine, something that would pull the rope up the hill.

It was my lucky day! Walking home from school someone had put out an old clothes washer for the trash. Apparently it leaked water and was not good anymore, but it still worked. There was a motor inside that turned a drum, where the clothes go. I got to thinking that I could wrap my rope around the drum and back down the hill around a pulley. I had a hypothesis; I just had to test it. Into the shop I went and a few days later my masterpiece was complete. Time to test it.

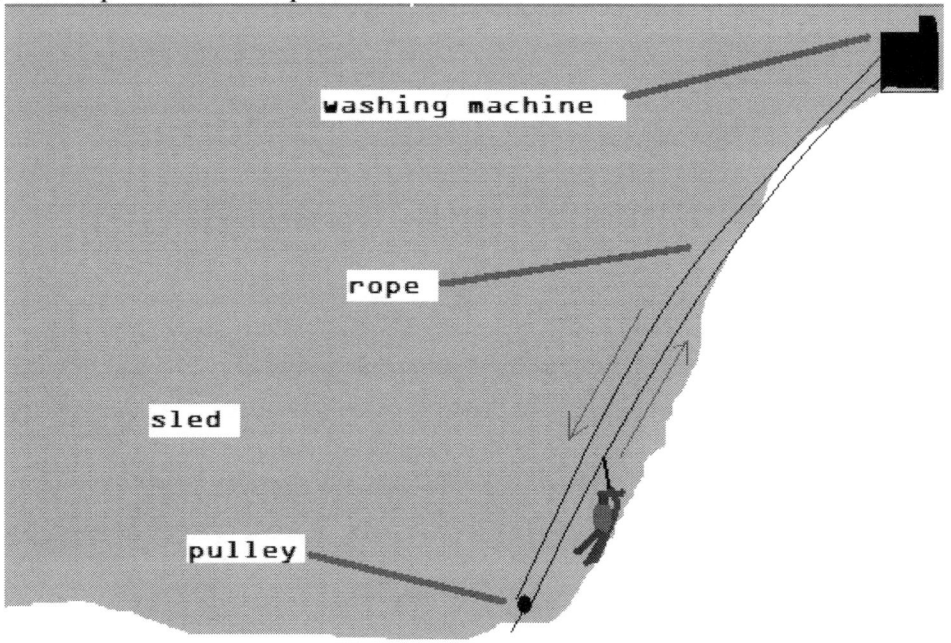

After my awesome ride down Terminator hill, I prepared for a just as fun a ride back up. The Washing machine was plugged in with a very, very long extension cord for electricity. The dial was set to the gentle cycle, the power button was pushed and the drum started to rotate. The rope that was wrapped around the drum began to move. My sled and I started to slowly go up the hill. It was a success! When I reached the top I just unhooked from the rope and back down I went. Everyone was impressed, at least at first, on my third trip up the hill something changed. The washing machine had reached the SPIN CYCLE! The spin cycle makes the motor spin the drum superfast, and acts like a centrifuge, to spin the water out of wet clothes. My ski lift was now in overdrive. My velocity instantly changed, I was accelerating at a rapid pace. The rope tightened up and hummed like a guitar string, the sled shot up the hill like a rocket. I hung on for all I was worth. When I reached the top of the hill I rolled off the sled and slid for half a mile. With the reduced weight, my sled shot high up into the air when the rope snapped, its trajectory carrying it over my town. As it cut through the air, it made a roaring sound. It was gone in a second. I never saw my sled again. But other people did. They reported an object streaking over the town faster than any airplane. The blurry object zipped by everyone. There was a loose piece of wood on the sled vibrating, making a humming sound. The police were called, the Air Force, the Army, my town had a UFO sighting! The newspaper reporters came and we were all interviewed. It was on the T.V. news.

We sold hot dogs to the many tourists that came looking for flying saucers, and we made a fortune.

No one ever explained that UFO sighting, some suspect a secret military weapon, and others swamp gas, or Russia, or some kind of government cover-up. The Internet was crazy with ideas. I hated to ruin everyone's fun, so I never told them about my sled.

Fun things to Google
Rocket sled
Crash test dummies
Car safety
Acceleration
Land speed record
Fastest man alive

Links
John Stapp: https://en.wikipedia.org/wiki/John_Stapp

Rocket sled: https://en.wikipedia.org/wiki/Rocket_sled

G force: https://en.wikipedia.org/wiki/G-force

A short video of some of Dr. Stapp's experiments:
https://www.youtube.com/watch?v=s4tuvOer_GI

Neil Armstrong talks about acceleration:
https://www.youtube.com/watch?v=7AJkC6H-fSg

Lincoln explains m/s/s to Washington:
https://www.youtube.com/watch?v=fJ8_zy0hFig

Crash test with and without a seat belt.
https://www.youtube.com/watch?v=d7iYZPp2zYY

The physics of seat belts. https://www.youtube.com/watch?v=04gD8dCVbys

How airbags work. https://www.youtube.com/watch?v=SSz6y-W-R_A

Unit 7 Forces

Stationary and movable pulleys. https://www.youtube.com/watch?v=ju7Wciv4RXk

Chapter 23
Force

Video of this power point lesson:
https://www.youtube.com/watch?v=QW5yEo_gj7A

The wonders of science
Others had thought about it, and some had come close, but it took the genius of Sir Isaac Newton, to put it all together. He once said, *"If I have seen further than others, it is by standing upon the shoulders of giants."* What he meant by this is, others before him had made fantastic discoveries, which gave him a head start on his own discoveries. Newton came along at just the right moment in history where his knowledge and genius could be used best. He may have been the greatest scientist who ever lived, and those that came after him had his discoveries to work with.

Sir Isaac Newton is famous for many things, some of which you never even knew. He is probably most famous to you as the guy that discovered gravity when an apple fell on his head. Not exactly true, he did not *discover* gravity, he just figured it out. An apple did not fall on his head, but he did watch apples fall from a tree in front of his office window while he tried to figure out how gravity behaved.

One of his many amazing discoveries was the invention of calculus, a very complex form of math that was necessary for some of his calculations. He gets most of the credit for this, but another man named Leibniz, kind of invented it separately at the same time.

He went on to investigate optical prisms and how sunlight formed a rainbow when shined though it. Then he shined the colors of the rainbow back through the prism and made white light again. This is how we know white light contains all the colors of light mixed together (the colors of the rainbow - ROYGBIV). His discoveries of optics resulted in the first reflecting telescope and advancements in many other things such as mirrors, lenses, telescopes, microscopes, lasers, and fibre optics.

In his studies about gravity and Newton's law of universal gravitation, he explained not only why apples fall straight to the ground, and not sideways, but he also realized that the force of gravity did not stop at the apple but reached out everywhere; to the moon, the sun, the planets, everything that had mass. He realized everything that had mass had gravity of its own.

His most famous discoveries were about moving objects and his laws of motion. It was Sir Isaac himself who realized that all objects want to do whatever it is they are doing (staying still or moving at a constant velocity) and they will do it **forever** unless forces makes them change. Up to this point in time, everyone thought a force *kept* things moving, an easy mistake to make, when you forget about friction. Sir Isaac Newton introduces forces, https://www.youtube.com/watch?v=spd_SllGovQ.

What you need to remember
Back in chapter 23, I made a big deal about how acceleration is a change in velocity, and that acceleration can change three ways, by slowing down, speeding up, and turning in a new direction. The other thing you need to remember is from way back in chapter 19, when I made a big deal about how you cannot really tell if you are moving at a constant

velocity or if you are stationary. As far as science goes, they are the same thing. The example I used was that you are stationary in this room (compared to your desk) but moving at a constant 1000 miles per hour (compared to the sun) and you never realized it. That is because stationary and constant velocity is the same. Only acceleration is different. That is what this chapter is about, how to accelerate.

REMEMBER

- **ACCELERATION IS A CHANGE IN VELOCITY**
  - **FASTER, SLOWER, OR TURNING**
- ☆ **STATIONARY OBJECTS** AND MOVING AT A **CONSTANT VELOCITY** OBJECTS *ARE THE SAME*

Force

A force is a push or pull on matter. It could be any object hitting another object like kicking a ball; this would be a *contact force* where your foot actually touches the ball. Some forces act at a distance but still push or pull things, like the sun's gravity. This would be a *non-contact force*. All **forces** no matter what type *change an object's velocity*. That means a force makes the object accelerate in some way (faster, slower, or turning). I can't say this enough; *a force causes an object to accelerate*. **Never say a force causes an object to move**, because that is not true. Objects want to keep moving by themselves, but it takes force to *make* them change their movement.

FORCE

- A PUSH OR PULL ON MATTER
- UNIT = NEWTON (N)
- IT CAN CHANGE AN OBJECTS VELOCITY (SPEED OR DIRECTION)
- ☆ MAKES THINGS ACCELERATE (OR DECELERATE)
- MAKES THINGS TURN

Some types of forces

Contact forces are when two objects actually touch or run into each other. You are most familiar with these. Hitting a ball with a baseball bat is a contact force, so is

banging your head on something too low to walk under. *Friction* is a contact force and is always present between touching objects. Friction always makes moving objects slow down. Listen to a short video of Napoleon learning about friction:
https://www.youtube.com/watch?v=llZ1Zl57r1Q

Using friction to start a fire: https://www.youtube.com/watch?v=APCgA9uxMGI

Non-contact forces are almost like magic; the objects involved never touch but push or pull on each other just the same. You are familiar with gravity pulling you down to the earth when you jump off a step or a small hill. It is a good thing it does too, or you would just fly away into outer space if it didn't. Magnetism is a non-contact force also. The magnetic field of the earth makes a compass needle spin; you can levitate one magnet over another magnet as they push away. Static electricity can attract things from a few inches away, or repel them

Bill Nye The Science Guy on Static Electricity: https://www.youtube.com/watch?v=Z-77IzaXGcg

Static Electricity demonstrations : https://www.youtube.com/watch?v=QxZ6AWLpnUw

9 Awesome Science Tricks Using Static Electricity!: https://www.youtube.com/watch?v=ViZNgU-Yt-Y

hair-raising static electricity Van-de-Graaff: https://www.youtube.com/watch?v=jZEFuCxD7BE

Van de Graff Generator : https://www.youtube.com/watch?v=3Ptu07enIsY

Floating Magnets !: https://www.youtube.com/watch?v=4T2XfrsvZRw

A top levitating over a magnet https://www.youtube.com/watch?v=9zv7mBQaXPg

SOME TYPES OF FORCES

- CONTACT FORCE – WHEN OBJECTS TOUCH
  - FRICTION, A KICK
- NONCONTACT FORCES – PULL FROM FAR AWAY
  - GRAVITY, MAGNETISM
- GRAVITY – BETWEEN MASSES
- FRICTION – RESISTS MOTION

THE FORCE OF GRAVITY FROM THE MOON MAKES THE TIDES

Gravity from the moon causes the ocean tides
Napoleon explains force to Gandhi: https://www.youtube.com/watch?v=JOiorue6GO8

Combining forces

Sometimes (all the time) more than one force will push on an object at the same time. A book sitting on a desk is being pulled down with the exact same force that the desk is pushing it up. The result is the forces cancel out, as if there was no force at all. This is called the *net force*, or the result.

If the forces push in opposite directions, they may cancel out; if they push in the same direction they are combined.

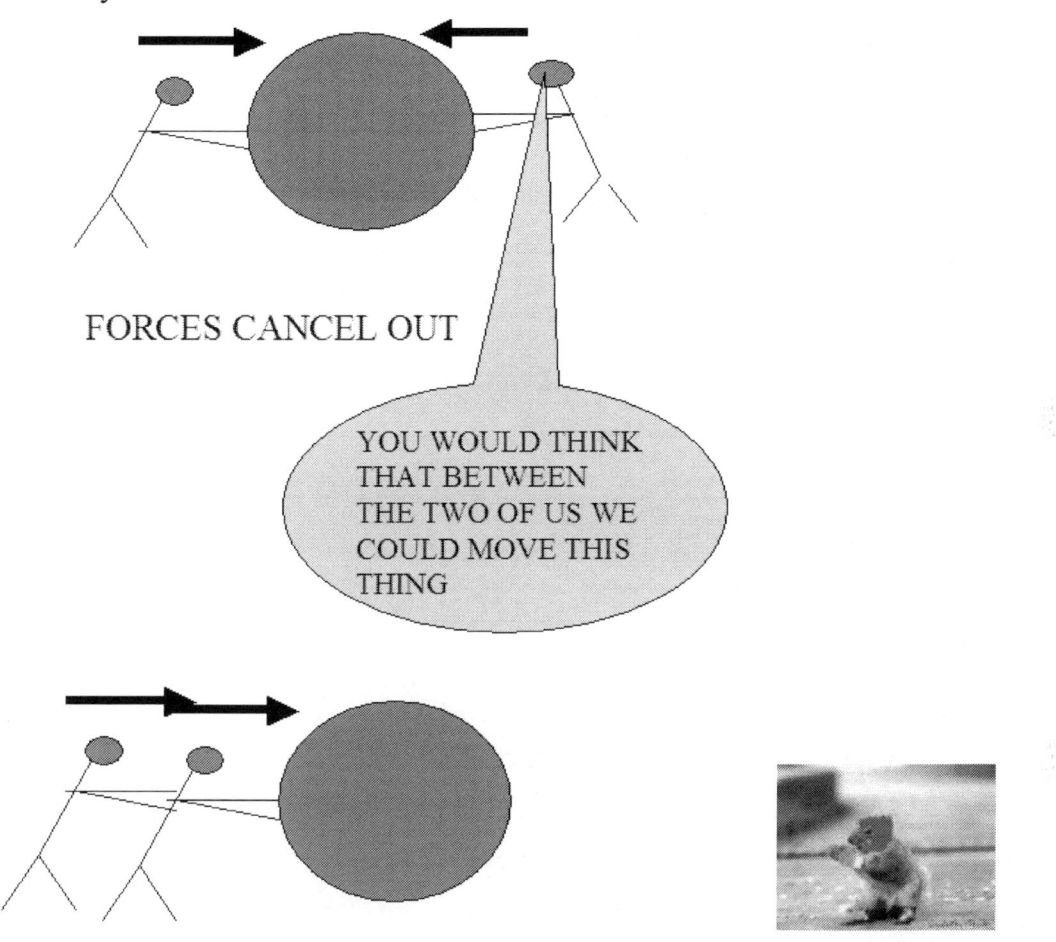

FORCES CANCEL OUT

YOU WOULD THINK THAT BETWEEN THE TWO OF US WE COULD MOVE THIS THING

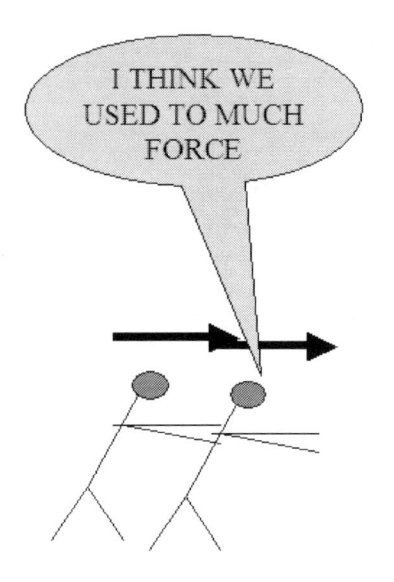

"I THINK WE USED TO MUCH FORCE"

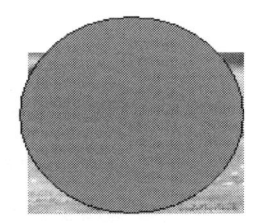

FORCES ARE ADDED

A cool video of indoor sky diving. The person's weight is cancelled out by the air blowing upward. https://www.youtube.com/watch?v=WHmmdW3szwk

COMBINING FORCES

- **OPPOSITE FORCES ARE SUBTRACTED**

- **SAME DIRECTION FORCES ARE ADDED**

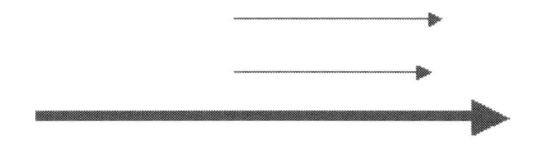

An example summing up force problems

EXAMPLE 1
WHAT IS THE NET FORCE ACTING ON A 10 N BOOK BEING HELD UP WITH A FORCE OF 4 N?

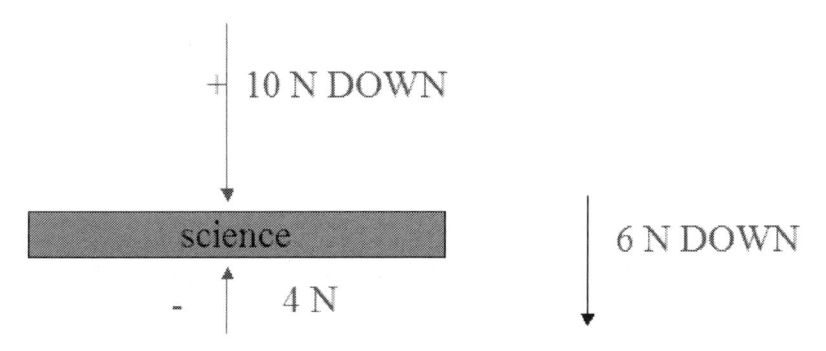

+ 10 N DOWN

science

- 4 N

6 N DOWN

Net force

Net force means *after you do the math*. In the above example, 10 n down (the weight of the book) minus 4 n up (air resistance) results in a net force of 6 n down. The book will be accelerating downward from this net force. In this example the book has been dropped and air friction is pushing it up, but not hard enough, so the book still accelerates.

NET FORCE

- SUM OF ALL FORCES ACTING ON AN OBJECT

- THE ANSWER

Now after you calculate the net force the answer will be one of two things. It can be zero, in which case the forces cancel out, and the motion of the object *stays the same* (no acceleration). The net force can also be any number except zero, in which case the object can only be accelerating.

Unbalanced Forces

This is when the net force is *not* zero. The book in the above example has unbalanced forces. The book does not care how many forces are pushing it, only that the *net force* is 6 n down. This unbalanced force will always cause the object to accelerate faster and faster until it lands.

UNBALANCED FORCES

- WHEN THE NET FORCE **IS NOT** ZERO

- OBJECT WILL **ALWAYS** ACCELERATE OR DECELERATE

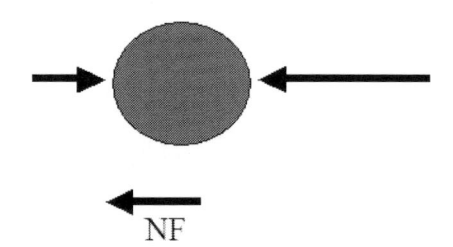

→ ● ←

← NF

Balanced forces
Balanced forces are when the net force *is* zero. No matter how many forces are pushing something, if they all cancel out the forces are balanced.

In the following example a book is sitting on a table, and the table pushing up cancels out the weight of the book pushing down. The book remains stationary, which is easy to understand.

EXAMPLE 2
WHAT IS THE NET FORCE ACTING ON A 10 N BOOK BEING HELD UP WITH A FORCE OF 10 N?

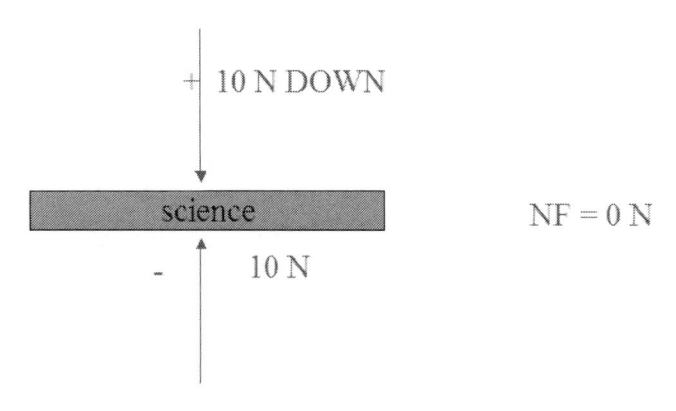

+ 10 N DOWN

science

− 10 N

NF = 0 N

But now for the part some of you will have trouble with. *Since in science stationary and constant velocity is the same thing,* the book could just as well be moving at a constant velocity (but not accelerating). In fact if you think about it, the book is moving at 1000 miles per hour to the east because of the Earth's rotation and we do not even notice.

68

BALANCED FORCES

- THE NET FORCE **IS** ZERO
- FORCES CANCEL EACH OTHER OUT
- THE OBJECT'S MOTION WILL NOT CHANGE
- STATIONARY OBJECTS
- OBJECTS MOVING AT A CONSTANT VELOCITY
- CAN NEVER ACCELERATE BUT THEY CAN MOVE AT A CONSTANT VELOCITY

Let's take this constant velocity idea one step farther. In a car moving at a *constant* 60 miles per hour, the engine is pushing the car, and if the driver took their foot off the gas, the car would slow down. We see this every day. What is really happening though is the force of the car's engine is canceling out the force of friction trying to slow it down. The forces are balanced with the engine on, resulting in a constant velocity, but when the driver took their foot off the gas, friction was stronger and decelerated the car.

THIS CAR HAS JUST STARTED ACCLERATING FROM A STOPPED POSITION, SINCE THE ENGINE IS STRONGER THAN THE AIR FRICTION THE CAR WILL INCREASE SPEED FORWARD

AIR RESISTANCE   ENGINE

Net force

When the drover hits the gas the engine accelerates, until it *can't accelerate anymore*. Once the car reaches its top speed, the engine is still pushing the car, but air resistance is trying to slow it down with the exact same force.

IS GOING AT A CONSTANT 60 MILES/HOUR,
IT IS NOT SPEEDING UP OR SLOWING DOWN,
IS IS MOVING BUT NOT ACCELERATING, THE
FORCES ARE BALANCED.

AIR RESISTANCE → ← ENGINE

Net force = 0

Now the driver will take his foot off the gas causing the force from the engine to be less. The forces are unbalanced and the car will decelerate.

THE CAR NOW HAS UNBALANCED FORCES
AND WILL DECELERATE FROM THE NET FORCE
PUSHING BACKWARDS.

AIR RESISTANCE → ← ENGINE

→
Net force

The same idea with a motor cycle: https://www.youtube.com/watch?v=bjxZ1JoPWl4

This is Sir Isaac Newton's explanation of net force on an airplane:
https://www.youtube.com/watch?v=sHLXK2APiwA

This is one of my demonstrations with a fan car that has a sail to cancel out the thrust, watch what happens. https://www.youtube.com/watch?v=pwIGcsitXBo

Here is an interesting video of a man (my cousin) named Dennis Rogers holding back a pair of motorcycles trying to rip him apart. The forces are balanced and he stays stationary. https://www.youtube.com/watch?v=blcgoUx2G-Y

COMPARE

BALANCED FORCES	UNBALANCED FORCES
NET FORCE IS ZERO	NF CAN BE ANYTHING **BUT** ZERO
STATIONARY OBJECTS	ARE ACCELERATING
MOVING AT A CONSTANT VELOCITY	
DOES NOT CHANGE MOTION	CHANGES AN OBJECTS MOTION
ACTS LIKE NO FORCE AT ALL	

Review

ALWAYS REMEMBER AND NEVER FORGET

- <u>FORCE</u> MAKES THINGS <u>ACCELERATE,</u> (NOT JUST MOVE)
- UNBALANCED = FORCE = N.F. <u>NOT</u> ZERO
- BALANCED = NO FORCE = NF = ZERO
- NO FORCE = STATIONARY, CONST. VEL.

A day in the life of Earwig Hickson III

I had to clean the garage today, mostly because of my sled construction destruction. It was a big job, and not just because of my mess. There were spiders in there! I looked at the mess and tried to come up with a way to get the job done with as little effort as possible. I had seen an old *Star Wars* movie the week before and figured it was worth a try. I realized that the "Force" was a non-contact force, kind of like gravity or magnetism but no matter what I did, it just did not work. I tried to use my mind to push away the spiders, but they just sat and stared at me, with their little beady spider eyes, all eight of them! The *force* was not with me. I had also watched an old Harry Potter movie, and tried some of the spells I learned from that, another non-contact kind of force. I was defiantly leaning toward the non-contact force way of doing things because that meant I did not have to touch the spider webs. The magic spells failed. I even tried to trick my friends with a method I learned in a book I read called *The Adventures of Tom Sawyer*, where he tricked his friends into paying him to do his chores. Failed again (never trust literature). I was getting worried, but then I realized the *Star Wars* idea was probably the best place to start. Not with "magic" since my teacher said there was no such thing, but with science, he said that was real. So I began my plan. I would use a contact force of some kind, but without touching the spiders. I would rewire the shop vacuum for extra power but not to suck up the spiders, but to blow them away. I replaced the puny little motor on the shop vacuum with a bigger one from the old washing machine from the ski lift project. I had built a hurricane wind generator. The spiders went flying, so far that my neighbors found them all over the walls of their house. The dust, dirt, sled parts, and assorted large metal pieces flew out of the garage like an explosion. The wind I made was a contact force, and it made everything it touched accelerate to such a degree that I never saw the pieces again, but other people did. My garage was spotless, even the workbench was gone, but the neighborhood looked awful. The trees were filled with assorted things that used to be in my garage. Lawns looked like junkyards. The force was with me, I had cleaned the garage!

My dad was happy until the calls from the neighbors came. I had to clean up the neighborhood, but with my "force generator," I just accelerated all the junk out of sight, and out of mind, at least until the next neighborhood complained.

Review of terms – quizlet: https://quizlet.com/124361149/chapter-5-volume-2-forces-flash-cards/?new

Fun things go Google
Sir Isaac Newton
Sky diving
Terminal velocity
Prism
Rainbow

Magnetic levitation
Static electricity

Links

Listen to a short video of Napoleon learning about friction: https://www.youtube.com/watch?v=llZ1Zl57r1Q

Napoleon explains force to Gandhi: https://www.youtube.com/watch?v=JOiorue6GO8

A cool video of indoor sky diving. The person's weight is cancelled out by the air blowing upward. https://www.youtube.com/watch?v=WHmmdW3szwk

The same idea with a motorcycle: https://www.youtube.com/watch?v=bjxZ1JoPWl4

This is Sir Isaac Newton's explanation of net force in an airplane: https://www.youtube.com/watch?v=sHLXK2APiwA

This is one of my demonstrations with a fan car that has a sail to cancel out the thrust, watch what happens. https://www.youtube.com/watch?v=pwIGcsitXBo

Here is an interesting video of a man (my cousin) named Dennis Rogers holding back a pair of motorcycles trying to rip him apart. The forces are balanced and he stays stationary. https://www.youtube.com/watch?v=blcgoUx2G-Y

A top levitating over a magnet https://www.youtube.com/watch?v=9zv7mBQaXPg

Magnetic levitation: https://www.youtube.com/watch?v=vA0_5BoVpDg

Using friction to start a fire: https://www.youtube.com/watch?v=APCgA9uxMGI

Bill Nye The Science Guy on Static Electricity: https://www.youtube.com/watch?v=Z-77IzaXGcg

Static Electricity demonstrations : https://www.youtube.com/watch?v=QxZ6AWLpnUw

9 Awesome Science Tricks Using Static Electricity!: https://www.youtube.com/watch?v=ViZNgU-Yt-Y

hair-raising static electricity Van-de-Graaff: https://www.youtube.com/watch?v=jZEFuCxD7BE

Van de Graff Generator : https://www.youtube.com/watch?v=3Ptu07enIsY

Floating Magnets !: https://www.youtube.com/watch?v=4T2XfrsvZRw

Chapter 24
Simple Machines and increasing force

Video of this power point lesson: https://www.youtube.com/watch?v=A08R-NXQxRY

What you need to remember
You should know that a force is a push or pull on matter. If you want to move (or accelerate) something to a new location, you need to use force. The less force you use, the easier it would be for you.

The wonders of science
They did it without bulldozers. They did it without power machines. They did it without steel. They did it without trucks. They did it 5000 years ago before any modern construction equipment. They did it by hand. They built the Great Pyramid of Giza along with two other great pyramids and The Great Sphinx, The Great Pyramid was made of rock, 5.9 million tons of rock, 2.3 million blocks of granite and limestone, the largest of these blocks weighed 80 tons! Not only were these blocks incredibly heavy and huge, they were moved to Egypt from up to 500 miles away! Not only that, these pyramids are not just piles of rocks. They have tunnels and rooms inside, with many secret tunnels recently discovered, and there may be undiscovered rooms left to be found.

The great pyramid is 481 feet tall (that is as tall as a 40 story sky scraper today). The base is 755 feet in each direction (that is more than two football fields), and they built it in less than 20 years. If this is not amazing enough, ancient people built pyramids all over the world; they built giant obelisks (think of the Washington Monument) from a single stone and lifted up into a vertical position. A short video of the Giza complex: https://www.youtube.com/watch?v=y-9q9RYa5CY

Even older are Stonehenge and Avebury, rings of rocks weighing up to 25 tons each and standing 30 feet tall, with other heavy rocks sitting on top! These rocks were quarried 20 miles away and moved to the monument! We do not even know for sure who built these monuments; they were so primitive they evidently had no written language. Truly unbelievable that such primitive people could build such structures, and many different cultures built similar structures all over the world. A video of Stonehenge featuring President Obama: https://www.youtube.com/watch?v=rB0un0AL5MA

How did they do it? How did they cut the stones with simple bronze and wooden hand tools? How did they transport such heavy rocks, so far, and after they transported them, how did they get them up on top of other rocks? Could we even do these things today with modern technology? Have you ever watched T.V. shows explaining that only aliens from another planet or a super advanced civilization could have made these? If so, I wonder why they used rocks?

The truth is much less exciting than some of the T.V. shows I have seen. You need to remember (as these shows seem to forget) that people 5000 years ago (or even 50,000 years ago) were not dumb; they were just as intelligent as we are. What they were was primitive, as far a technology is concerned. They did not have modern machines but they did have simple machines and creativity. The large rocks were cut with wooden wedges soaked in water so they would expand and break the rock into blocks. They were transported on rolling logs and sleds lubricated with water and sand; they probably used barges to float them long distances. Building dirt ramps and sliding the rocks up to the next layer stacked the blocks. A work force of 40,000 workers constructed the Giza pyramid complex. Not dumb peasants but skilled workers using levers, inclined planes, and wedges. They had *mechanical advantage* in their favor. There have been many modern experiments to test how this was done. There is no mystery, they are just amazing projects done by amazing people. Perhaps the science test next week does not sound so hard compared to this.

A theory about how the stones were cut:
https://www.youtube.com/watch?v=dGH93mt81BA

How the blocks were moved video:
https://www.youtube.com/watch?v=hRvtVrXcC3Y

A short animation of how the pyramids could have been constructed:
https://www.youtube.com/watch?v=nxP46jWYglI

Building a modern Stonehenge: https://www.youtube.com/watch?v=9yY7CVFTE1A

Raising an Obelisk: https://www.youtube.com/watch?v=BgekJnMeNiY

A man tries to recreate how Stonehenge was built:
https://www.youtube.com/watch?v=X4SkzBahfrQ

Simple machines are used to increase force

This chapter is about Simple machines and there are only six kinds. Their *only* function is to increase your force, in other words these machines make you stronger with no extra effort, and that is a very useful thing! Imagine a little machine that can make you 10 times stronger or 1000 times stronger! They do not help you to do *more work*, or to make you work *faster*, but they do make the work **easier**! That is always my goal, to do the same work with less effort. Simple machines are great!

Leonardo Da Vinci applies for a job with Napoleon:
https://www.youtube.com/watch?v=k50muw33Nlg

Work

To understand the power of simple machines we need to know exactly what *work* is. It is not what you think it is; it means to *move a mass a distance*. If you move a book 10 meters, you did work. If you sit and study your notes, you did not do any work. You have to move something a distance (in science) to do work. So if your teacher says, "get to work" you could move a book a few feet, and you did work! I think she meant something else though when she said that. So, work is moving a mass for a distance. It is not studying or reading, you have to *move something* (at least in science). When you are studying or writing the only work you actually do is to move your pen to write, turn a page, or pick your nose, if that is what you like to do, since you are moving a mass for a distance.

Let's imagine a giant snowstorm has deposited three feet of snow on my driveway. I order my kids to clear my driveway, since I am not about to do it. They go out and move the snow one handful at a time until the driveway is clear. I pay them one dollar. They did work; they moved a mass of snow off the driveway to somewhere else (a distance). They earned their money. The next snowstorm hits with three more feet of snow and I give them a shovel, they do the *same* amount of work, but do it with less effort and a lot

77

faster than when they used only their hands. It may have taken a week to clear the driveway with just their hands, but with the shovel, they could do it in a few hours. They did the *same* amount of work (mass of the snow times the distance they moved it) but it took less effort, or sweat and complaining. They still get a dollar, since they did the *same* amount of work, or got the same thing accomplished.

After shoveling the snow they need to study for their next science test. They sit there looking at their notes and do NO WORK; they moved no mass for a distance, they still complain, even though they are doing NO WORK. Kids.

What is work, a short video: https://www.youtube.com/watch?v=b5phRAbVdn0

WORK

- MOVING A MASS A DISTANCE
- WORK = MASS X DISTANCE
- IF YOU MOVE A BOOK, 10 M YOU DID WORK
- IF YOU STUDY YOU DID NOT
- EXCEPT FOR MOVING YOU PENCIL
- OR PICKING YOUR NOSE

Power

Now my kids are not dumb, they just have a mean father. They realized that using a shovel (a simple machine) made the job of clearing the driveway *easier*. They liked the shovels because they got the job done *faster*. Getting the *same job done faster is called power*. Power tools, like an electric drill or electric screwdriver, are not called power tools because they have motors; it is because they do the *same job faster*. **Power is how fast you do work**. You still do the *same* amount of work, but you do it faster. If I gave them a snowblower, they would get the *same* job done even faster, resulting in more power! I am still only paying them a dollar though. The same amount of work is worth the same pay, even if it was harder or easier.

POWER

- HOW FAST YOU DO WORK
- IF YOU SHOVEL A DRIVE WAY YOU MOVED SNOW AND DID WORK
- IF YOU USE A SNOWBLOWER, YOU DO THE SAME AMOUNT OF WORK BUT DO IT FASTER
- THIS MEANS YOU HAVE MORE POWER

LOW POWER WORK

WORK = MASS X DISTANCE

HIGH POWER – SAME AMOUNT OF WORK

WORK = MASS X DISTANCE

Each does the same amount of work; the snowblower just does it faster (more power).

Force can be magnified

Making you stronger than you really are is called <u>mechanical advantage</u>. Simple machines give you this. There are only six simple machines, but they all do the same thing, they increase your force without increasing your effort.

FORCE CAN BE MAGNIFIED

- WITH SIMPLE MACHINES (6) 
- 1. PULLEY – BLOCK AND TACKLE
- 2. LEVER – CROW BAR, HAMMER 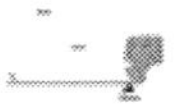
- 3. WHEEL AND AXLE – DOOR KNOB 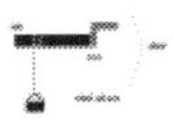
- 4. SCREW – CAR JACK 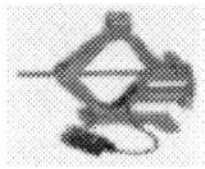
- 5. WEDGE - KNIFE 
- 6. INCLINED PLANE - RAMP

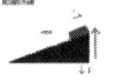

Mechanical advantage

Mechanical advantage is how many times stronger a simple machine makes you, or how many times your force is increased. You probably call it leverage. Nothing is free though. When you have a lot of mechanical advantage making your work easier, you have to sacrifice something, and what you sacrifice is distance. It is a kind of trade off, you get to move the easier force but you need to move it a lot farther. An easy example of this is with an inclined plane (a ramp). Your goal might be to lift a big rock into the back of a pickup truck, which is only three feet off the ground. Using a long ramp you still get the rock into the truck (with less effort) but the ramp may be 15 feet long, so the distance you move the rock is greater.

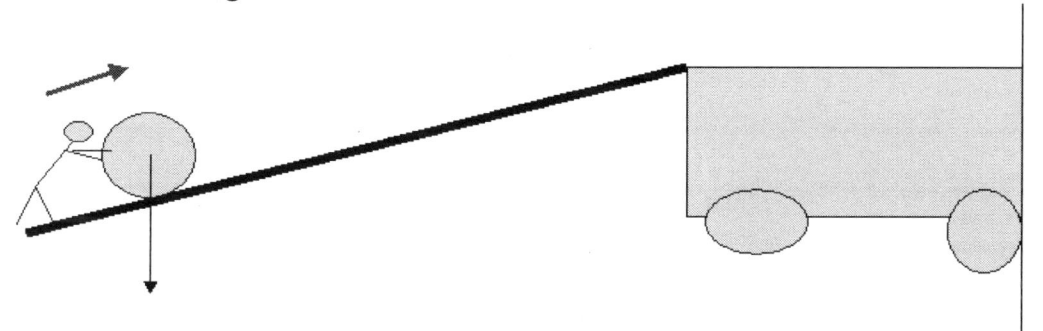

MECHANICAL ADVANTAGE
- HOW MANY TIMES STRONGER A MACHINE MAKES A FORCE
- LEVERAGE

Calculating mechanical advantage does have a formula, but since we already learned, how to use a formula, it should be no problem. Mechanical advantage is just a ratio between how much force (effort) you use compared to how much force you can lift (resistance force). So if you can lift 200 pounds with 100 pounds of force, your mechanical advantage is 2 (200 divided by 100).

TO FIND MECHANICAL ADVANTAGE

$$MA = \frac{\text{RESISTANCE FORCE}}{\text{EFFORT FORCE}}$$

$$MA = \frac{F_R}{F_E}$$

If you had a lever (or any other simple machine) and you could lift 100 Newtons with 5 Newtons of force, your mechanical advantage would be 20. You would be 20 times stronger than you really are!

Da Vinci explains mechanical advantage to Sacagawea:
https://www.youtube.com/watch?v=TjIUdLMf5wM

WHAT IS THE MA OF A LEVER THAT CAN LIFT 100 N WITH A FORCE OF 5 N?

$$MA = \frac{F_R}{F_E}$$

$$MA = \frac{100 \cancel{N}}{5 \cancel{N}}$$

$$MA = 20$$

The lever group of simple machines

Even though there are six kinds of simple machines, there are only two groups. The lever group includes levers, wheel and axle and pulleys. They are placed together because they are all really different shaped levers that act a little different from each other. A pulley is like a lever that goes in a circle (one side goes up as the other side goes down, just like a lever) and a wheel and axle is like a round lever (a handle).

Levers

A Lever is nothing more than a stick, a long stick. It is like a seesaw. One end of the lever is longer than the other (the easy side). There is a fixed point (a spot that will not move) called the fulcrum, or a pivot point. Levers are usually used to move a heavy object with a smaller effort force. Using a hammer to pull out a nail is a good example of this.

A short video about how levers work by TEDed:
https://www.youtube.com/watch?v=YlYEi0PgG1g

Levers can also be used kind of backwards to *lose* mechanical advantage on purpose. The reason for this is because of the distance sacrificed to gain mechanical advantage in the first place. A catapult takes advantage of this increased distance and uses it to launch something very far.
Check out this video of catapults. The Physics of "Pumpkin Chucking"
https://www.youtube.com/watch?v=sXuQvAPwcOE

Some more pumpkin Chucking: https://www.youtube.com/watch?v=qC6RJxFEMfY

How to use a lever, Red Green explains:
https://www.youtube.com/watch?v=020mtTI9pGA

LEVER

LONGER HANDLE INCREASES THE FORCE

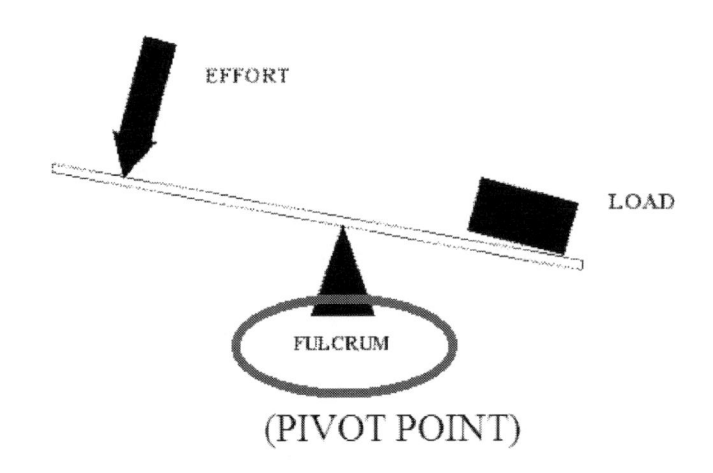

EFFORT

LOAD

FULCRUM

(PIVOT POINT)

With the proper lever, a mouse could lift an elephant!

F_E

lever

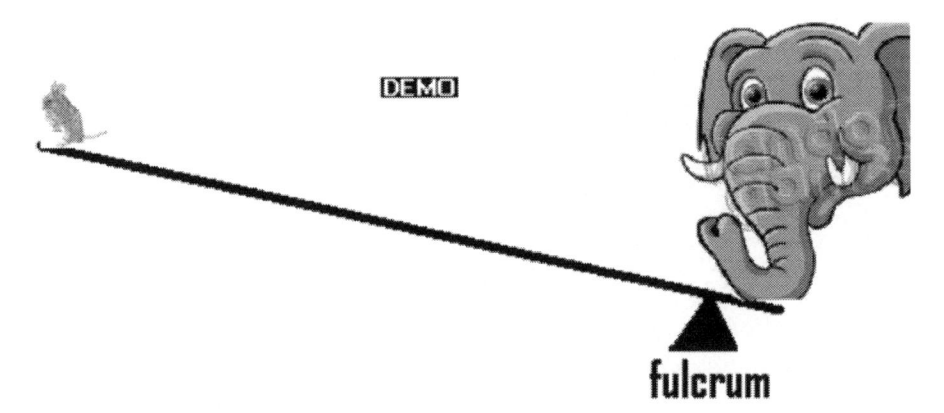

F_R

DEMO

fulcrum

The Lever, a Simple Machine, a video by Eureka:
https://www.youtube.com/watch?v=Us2KfO_yrPA

TYPES OF LEVERS

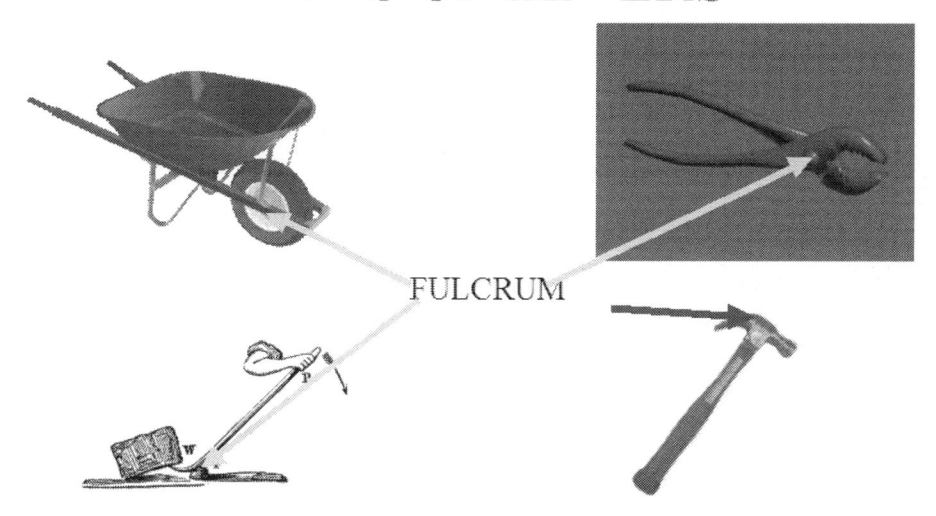

FULCRUM

Play with levers with this simulation by Phet:
https://phet.colorado.edu/en/simulation/balancing-act

Wheel and Axle

A Wheel and axle is one solid piece. The axle is attached to the wheel so that when one turns, so does the other. A wheel and axle is a handle, and has almost nothing to do with a *wheel*.

WHEELS AND AXLES

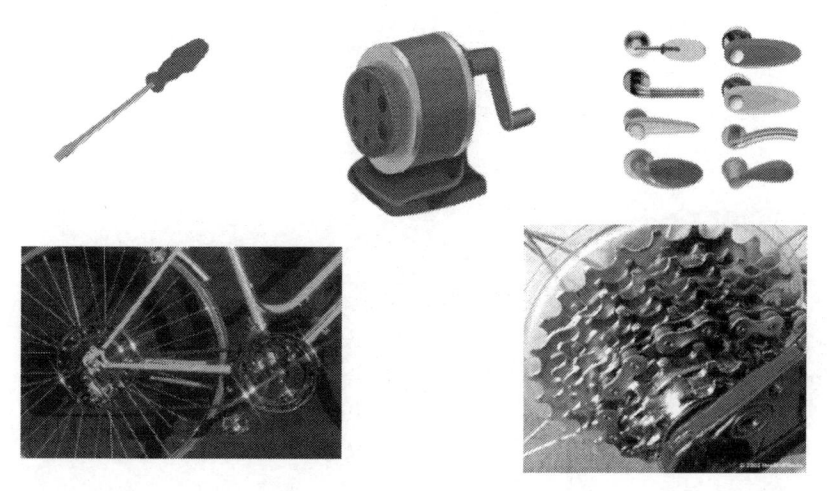

A wheel is just a wheel; it spins **ON** an axle to reduce friction, and is not the same as a wheel AND axle. A wheel is just a wheel.

Newton explains a wheel and axle to Napoleon:
https://www.youtube.com/watch?v=7QGJqFcWuwI

> I WANT TO BE A WHEEL AND AXLE TOO

> YOU HAVE TO BE ATTACHED TO AN AXLE

A wheel and axle is a *handle*. When you turn the handle it is attached to an axle that moves also. Think of a crank like on a wishing well, called a *windlass*.

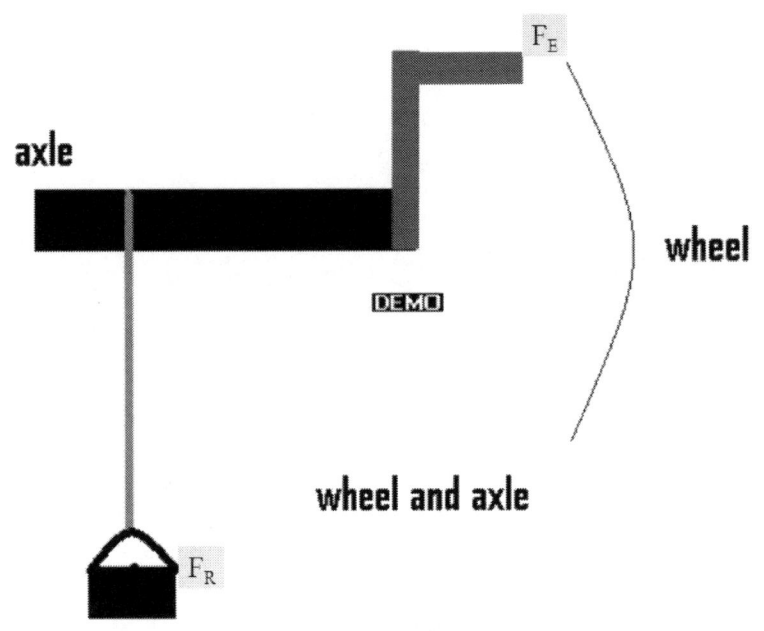

axle — F_E

wheel

wheel and axle

F_R

The handle turns in a big circle, and someone somewhere thought this looked like a wheel shape, so they gave it that name. The bigger the "wheel" diameter, the greater the mechanical advantage becomes. This is why big handles are easier to turn than small handles.

Wheels and axles include gears on the back of your bike, which are attached to the wheel, so the back tire of a bike is a wheel and axle but the front tire is just a wheel. When you pedal a bike the chain turns the gears, which turn the back tire, giving you mechanical advantage. The front tire just spins.

A screwdriver is also a wheel and axle because the larger diameter of the handle makes it easy to turn a screw; the shaft of the screwdriver is the axle.

A simple wheel and axle demonstration:
https://www.youtube.com/watch?v=nzXdKRFxU7U

WHEEL AND AXLE

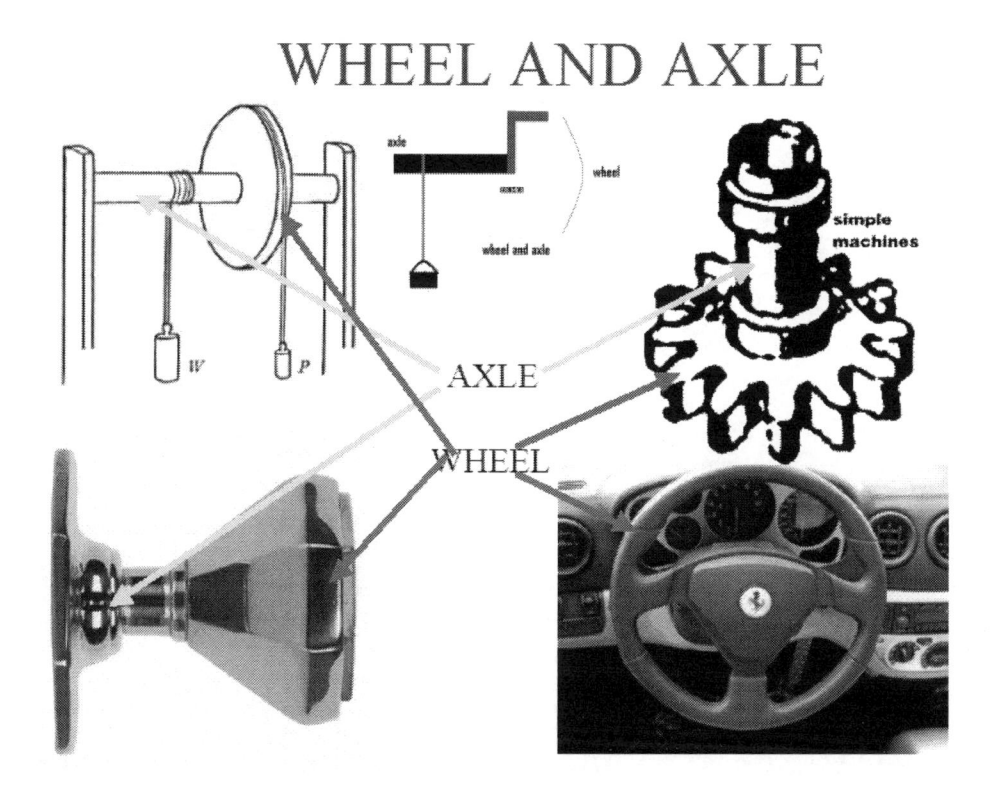

Pulleys

There are two kinds of Pulleys, stationary and moveable. Often times the two types work together to form a block and tackle or a pulley system.

Stationary pulley

This is a pulley that does not move when you use it. It still spins but it does not move up or down, it just stays put. The pulley at the top of a flagpole to help raise the flag is a stationary pulley. It does *not increase force*; its job is to *change the direction of the force* (or rope). Can you imagine having to climb a flagpole every morning to put up the flag? It might be fun the first time, but it will get old quick. A stationary pulley lets you raise the flag from the ground.

STATIONARY PULLEY

THIS DISTANCE DOES NOT CHANGE

NO MECHANICAL ADVANTAGE

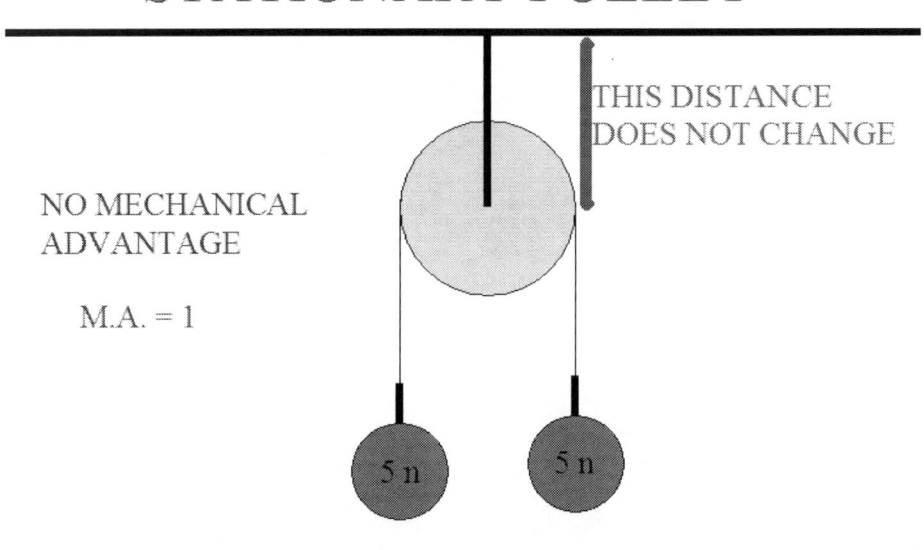

M.A. = 1

5 n   5 n

A *moveable pulley* does actually move, it goes up or down. It also increases your force by a factor of two; a moveable pulley makes you two times stronger.

MOVABLE PULLEY

Stationary pulley

THIS DISTANCE DOES CHANGE

MECHANICAL ADVANTAGE IS 2 (TIMES STRONGER)

M.A. = 2

5 n   10 n

F_E F_R

Notice how the moveable pulley goes up in the next picture.

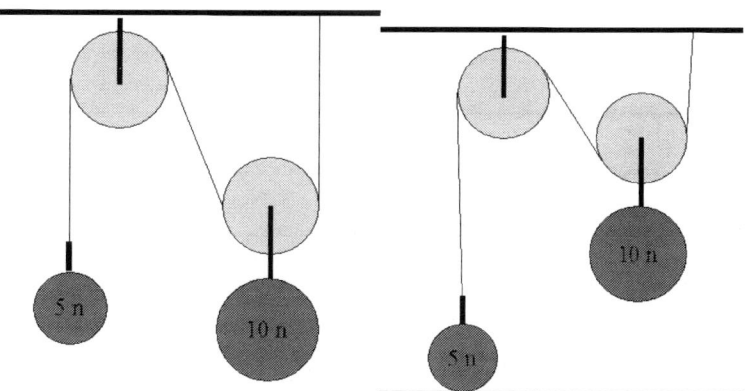

A short video showing stationary and moveable pulleys:
https://www.youtube.com/watch?v=ju7Wciv4RXk

When pulleys work in a system it is called a block and tackle and **each** moveable pulley has a mechanical advantage of two.

A short video about how pulleys work:
https://www.youtube.com/watch?v=9T7tGosXM58

4 PULLEYS

M.A. = 2 x 4 = 8

4 PULLEYS

1.25
F_R

10 n
F_E

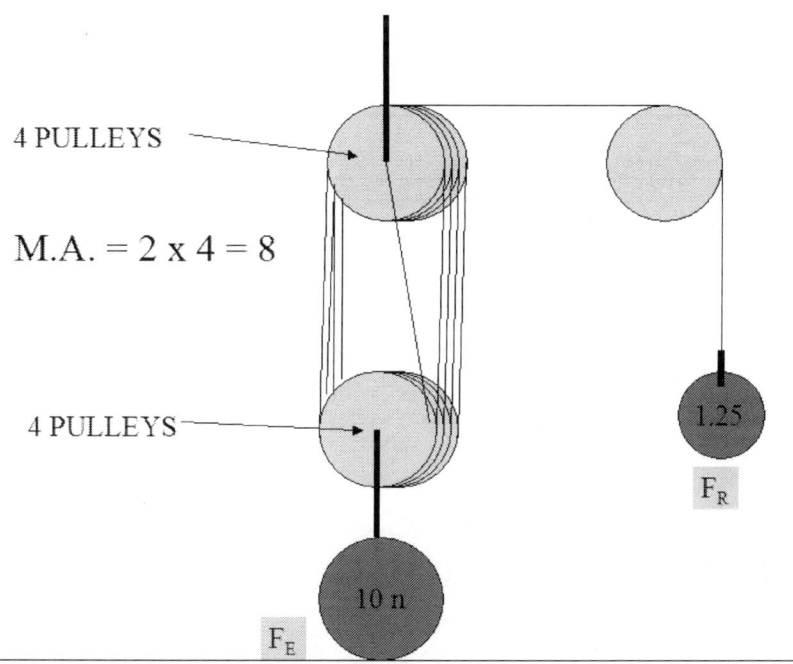

A block and tackle used to lift a bowling ball:
https://www.youtube.com/watch?v=IquTgpr7nvc

Notice in the following pictures how a block and tackle works to pull Einstein toward me.

89

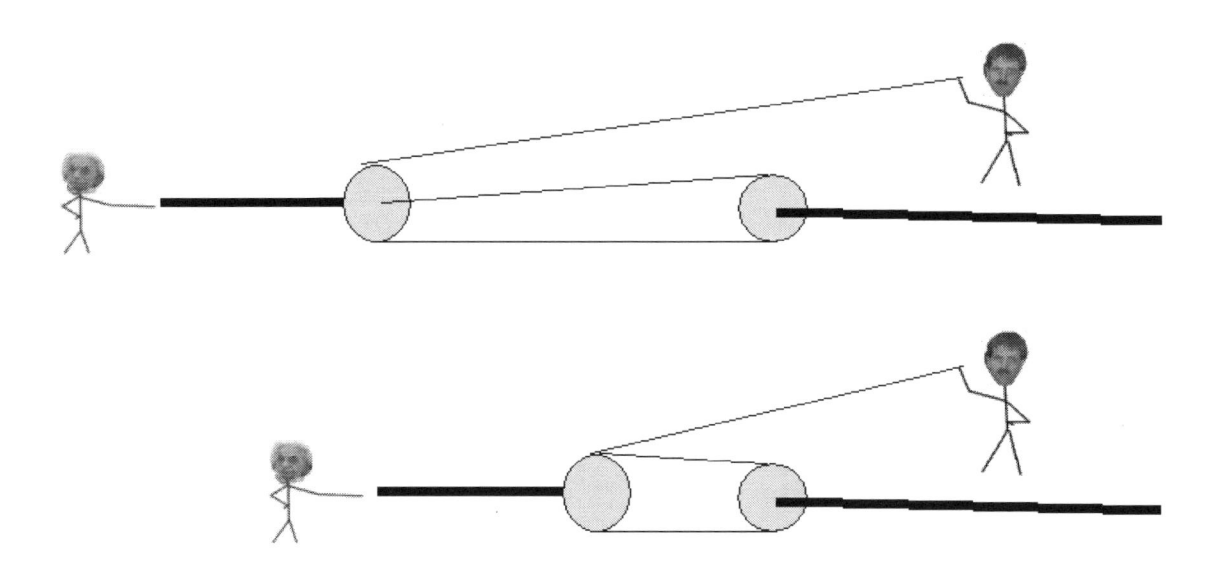

Tug of war using pulleys: https://www.youtube.com/watch?v=ELoqD2K9gs8
Using a pulley to make an arm wrestling machine: https://www.youtube.com/watch?v=4RSeMExrN4c

Inclined Plane group

This brings us to the second group of simple machines, the incline plane group. These all look like triangles (a pointy end); they are used differently and include the incline plane, the wedge and the screw.

Inclined plane

INCLINED PLANE

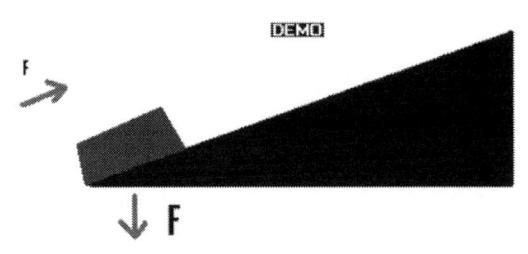

The Inclined plane is a ramp. It is a sloped object used to gain height. A common use would be a ramp to push something up into a truck, but it could also be a set of stairs, or even a sloped pile of dirt. The job of an inclined plane is to make it easier to gain *height*. Stairs are an inclined plane with steps; it is kind of hard to get to the upper floor in my school without a flight of stairs.

The important thing about an inclined plane is the longer the ramp (more gentle the slope) the easier it is to use. A steep ramp makes you push the object harder. Long ramps have more mechanical advantage.

90

HARDER – MA = 2

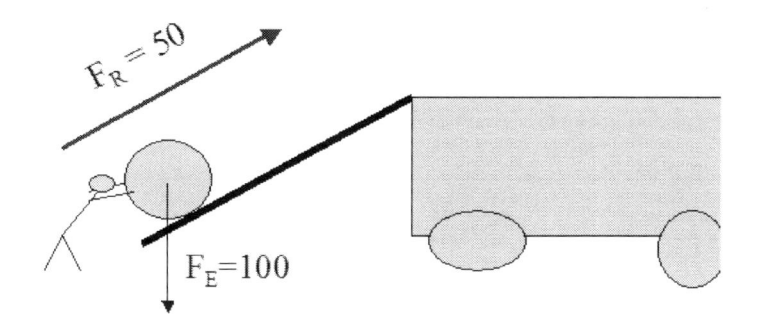

$F_R = 50$
$F_E = 100$

EASY – MA = 4

THE LONGER THE RAMP THE MORE MA.

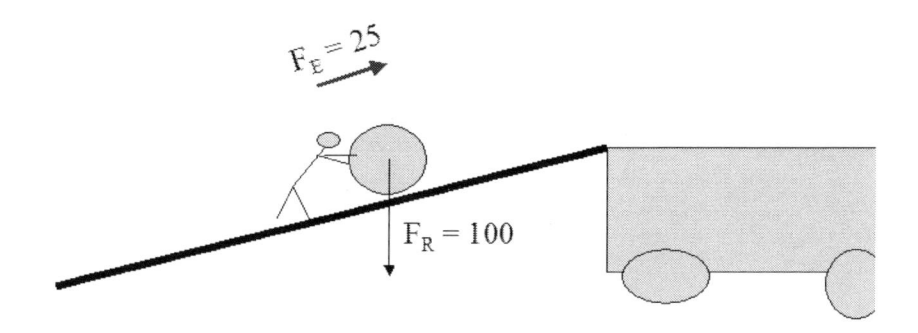

$F_E = 25$
$F_R = 100$

Screw

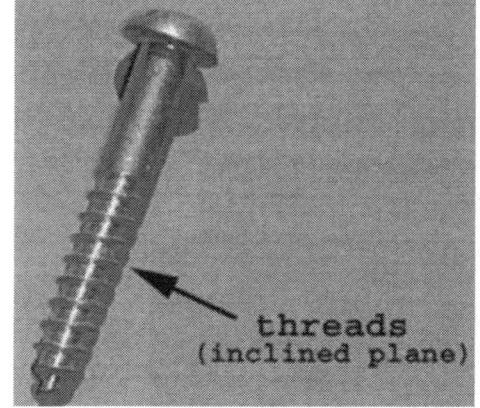

threads (inclined plane)

A Screw is really an incline plane wrapped around a cylinder. This is like a spiral staircase; you still go up a gradual slope but you spiral your way up. A screw for holding wood together is a screw. Did you ever wonder how many threads there are on a screw? Well, the answer is one. The one thread spirals up the shaft just like a spiral staircase. In the case of a wood screw, as you turn it, it goes down into the wood. Screws can also be used to make things go up, like a carjack.

Now the closer the treads of a screw are to each other, the more gentle the slope, and the easier the screw is to turn, which means more mechanical advantage.

SCREW
CLOSER THREADS = MORE M.A.

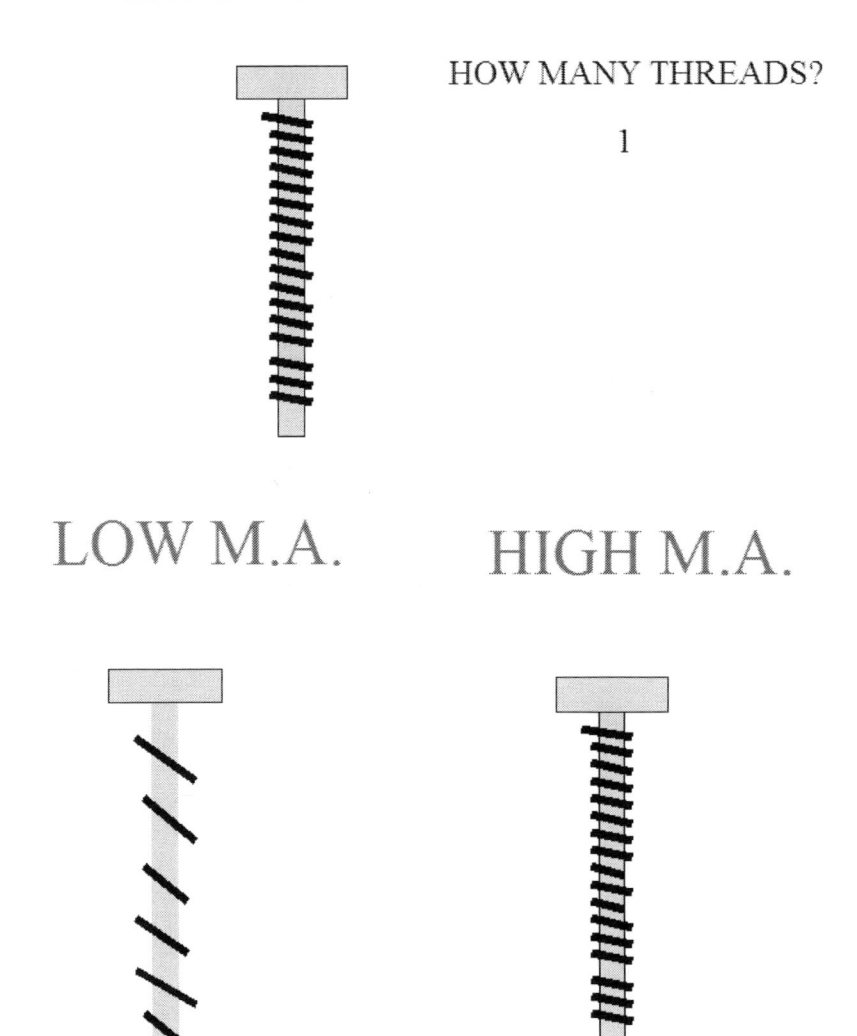

HOW MANY THREADS?

1

LOW M.A.   HIGH M.A.

Newton explains a screw to Napoleon:
https://www.youtube.com/watch?v=UL_YAQd7t_w

Wedge

A <u>Wedge</u> is an inclined plane (at least the shape) that moves. Anything with a point or a blade that goes through a material and pushes it apart is a wedge. An axe is a wedge. When it goes into a piece of wood, it pushes it apart sideways.

Wedges can be used to lift things up too. When a wedge is placed under a door it pushes it up and "wedges" it, so it will not open.

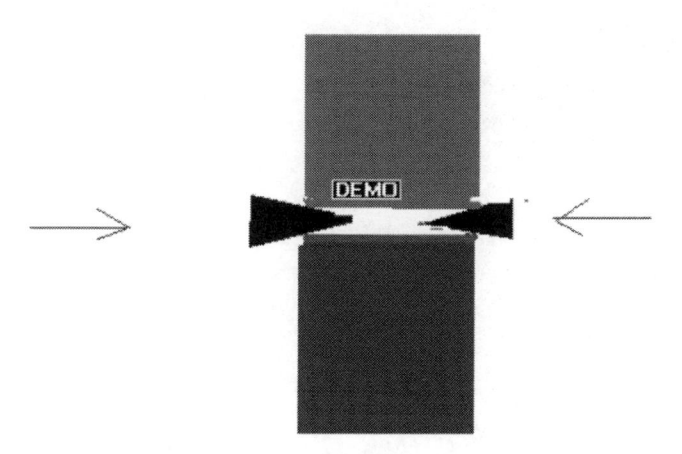

WEDGE

Newton explains a wedge to Napoleon: https://www.youtube.com/watch?v=-SuECgnhoLw

A day in the life of Earwig Hickson III

It was meant to be just a snowman. A simple snowman. That is how it started. We had a snow day! A whole day of no school and the best creative art medium ever invented – snow, and a lot of it. I found a nice field in the local playground and started rolling a big snowball for my little snowman. I soon got to the point where I could not push it anymore. "Here," I declared, "is where my snowman will stand." Some kids were watching me and came over and began pushing the big ball farther, making it bigger, but soon we could push no farther. "Here," I declared, "is where the snowman will stand!" A pair of old men were watching us and came over and after discussing something among themselves one left and came back with a long board. A lever he called it and he placed it under the giant snowball while we kids lifted it up pushing the snowball forward, farther, and larger. We pushed it to the edge of the hill that went down quite far. Over the side it went rolling and growing to the bottom. It came to rest. "Here," I declared, "is where the snowman will stand." And it did. The bottom of the snowman was 15 feet high, taller than any three kids. We began rolling the snowman's middle. Using the lever and about 20 kids who had joined us we made the giant middle. A ramp was built, an inclined plane to the top of the snowman bottom, of snow, ice and wood planks. Twenty kids with levers pushed the middle ball up the ramp and onto the snowman. He stood 25 feet tall! The head came next, up the new and improved inclined plane it went. The snowman stood 30 feet tall! Taller than a three-story house, he was a giant. Trashcan lids were added for his buttons and eyes. For the hat we used a 55 gallon drum, the nose was an orange pool noodle, a broom was made from a telephone pole, the mouth was seven mailboxes painted red. He was awesome. At the last minute we decided to decorate him to look like an alien. He was sprayed with green food dye, and two antennae made of golf clubs were added. His eyes were slanted like giant almonds, and his ears were pointed. We decorated him with weird symbols. We added a giant crop circle (a snow circle) in the field beside him. We made big 11-toed footprints, from the circle to the snowman. I poured the juice from a glow stick into each footprint. We removed all traces of the ramp and went home exhausted, for some sleep.

The next morning was like a circus. All the T.V. stations were there. "Experts" were called in from the local university, the FBI, CIA, the Army, Air Force, Homeland Security, and the History Channel. It was considered a mystery. The 8th wonder of the world, the local T.V. stations investigated. Who built this they all asked, why did they build it? How did they build it? Was it a message from an advanced alien race? The area was posted *no trespassing, US Government*, and an official investigation was conducted. It made all the papers. For the next few months, we kids were scared to death and silent. Finally spring came and the giant snowman alien melted into mush. Samples were carted off, never to be seen again. The investigators left and my town was finally left alone.

Somebody opened a restaurant called Alien Snowman and a museum was established. People come from all over the world to see it. We are a tourist attraction, just like Roswell NM. I did my part for the local economy. Sometimes I see my snowman on Cable T.V. shows. They say humans could not have built it, I just smiled.

Flashcards for this chapter on Quizlet: https://quizlet.com/128306688/chapter-6-volume-2-simple-machines-flash-cards/ and https://quizlet.com/126736481/chapter-24b-more-about-simple-machines-flash-cards/

Links

A theory about how the stones were cut:
https://www.youtube.com/watch?v=dGH93mt81BA

How the blocks were moved video: https://www.youtube.com/watch?v=hRvtVrXcC3Y

A short animation of how the pyramids could have been constructed:
https://www.youtube.com/watch?v=nxP46jWYglI

Building a modern Stonehenge: https://www.youtube.com/watch?v=9yY7CVFTE1A

Raising an Obelisk: https://www.youtube.com/watch?v=BgekJnMeNiY

A man tries to recreate how Stonehenge was built:
https://www.youtube.com/watch?v=X4SkzBahfrQ

A short video of the Giza complex: https://www.youtube.com/watch?v=y-9q9RYa5CY

A video of Stonehenge featuring President Obama:
https://www.youtube.com/watch?v=rB0un0AL5MA

Leonardo Da Vinci applies for a job with Napoleon:
https://www.youtube.com/watch?v=k50muw33Nlg

What is work, a short video: https://www.youtube.com/watch?v=b5phRAbVdn0

A short video about how levers work by TEDed:
https://www.youtube.com/watch?v=YlYEi0PgG1g

Da Vinci explains mechanical advantage to Sacagawea (who only speaks French):
https://www.youtube.com/watch?v=TjIUdLMf5wM

Check out this video of catapults. The Physics of "Pumpkin Chucking"
https://www.youtube.com/watch?v=sXuQvAPwcOE

Some more pumpkin Chucking: https://www.youtube.com/watch?v=qC6RJxFEMfY

How to use a lever, Red Green explains:
https://www.youtube.com/watch?v=020mtTI9pGA

The Lever, a Simple Machine, a video by Eureka:
https://www.youtube.com/watch?v=Us2KfO_yrPA

Play with levers with this simulation by Phet:
https://phet.colorado.edu/en/simulation/balancing-act

Newton explains a wheel and axle to Napoleon:
https://www.youtube.com/watch?v=7QGJqFcWuwI

Newton explains a screw to Napoleon:
https://www.youtube.com/watch?v=UL_YAQd7t_w

A short video about how pulleys work:
https://www.youtube.com/watch?v=9T7tGosXM58

Newton explains a wedge to Napoleon: https://www.youtube.com/watch?v=-SuECgnhoLw

A simple wheel and axle demonstration:
https://www.youtube.com/watch?v=nzXdKRFxU7U

A short video showing stationary and moveable pulleys:
https://www.youtube.com/watch?v=ju7Wciv4RXk

A block and tackle used to lift a bowling ball:
https://www.youtube.com/watch?v=IquTgpr7nvc

Tug of war using pulleys: https://www.youtube.com/watch?v=ELoqD2K9gs8

Using a pulley to make an arm wrestling machine:
https://www.youtube.com/watch?v=4RSeMExrN4c

Fun things to Google
    How were the pyramids made?
    How was Stonehenge made?
    Eureka simple machines?
    The great pyramid of Giza
    Archimedes simple machines
    Pumpkin Chucking
    Medieval catapults

Chapter 25
Newton's First law of motion

An introduction video of the first law of motion (funny):
https://www.youtube.com/watch?v=Bg34At4-2K4

Video of this lesson's power point:
https://www.youtube.com/watch?v=Xyuds1h4Dv4

The wonders of science
It was the age of knights. They wore heavy suits of iron armor. Not just for protection but to stay on their horses when someone hit them. They were the tanks of the 15th century, before gunpowder and rifles made them obsolete. A knight wore heavy armor of iron and tried to knock other knights off their horse with a long lance or spear. The armor was not so much for protection (although it did offer a lot), but for stability. A knight was fearsome while on a horse, but useless if he fell off. The heavy armor made him unable to move so he was helpless on the ground. The goal was to knock him off the horse. But a lot of iron armor made him hard to knock off. A large mass of iron had a lot of inertia, making it hard to change his position (on the horse to off the horse); the knight was the tank of the middle ages. Armies, who had no knights, won few battles. The heavier the knights armor, the more successful the knight, the more successful the knight, the better the chance of winning the battle. Mass is everything in science and warfare. Mass means inertia, and inertia is the resistance to changes in motion.
Modern jousting for fun! https://www.youtube.com/watch?v=tWVZgp-eQG8

What you need to remember
Remember back in chapter 19, when I said *stationary objects and object moving at a constant velocity* are the same? That is important now. I also said that a *force causes objects to accelerate*, and not just move, in chapter 23. That is important too. The other thing to remember is about how unbalanced forces change an object's motion, it makes them accelerate.
This is a good time to review forces with a cool simulation by Phet about forces and motion: https://phet.colorado.edu/en/simulation/forces-and-motion-basics

Newton's First law of motion
Newton introduces himself: https://www.youtube.com/watch?v=UYk0tQrstrc
Newton himself explains the first law of motion:
https://www.youtube.com/watch?v=yfeiD1nq-v4
Newton named everything after the smartest person he knew, which was himself. Other people call this **The Law of Inertia**. Inertia comes from a Latin word meaning *idle or lazy*, so if someone tells you, you have a lot of *inertia*, it is not a compliment. If you tell your mom, she has a lot of inertia, make sure she does not know Latin first. In science, inertia means *to keep doing whatever you are doing*. If an object is stationary, that is what it wants to do, if it is moving at a constant velocity, that is what it wants to do. Objects in science are kind of stubborn; *they do not like to change their motion*. In fact, the only way to change an object's motion is with a force. A force causes things to

accelerate, which is not stationary or a constant velocity. If there is no force applied, all objects will continue doing whatever it is they are doing. It is easy to imagine a stationary object, like a rock, staying stationary forever, until some force comes along to accelerate it. It is much harder to imagine an object moving at a constant velocity, continuing to move at a constant velocity forever. We just do not see this. Friction (which we forget about) is a force that slows down moving objects all the time. But, what about the planets? Why don't they slow down and stop? The answer is, because, since in the vacuum of space there is very very very little friction, the planets keep doing whatever it is they were doing, which is to move at a constant velocity. So we see them just keep on moving like they were before. This is because they have inertia. And since they have **A LOT** of mass, they have a **lot of inertia**. Inertia is much related to mass. Heavy things keep doing what they are doing a lot more than light things do.

I have a bowling ball that has a mass of 6 Kg, a bit over 12 pounds. It **REALY** wants to keep doing whatever it is it is doing. It is hard to move from a stationary position, but it is hard to stop once it is moving. I also have a Styrofoam rock, it has very little mass and is *easy to accelerate*, and *just as easy to decelerate*. Not much inertia in the foam rock, since there is not much mass.

Newton's first law of motion has three parts:

NEWTON'S FIRST LAW OF MOTION (LAW OF INERTIA)

- AN OBJECT AT REST WILL REMAIN AT REST
- AND AN OBJECT IN MOTION WILL STAY IN MOTION AT A **CONSTANT** VELOCITY
- UNLESS ACTED UPON BY AN OUTSIDE **UNBALANCED** FORCE (FRICTION OR OTHER)

Or more simply, this is what you must memorize:

IN ENGLISH

- OBJECTS AT REST ……….. STAY
- OBJECTS IN MOTION ……. STAY
- UNLESS FORCE ACTS

In other words, *objects at rest (stationary) wish to remain at rest*, and *objects moving at a constant velocity (with balanced forces) want to remain moving at a constant velocity* (with balanced forces) *unless an outside force (any force) acts on them.* Remember, a force changes motion and causes an object to accelerate.

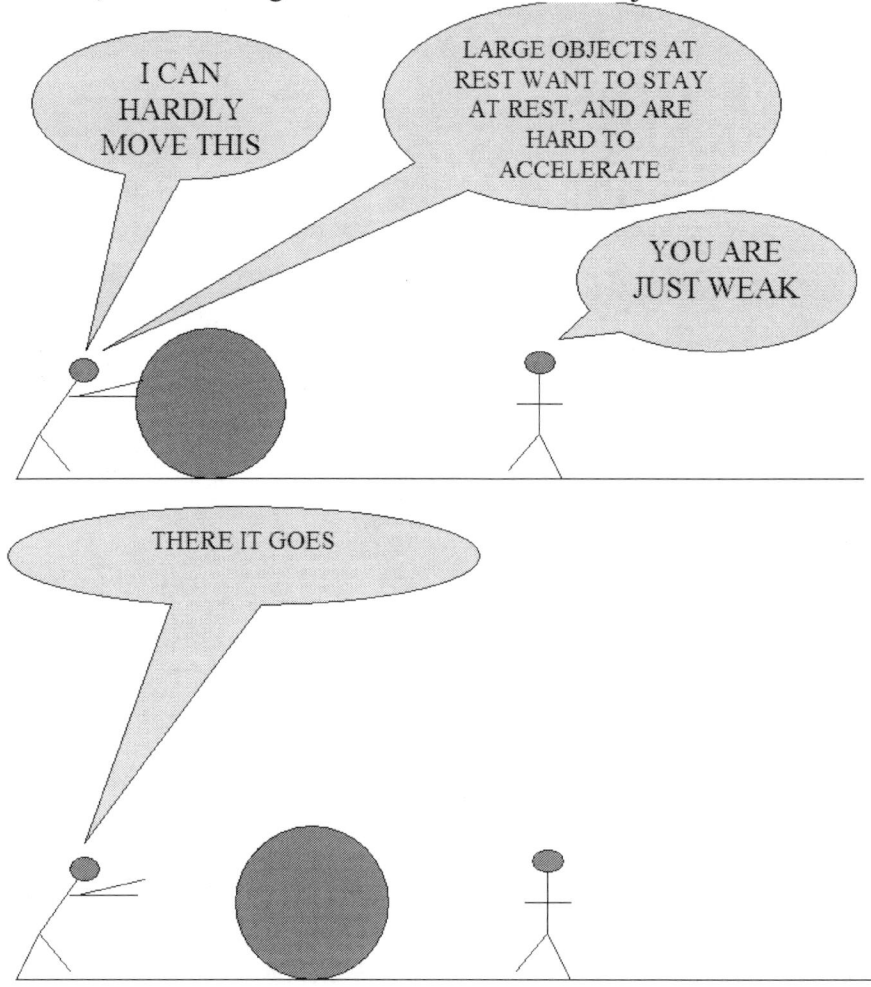

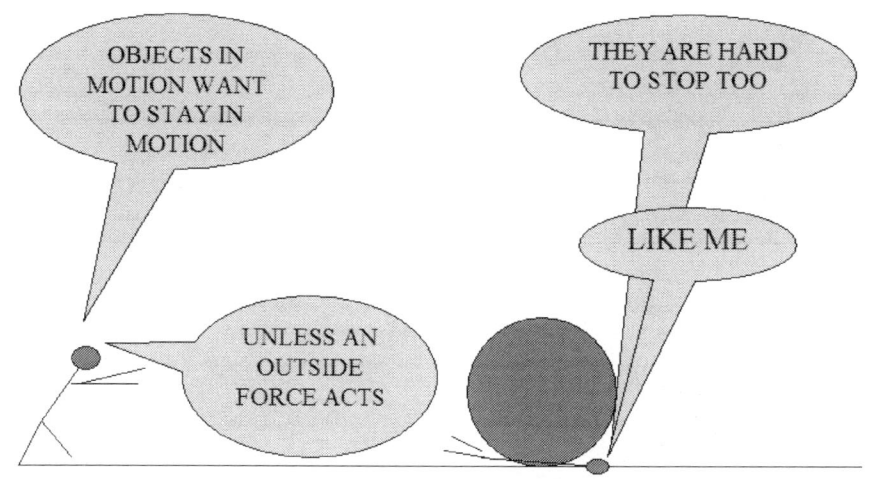

[Speech bubbles around illustration:]
- OBJECTS IN MOTION WANT TO STAY IN MOTION
- THEY ARE HARD TO STOP TOO
- LIKE ME
- UNLESS AN OUTSIDE FORCE ACTS

What does a helium balloon do in a moving car? It is cool and unexpected: https://www.youtube.com/watch?v=xHTSTuwVF1I

Something to think about – here is a video of crash test dummies NOT using a seat belt (or an air bag): https://www.youtube.com/watch?v=PY_-A9-KNNA

Inertia

So if this is the law of Inertia, what is inertia? Well inertia is simply, how much an object wants to do *whatever it is doing*. It is based on mass. The more mass, the more inertia, the more it wants to resist changes in motion. Big massive things resist changes in motion, more than little tiny, less massive things. The more mass an object has, the more inertia it has.

Tablecloth trick using Newton's law of Inertia: https://www.youtube.com/watch?v=FTxm_od9kVQ

Using the inertia of air to break a wooden board: https://www.youtube.com/watch?v=sOe2Pk_FjEQ

INERTIA

- THE TENDENCY FOR AN OBJECT TO RESIST CHANGES IN MOTION
- A STATIONARY ROCK IS HARD TO MOVE
- AND IT IS HARD TO STOP ONCE IT IS MOVING

THE MORE MASS AN OBJECT HAS
THE MORE INERTIA IT HAS
INERTIA IS EQUIVALENT TO MASS

A short video explaining the fist law of motion:
https://www.youtube.com/watch?v=OHw80HXSuAQ
This is a great video from
Jules Sumner Miller (a great teacher) of some demonstrations about the first law of motion https://www.youtube.com/watch?v=Za3DGUEpW2U
Remember Hero's engine? Here it is used to accelerate an object:
https://www.youtube.com/watch?v=pxWHWOYVov4
How do you stay safe in a car crash?
https://www.youtube.com/watch?v=8zsE3mpZ6Hw

Review
The first law of motion says three things. Objects at rest stay at rest, objects in motion stay in motion, unless an outside force is applied. Inertia is the tendency for objects to resist changes in motion and is proportional to the objects mass.

Flash cards on this chapter on quizlet: https://quizlet.com/128305265/chapter-7-volume-two-newtons-first-law-of-motion-flash-cards/

A day in the life of Earwig Hickson III
It did not seem that bad. My science teachers asked for a *brave volunteer*, someone who is *expendable*. I did not know what expendable meant, but I found out later, it meant someone who no one cares about if something bad happened to him or her. I was the expendable volunteer! The first thing my teacher did was to bring out a big board full of hundreds of sharp nails. He called it the bed of nails and made me lie down on it. I was scared but he said it was all right because he was a science teacher and knew everything. So I lay on the bed of nails! It was actually quite comfortable because my weight was distributed evenly over all those nails, no one nail could 'get" me. In fact there was only a fraction of a pound on each nail. I could have slept on the bed of nails, especially since I was still sleepy from math class. I thought my torture was done but then he placed a 50-pound barbell weight on my chest! I was getting worried but when he put it on me, I actually still felt comfortable; it made no difference at all! But then he got mean; he took out a big hammer! He raised the hammer as if he were going to hit the weight and drive

me into the bed of nails. I begged for forgiveness for whatever crime I had committed. I promised to do my homework, I promised to stop drawing cartoons of him with devil horns. Nothing worked; he swung the hammer onto the weight on my chest. I braced for the worst. Nothing happened!? I felt nothing. The hammer did not hurt, the nails did not hurt, and I was fine.

But how I wondered, did I not feel pain? My teacher said that since there were so many nails, there was hardly any force on any one nail. The mass of the 50-pound weight wanted to stay still because of inertia; the little hammer (which was in motion and wanted to smash me) did not have enough mass, or inertia, to overcome the inertia of the giant weight. The 50-pound weight was at rest and that is what it wanted to do. The weight saved me! It was actually rather impressive. He thanked me for promising to do my homework and stop the silly cartoons, though I still draw the cartoons.

A short video of me on the bed of nails getting whacked with a hammer: https://www.youtube.com/watch?v=HrUFHn0Ygec

Links

A video of the first law of motion (funny): https://www.youtube.com/watch?v=Bg34At4-2K4

Bed of nails demonstration: https://www.youtube.com/watch?v=HrUFHn0Ygec

Phet forces and motion: https://phet.colorado.edu/en/simulation/forces-and-motion-basics

A documentary about Sir Isaac Newton by NOVA.
https://www.youtube.com/watch?v=P2UF0tX-45M

Newton introduces himself: https://www.youtube.com/watch?v=UYk0tQrstrc

Newton himself explains the first law of motion:
https://www.youtube.com/watch?v=yfeiD1nq-v4

What does a helium balloon do in a moving car? It is cool and unexpected:
https://www.youtube.com/watch?v=xHTSTuwVF1I

Something to think about – here is a video of crash test dummies NOT using a seat belt (or an air bag): https://www.youtube.com/watch?v=PY_-A9-KNNA

A short video explaining the fist law of motion:
https://www.youtube.com/watch?v=OHw80HXSuAQ

This is a great video from Jules Sumner Miller of some demonstrations about the first law of motion https://www.youtube.com/watch?v=Za3DGUEpW2U

Remember hero's engine? Here it is used to accelerate explain how a rocket gets into space: https://www.youtube.com/watch?v=pxWHWOYVov4

How do you stay safe in a car accident?
https://www.youtube.com/watch?v=8zsE3mpZ6Hw

A cool inertia experiment you can do at home:
https://www.youtube.com/watch?v=6gzCeXDhUAA

Newton's First Law Inertia - Science Theater 14:
https://www.youtube.com/watch?v=iiUzRKW_4lw

Using Newton's first law of motion for a neat magic trick:
https://www.youtube.com/watch?v=S_mhVq9iy4k

Tablecloth trick using Newton's law of Inertia:
https://www.youtube.com/watch?v=FTxm_od9kVQ

Using the inertia of air to break a wooden board:
https://www.youtube.com/watch?v=sOe2Pk_FjEQ

The bed of nails explained: https://www.youtube.com/watch?v=cM3lOjR5n-I

Newton's First Law of Motion - Science of NFL Football, for you football fans:
https://www.youtube.com/watch?v=08BFCZJDn9w

Fun things to Google
Inertia
First law of motion
Law of inertia
Bed of nails

Medieval knights
Jousting
Crash test dummies

Chapter 26
Newton's second law of motion

Video of this chapter's power point lesson: **https://www.youtube.com/watch?v=-mbbt7j_krM**

The wonders of science
It was 1962. President Kennedy gave a speech at Rice University, forever to be known as the We choose to go to the Moon, speech, in which he challenged the Soviet Union to a "moon race". It was a way to challenge our enemy in a non-violent way (at this time the cold war was active and the Soviet Union (Russia) was our enemy). We would land a human on the moon and return him safely to Earth. The U.S.A would do this first, before the Russians, we would show our scientific superiority to the world. It was a tall order for the United States Space agency (NASA), how could they do this? The Russians were far ahead of us in space technology. They were winning; it was not even close at that time.

NASA went back to Newton's Laws of Motion, specifically the second law, The Law of Acceleration, and this would fill the bill nicely.

The moon is ¼ of a million miles away. How do you get a spacecraft to the moon and back again? And land on the moon too? How do you carry enough fuel? No one knew. Now in space there is almost no friction. This is because there is no air. So as long as your engine is on, you will accelerate. Then when you are going fast enough, turn off the engine and you will continue at a constant velocity, with no friction to slow you down. This is what NASA did. They accelerated the spacecraft up to speed, and let it 'coast' to the moon. The craft orbited the moon, a smaller spaceship separated and landed on the moon, then returned to the mother ship and everyone came back home. To this day America is the only country to land men on the moon. Quite an achievement.

What you need to remember
Remember back in chapter 23, where I made a big deal about how force causes a mass to *accelerate* (and not just move)? I also made a big deal about how *constant velocity* had *balanced forces* and did not slow down or speed up? I even said that objects moving at a *constant velocity and objects that were stationary were kind of the same thing*? Now is when these things are important.

Newton's second law of motion
Newton's second law of motion, also called The Law of Acceleration, is easy. You already know it, but do not know that yet. I used to say '*A force causes an **object** to accelerate.*' I am now going to slightly change that to, 'A force causes a **mass** to accelerate.' The same thing, but that is what Sir Isaac discovered.

Objects either wish to be *stationary* or moving at a *constant velocity*, the only way to change that motion is with a *force*. Since a force *changes* motion, it causes things to accelerate.

So Newton's second law of motion says: **A force causes a mass to accelerate**. It is that simple, it explains why objects (masses) accelerate.

A short video explaining Newton's Second Law: https://www.youtube.com/watch?v=nO7XeYPi2FU

NEWTON'S SECOND LAW OF MOTION (LAW OF ACCELERATION)

- A **FORCE** CAUSES A **MASS** TO **ACCELERATE**
- OBJECTS DO NOT JUST MOVE WHEN A FORCE PUSHES THEM, THEY **ACCELERATE**
- BIG THINGS TAKE MORE FORCE
- REMEMBER TURNING IS ACCELERATION TOO

A spacecraft goes faster (accelerates) because the rocket engine is a force, the rocket is a mass, the result, is the rocket accelerates. Quite simple, actually.

A **FORCE** CAUSES A **MASS** TO **ACCELERATE**

GEORGES ROCKET IS PUSHED BY A FORCE

WHICH MAKES IT ACCELERATE

The second law of motion formula
There is a formula that goes with the second law of motion. It is an easy formula and gives a lot of useful information. When a car company designs a new car, how do they

decide how strong the engine should be? They use the formula for the second law of motion to calculate it.

The formula is just as the law states: A **force** causes a **mass** to **accelerate**.

The formula then is:

Force = Mass x Acceleration

Or

F = ma

THE SECOND LAW
FORMULA

FORCE = MASS X ACCELERATION

F = ma

UNITS = NEWTON'S (N)

The units are Newtons (N) as Isaac would want, but you will get different units if you use the formula set up correctly. Both units are correct.

NEWTON

- UNIT FOR FORCE
- 1 N = 1 Kg-M/S/S
- ONE NEWTON IS THE FORCE REQUIRED TO ACCELERATE A 1 KG MASS AT 1 M/S/S

Notice that 1 Newton is the same as 1 Kg*m/s/s. You can use either unit, both are correct.

EXAMPLE

WHAT IS THE **FORCE** REQUIRED TO ACCELERATE A **30 KG** CAR AT **4 M/S/S**?

Start with the formula and set up, just like before:

EXAMPLE

WHAT IS THE **FORCE** REQUIRED TO ACCELERATE A **30 KG** CAR AT **4 M/S/S**?

F = MA

F =   30 KG   x   4 M/S/S

Make sure you put the correct numbers and units in the set up:

EXAMPLE

WHAT IS THE **FORCE** REQUIRED TO ACCELERATE A **30 KG** CAR AT **4 M/S/S**?

F = MA

F =   30 KG   x   4 M/S/S

F = 120   KG-M/S/S

The units for your answer can be found in the set up, or you can decide to replace them with Newtons. Both are correct.

EXAMPLE

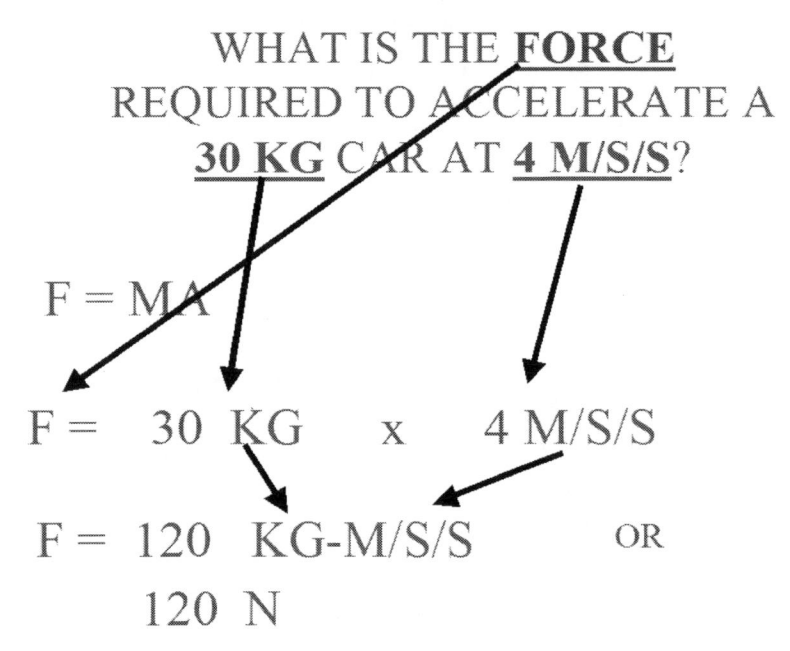

WHAT IS THE **FORCE** REQUIRED TO ACCELERATE A **30 KG** CAR AT **4 M/S/S**?

$F = MA$

$F = 30 \text{ KG} \times 4 \text{ M/S/S}$

$F = 120 \text{ KG-M/S/S}$ OR 120 N

So the final answer is 120 kg*m/s/s or 120 N.

Centripetal Force

Centripetal force is an INWARD force, which makes something go in a circle. You may have heard of centrifugal force pushing things outward, but that does not exist. Centrifugal force is actually a velocity and not a force at all. The only force present pushes *inward* and is called centripetal force.

Imagine someone swinging a bucket full of water in a vertical circle (over their head). Most people have done this when returning from the beach. What keeps the water in the bucket? It is centripetal force. This is what it looks like.

CENTRIPETAL FORCE PUSHES INWARD

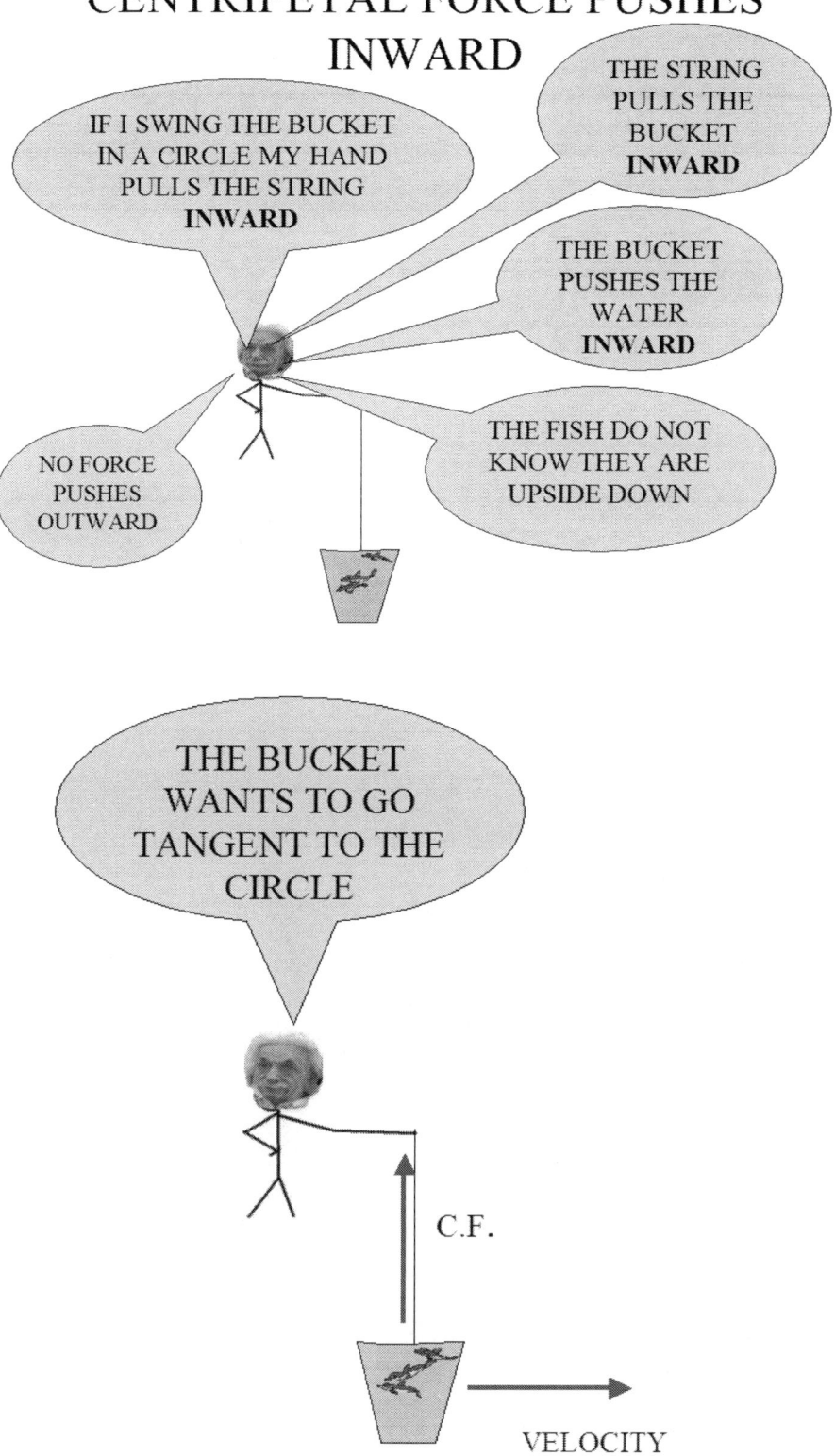

- IF I SWING THE BUCKET IN A CIRCLE MY HAND PULLS THE STRING **INWARD**
- THE STRING PULLS THE BUCKET **INWARD**
- THE BUCKET PUSHES THE WATER **INWARD**
- NO FORCE PUSHES OUTWARD
- THE FISH DO NOT KNOW THEY ARE UPSIDE DOWN

THE BUCKET WANTS TO GO TANGENT TO THE CIRCLE

C.F.

VELOCITY

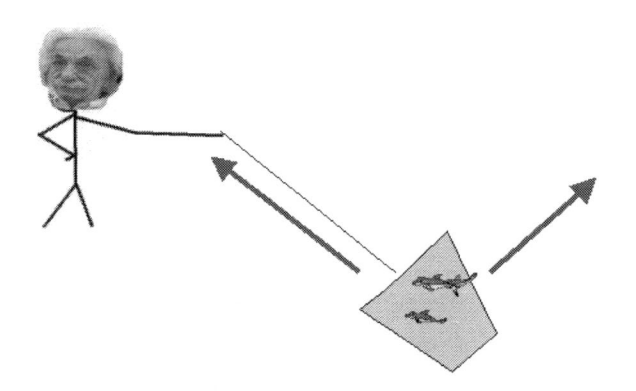

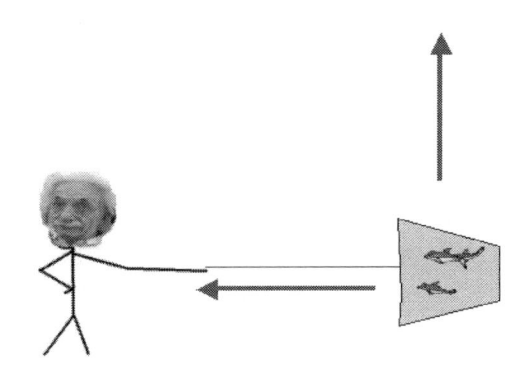

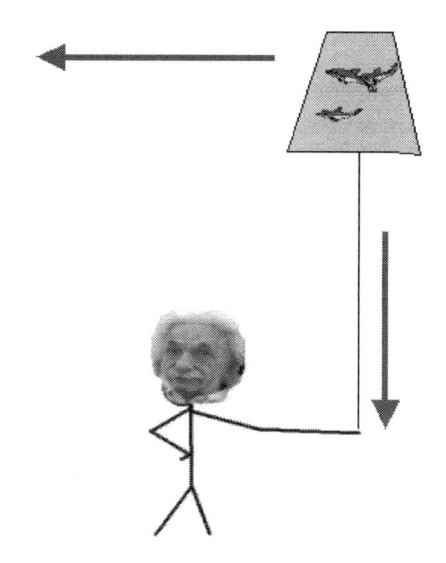

Notice the fish in the bucket do not know they are upside down.

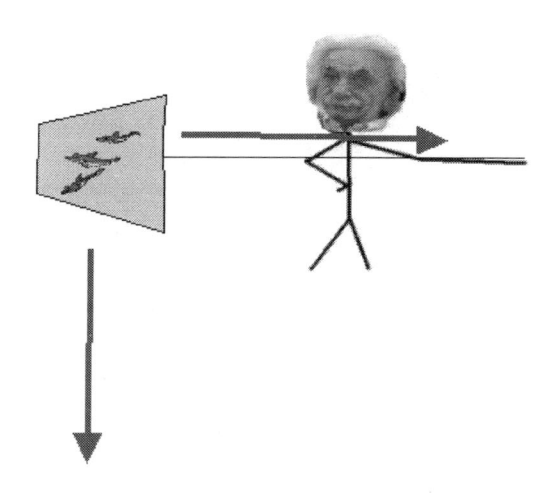

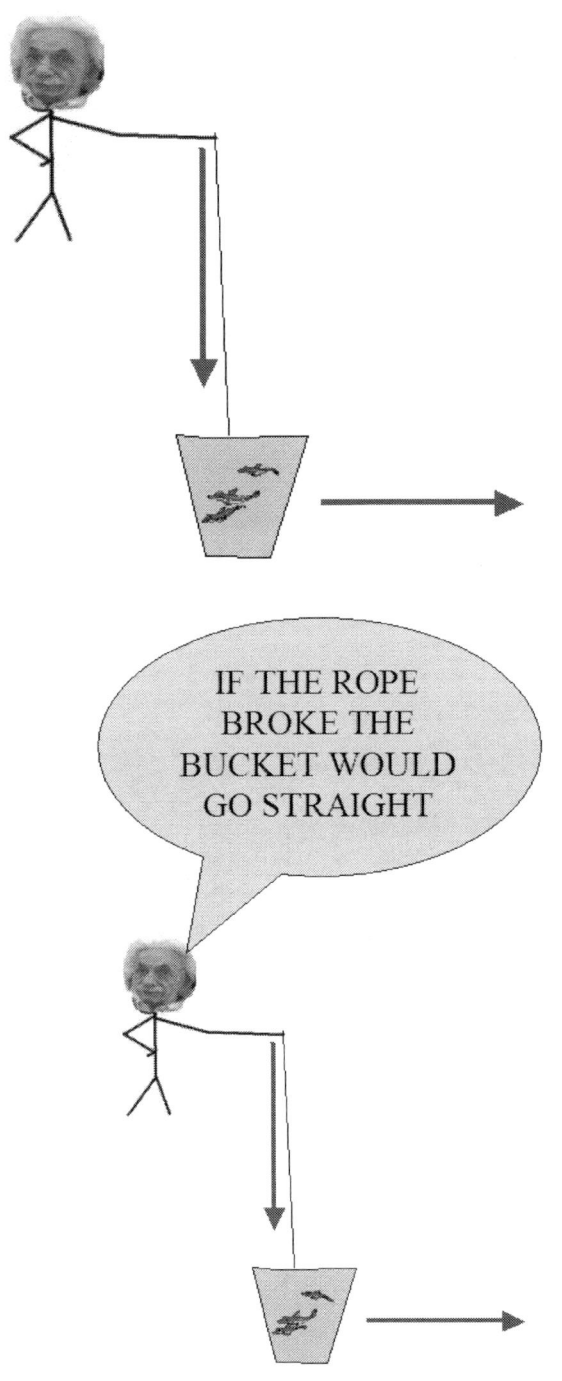

Since the velocity of the bucket is along the floor that is where it 'wants" to go, and if the string breaks….

CENTRIPETAL FORCE STOPS PULLING THE BUCKET INWARD

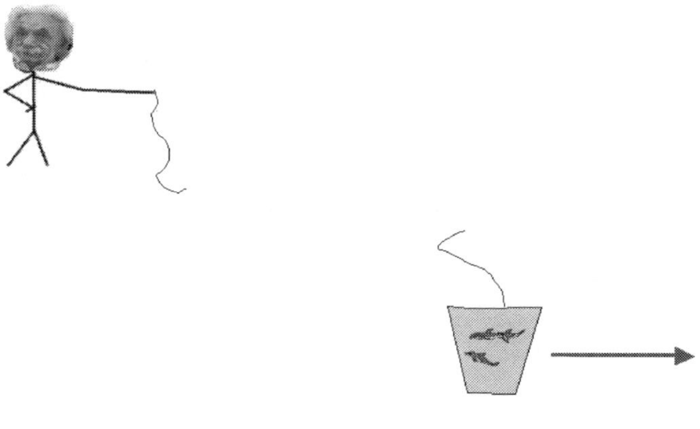

> HEY, I WAS SLEEPING!

The bucket slides along the floor until it hits something, like a sleeping student.

A day in the life of Earwig Hickson III

One day I was taking my goldfish for a walk and giving them some exercise. My Dad says I cannot have dog until I can show enough responsibility to take care of a pet. I got two goldfish to start with (Bubba and Nemo); just like a dog I figured I would have to take my goldfish for a walk. I put their bowl on a rope and carried them outside in the sunshine. They seemed to like the freedom, especially when I swung them in a circle parallel to the ground faster and faster until they threw up. I did not know goldfish could throw up. I knew they liked it because if I was a goldfish I would like to be spun in a circle too. The goldfish and water stayed in the bowl because of centripetal force, and what the fish were experiencing were g-forces. It had to be fun. I was on goldfish Bubba 6 and Nemo 6; gold fish have extremely short lives.

I had seen on T.V. that NASA likes to give potential astronauts rides in what they called human centrifuges, for high-G training. It was one of the main reasons I wanted to be an astronaut, when I grew up, to ride that thing. I had ridden the Gravitron ride at the local Fair, but the operator would only let me get to 3 g's, something about lawsuits. I tried to bribe him with a yo-yo, but he would not have any of that. I needed to build my own human centrifuge, right after I got some new goldfish.

Building the Beast (as it came to be called) was simple enough, an old lawn mower engine attached to a long PVC pipe with a lawn chair tied on the end. I would sit in the chair while my friend, Otis, started the engine. I waited for the perfect day to try it, the sun was shining, it was not too warm and most importantly, my parents were away for the day. I put on my bike helmet.

At first it was not too bad, the engine started slowly as it labored trying to move my mass. Eventually it got up to speed; I was flying around in a big circle and could feel the chair pressing up against my body. The bathroom scale I was sitting on to record my weight read two times my normal weight, then three times, I was at 3 g's, but then it read four times my weight and I was still gaining velocity. I called to Otis to slow down but he was long gone! When he saw me accelerating he got scared and ran home to hide (he had been a victim of my experiments before) so he did not get grounded again. I was on my own. My scale read seven g's and my face was sliding off my head. At eight g's my eyelids would not close and I decided I was not going to be friends with Otis anymore. At nine g's I began to pass out as the blood rushed out of my head. I could not move, I was dizzy, I was sick, and that is when the lawn chair gave out. I shot across the yard and if it were not for the shed, would surely still be going.

When my parents got home and asked what I had done with myself, I told them I made a machine for the next pumpkin chucking contest. I won that year.

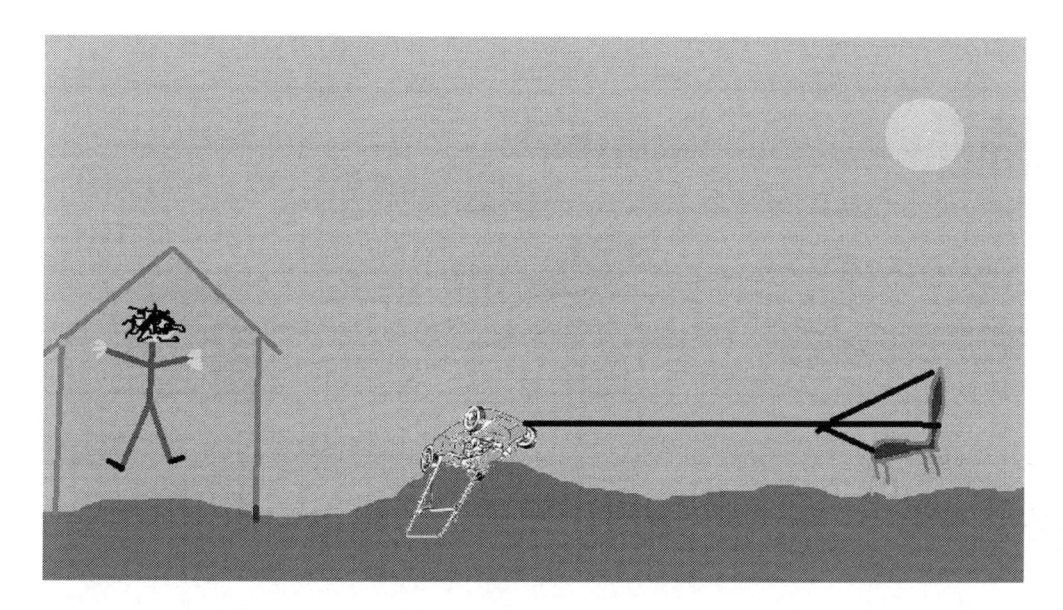

Flashcards for this chapter at Quizlet: https://quizlet.com/128329824/chapter-8-volume-2-newtons-second-law-of-motion-flash-cards/

Links

This is one of my demonstrations where I spin a cup of water in a vertical circle (on a platform) and don't spill a drop: https://www.youtube.com/watch?v=G79frInQcTw

Another one where a circular track is used to move a ball in a circle: https://www.youtube.com/watch?v=FjnDUuW4YDk

A harder one, where a penny is balanced on a coat hanger and spun in a vertical circle: https://www.youtube.com/watch?v=aVitgQbJ-KM

An even harder coat hanger and penny trick, you could do at home: https://www.youtube.com/watch?v=4OnsyUlhxh0

A Bola is like a lasso, it is used to catch cattle in place of a rope. This is fun, how to throw a bola, in this case a bola smurf to catch a bad guy:
https://www.youtube.com/watch?v=-GoIAuzvSrY

Newton's Second Law of Motion: https://www.youtube.com/watch?v=iwP4heWDhvw

A simplified explanation of Newton's second law:
https://www.youtube.com/watch?v=nO7XeYPi2FU

Newton's Second Law of Motion - Science of NFL Football, for you football fans:
https://www.youtube.com/watch?v=qu_P4lbmV_I

Fun thins to Google
Space program
Mercury program
Gemini program
Apollo program
Bola
Graviton ride
Centrifuge
Newton's second law demonstrations

Chapter 27
Newton's third law of motion

Video of the power point lesson on this chapter:
https://www.youtube.com/watch?v=kuQsHPKjcNY

The wonders of science

It was back in 1966. NASA had spaceships, they had astronauts, but they had no idea what they were doing. It was not NASA's fault; they were doing things no one had ever done before. They were experimenting in the unknown.

The Russians were the USA's competition. A competition for scientific dominance. At that time there were many small countries who we (the USA) wanted on our side, the Russians wanted the same thing. It was the Cold War. The country with the most allies would be the leader. It was between the United States of America and the Russians. Who would win? The battleground was not a war, but a competition in space. Who had the best technology? Who should a small country follow? Who was the superpower of the world? These were serious questions at the time. The most scientifically advanced country would gain allies.

The United States was behind. Russia had put the first man in Earth orbit (Yuri Gagarin). They had earlier put the first satellite in orbit, Sputnik 1. They even successfully had an astronaut make the first space walk, or extravehicular activity. This was the problem, at the time, we were losing.

Soon America let Edward H. White, II do the first American space walk. There were no problems, and we were catching up. Soon The United States had an astronaut do a spacewalk where he actually had to do more than float around. This proved to be a failure. Eugene Cernan was the guy. He was asked to actually do "work" in weightlessness; a simple task it was thought but it was a disaster, so much of a disaster than he almost died.

Everyone, it turned out, had forgotten Newton's third law of motion. Gene began by trying to turn a bolt; instead of the bolt turning, he turned. Soon he was spinning, totally out of control, his lifeline (the one that supplies his oxygen) from the Gemini spacecraft twisting. His commander, Thomas Stafford, tried to get him back into the capsule. Every time he touched the spaceship, it pushed him away, causing him to lose more control. He was in serious trouble.

Eventually, Stafford got him back in the small capsule. In the two-hour event, Cernan had lost 10 pounds and sweated over 3 pounds of water into his space suit.

It was all because of Newton's Third Law of motion, for every force (action) there is an equal and opposite force (reaction). Gene had nothing to hold onto, no gravity to help him, no leverage to help him. He was helpless to this law. Eventually NASA added foot and handhold to the outside the spacecraft, solving the problem of spacewalks. To this day NASA does not like when astronauts take spacewalks. Personally, I would like them, but they are dangerous.

What you need to remember

So far you know that objects want to continue in their state of motion (first law) and that a force causes a mass to accelerate (second law). You also know that the unit for force is called a Newton.

Newton's Third Law of Motion

The Third law of motion explains how forces "hit back"; it is also called the Law of Reciprocal Actions. It simply means that when a force is applied to an object, the object hits back with an equal force.

Let's let Sir Isaac Newton himself describe his Third Law: https://www.youtube.com/watch?v=a_QDQHidTSE

Did you ever kick a soccer ball and notice your foot hurts a bit? Or worse yet a rock and your foot hurts a lot? This is Newton's Third Law. These objects hit you back. Newton realized that forces do not exist alone, they came in pairs.

(Pear 1: "FORCES COME IN PAIRS")
(Pear 2: "SO BE CAREFUL WHEN YOU BITE PEARS")

The Third law of motion is simple. It simply states, *"For every Action there is an Equal and opposite Reaction"*. That is it; there is no more, very simple.

120

NEWTON'S THIRD LAW OF MOTION

1. FOR EVERY **ACTION**
2. THERE IS AN **EQUAL**
3. AND **OPPOSITE**
4. **REACTION**

Or even more simple.

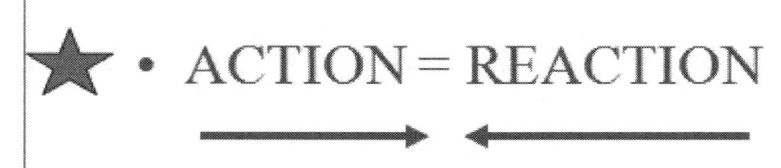

- ACTION = REACTION

If you were to run full speed into a brick wall, you are applying a force to it. You may notice the pain involved that is the wall hitting you with the same amount of force. The exact same amount. This is because the wall cannot accelerate (second law) and that force has to go somewhere, so it goes back into you in the opposite direction. The same thing happens when you are angry and punch a wall, so don't do that.

- WHEN AN OBJECT APPLIES A FORCE TO AN OBJECT THE SECOND OBJECT APPLIES AN **EQUAL** FORCE **BACK**

WALLS ALWAYS FIGHT FAIR!!!!

third law of motion - the hard way

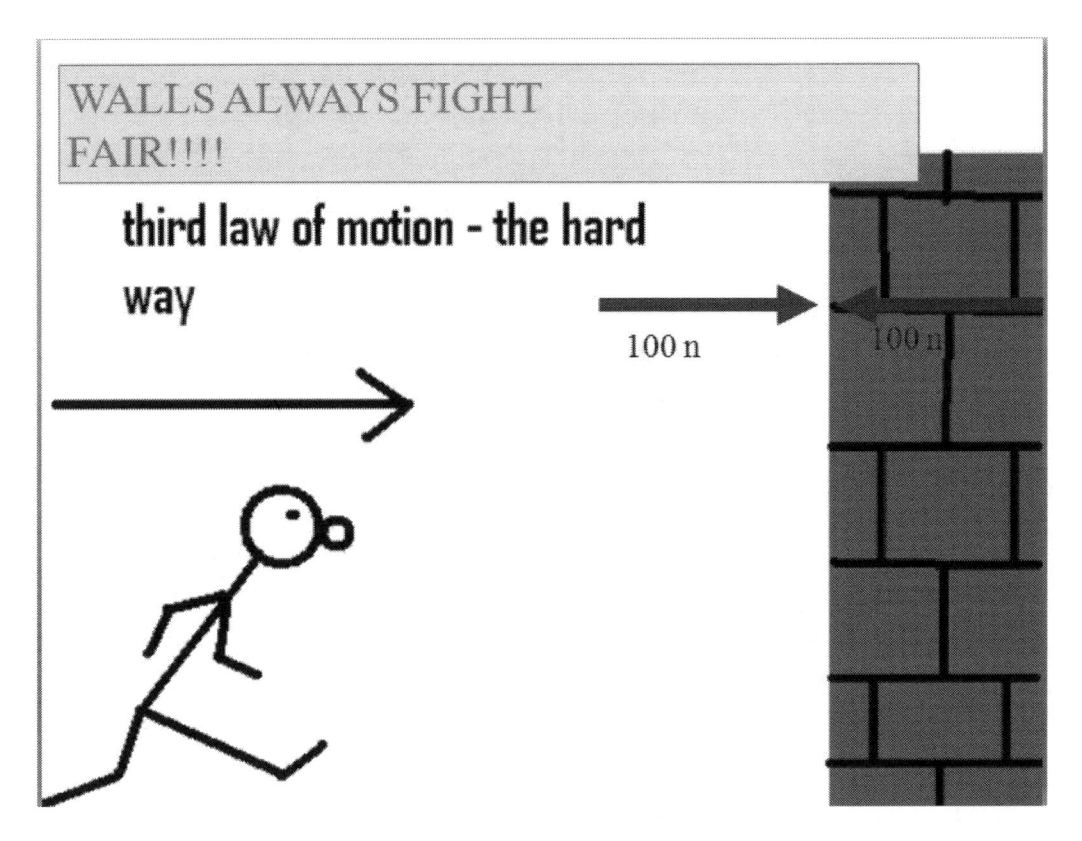

100 n   100 n

A rocket works the same way. The exhaust coming out the back is the *action force*, and if you were to be standing in front of that force you would be blown back (and burned to a crisp). The force from the exhaust pushing back applies an *equal* force in the *opposite* direction and the rocket accelerates forward. By the way, you do not want to be standing in front of the rocket either. Rocket lift off:
https://www.youtube.com/watch?v=vL1eXdVjN74

The Apollo 8 Saturn 5 rocket, https://www.youtube.com/watch?v=FzCsDVfPQqk

action ←   Reaction (accelerations) →

The Third Law is how flyboards allow a person to fly:
https://www.youtube.com/watch?v=LHL16av4C9k

For those of you who shoot guns (called projectile accelerators, in school because some people flip out over the word "gun") you may have experienced a "kick" or recoil. This is the reaction force of the projectile pushing back on the projectile accelerator.

It is very important to hold the projectile accelerator tight against your shoulder so it does not have a chance to accelerate into you. These can be dangerous, treat them with respect and responsibility.

https://www.youtube.com/watch?v=q7MaiL9ejW4

Here is my methanol rocket, notice the recoil, a lot when I am not holding it, less when I am holding it because my mass absorbs the action force.
https://www.youtube.com/watch?v=jbp0rKLv5tY

A larger methanol rocket with a minor malfunction:
https://www.youtube.com/watch?v=TQfr4BL_vAE

Jet engines force a plane forward for the same reason:
https://www.youtube.com/watch?v=_Si6o-k2iCI

What can happen to a school bus in the exhaust of a large jet:
https://www.youtube.com/watch?v=CpX1riSTeJc

Some of this may sound a bit confusing but will make much more sense in the next chapter (Ch 28).

Review
Newton's Third law of motion, also called the Law of reciprocal actions, states that for every action there is an equal and opposite reaction.

Flashcard review (quizlet): https://quizlet.com/143849821/chapter-27-newtons-third-law-of-motion-flash-cards/

A day in the life of Earwig Hickson III
It was such a simple idea. One that many people, I am sure, also had. I had the equipment, I had the knowledge. All I needed was a volunteer, and my friend Otis fit the bill nicely. I had built a rocket chair! The rolling chair from my Dad's office was the cockpit, four large fire extinguishers were duck taped to the sides. The triggers for the fire extinguishers were modified and attached to a lever, which turned them all on at once. Otis was securely duck taped into the chair and I was ready to do my experiment.

I had planned to take the trip myself, but could not, due to a signed contract my father made me sign before he left for the day. It said I could not do anything stupid while he was gone and included a 5 page list of "stupid things," one of them was about riding rockets. So Otis was enlisted for the fun part.

The chair sat in the middle of my street. Otis was strapped in but not quite ready. He changed his mind and was trying to get out of the chair. I knew he would regret chickening out so I did him a favor. I hit the starting lever. The lever snapped off, Otis began to accelerate; the force coming out the back knocked me down and shot him forward. It was actually a nice ride; Otis never went faster than about 30 miles/hour and was actually smiling and screaming "charge." The fire extinguishers ran out of power and the thrust ended. Otis almost had time to decelerate. Almost, I forgot about the big hill up ahead. Otis knew about the hill suddenly and his "charge" changed into a different word and down he went.

It was a long time before Otis spoke to me again. I tried to visit him in the hospital but I was not allowed in. I was also not allowed out of my house after my Dad found the pieces of his office chair piled up in his office. Although there was no evidence, I was blamed and punished for chairs destruction.

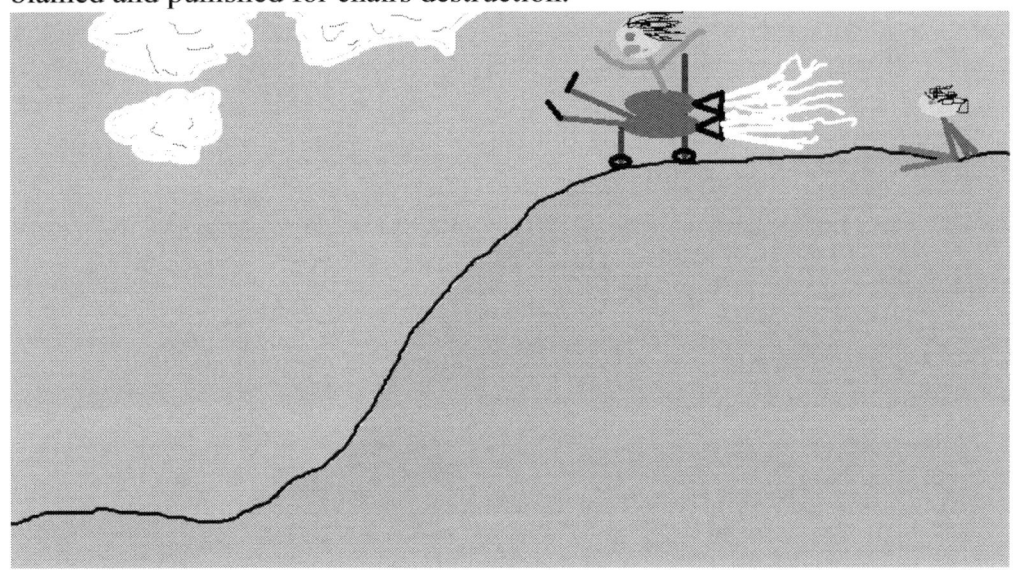

Fun things to Google
Third law of motion
Flyboards
Cool jet flyers
Rocket launch
Estes rockets
Recoil fails
Gun fails
Space walk
Gene Cernan
Fire extinguisher chair

Links

Sir Isaac Newton himself describe his Third Law:
https://www.youtube.com/watch?v=a_QDQHidTSE

Rocket blast off: https://www.youtube.com/watch?v=vL1eXdVjN74

The Apollo 8 Saturn 5 rocket, https://www.youtube.com/watch?v=FzCsDVfPQqk

The Third Law is how flyboards allow a person to fly:
https://www.youtube.com/watch?v=LHL16av4C9k

Recoil fails: https://www.youtube.com/watch?v=q7MaiL9ejW4

My methanol rocket. https://www.youtube.com/watch?v=jbp0rKLv5tY

A large methanol rocket with a minor malfunction: https://www.youtube.com/watch?v=TQfr4BL_vAE

Jets taking off: https://www.youtube.com/watch?v=_Si6o-k2iCI

Mythbusters see if a taxi can be flipped by jet exhaust: https://www.youtube.com/watch?v=rsCpX1riSTeJc

Chapter 28
The Law of Conservation of Momentum

Video of this power point lesson: https://www.youtube.com/watch?v=BgoT5sfNBaA

The wonders of science
It stood 365 feet tall (a 30 story building), weighed 6,478,000 pounds. It carried over 1.5 million gallons of fuel, weighing 5.6 million pounds. When the engine was lit it burned this fuel at a rate of 15 tons per second. It only burned for 11 minutes. All this to send a tiny capsule, a tiny module and three men on a long dangerous trip. It was July 16, 1969, and The United States of America was going to land men on the moon. It was a monster of a rocket, 60 feet taller than the Statue of Liberty. The three astronauts were sitting on a 30 story bomb. On July 20, 1969, only four days later, Neil Armstrong and Buzz Aldrin became the first humans to walk on the moon.
Photos of this amazing rocket:

The engine force of this amazing machine spit out 670 gallons of fuel molecules per second out the back. The velocity of this gas produced a thrust of 1,522,000 lb/ft. The combination of the mass of the fuel and its velocity resulted in a powerful force coming out the back. The reaction force from all this fuel pushed the rocket into space at a velocity of 7 miles per second, fast enough to cover the ¼ of a million miles to the moon. Only 12 men walked on the moon and no one has been back since 1972.
Video of the Saturn V lift off: https://www.youtube.com/watch?v=FzCsDVfPQqk

What you need to remember
You should know that Newton's Third Law of motion described how forces come in pairs, each going in opposite directions and equal in strength. Although the forces are the same, the objects do not behave the same; one usually goes faster. Why?
Napoleon learns about how the Third law is similar to momentum:
https://www.youtube.com/watch?v=LoNiSfaKF64

PROJECTILE ACCELERATOR

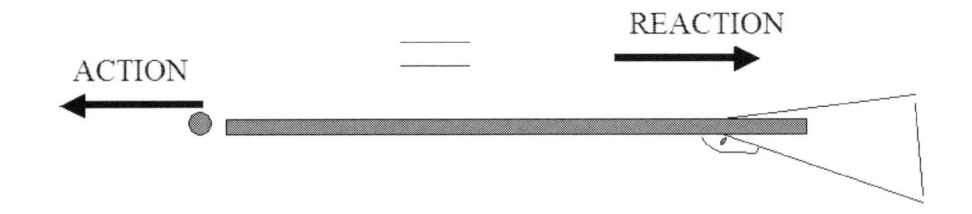

ACTION ← ● ══════════════ → REACTION

```
┌─────────────────────────────────────────────┐
│           PROJECTILE ACCELERATOR            │
│                                             │
│                          REACTION           │
│    ACTION                  ──→              │
│     ←──                                     │
│      •         BOOM    ▬▬▬▬▬▬▬▬▬            │
│                                             │
│    SMALL MASS GOES    LARGE MASS GOES       │
│    FASTER             SLOWER                │
│                                             │
└─────────────────────────────────────────────┘
```

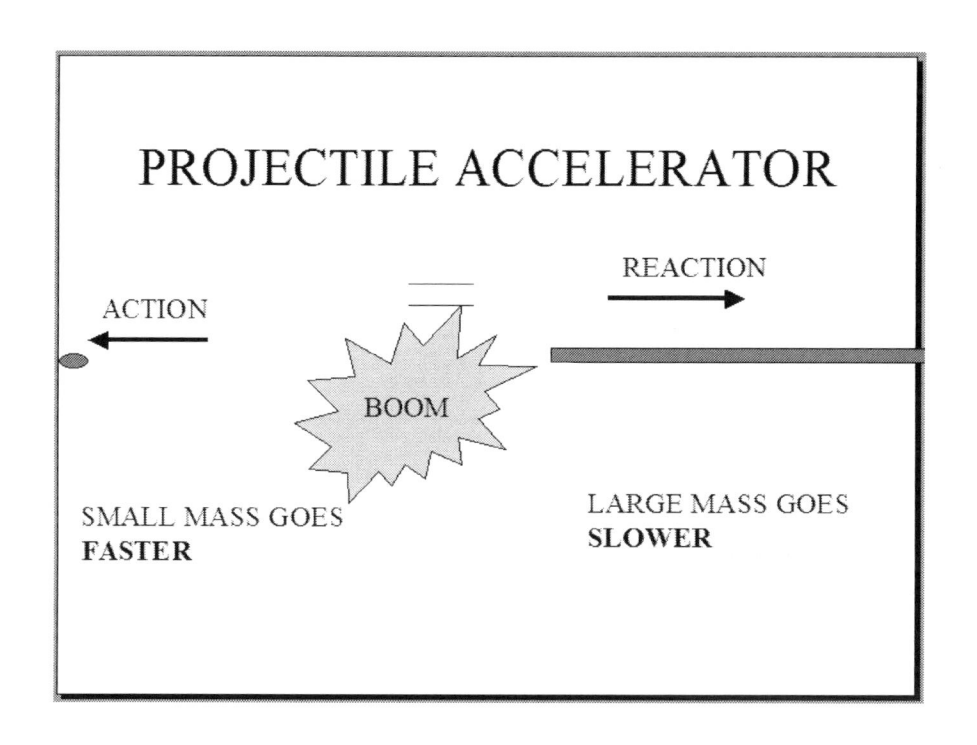

Momentum
Einstein explains momentum: https://www.youtube.com/watch?v=f6EYrILYeW8

Some objects hurt more than others when they hit you. Objects with a lot of *momentum* hurt more than those with little. *Momentum is the strength of a moving object.* Big massive things hurt more than light less massive things; this is because the mass of an object affects its momentum. This is why a feather does not hurt when it hits you but a bowling ball does. Fast velocity objects hurt more too, while slow things do not. If someone throws a bowling ball at you it can hurt a lot, while the same ball being handed to you does not. This is because faster objects have more strength and therefore more momentum than slow ones. So momentum is a combination of the mass of something and how fast it is moving, an object needs both to have momentum. A moving car is much more of a threat than a parked one (no momentum).

Momentum formula
Since momentum is a combination of an object's mass and its velocity, you can calculate it. It is a rather easy formula to use, just multiply the mass by the velocity. The units after the answer are weird but that will clarify shortly.

MOMENTUM FORMULA

MOMENTUM =   MASS   X   VELOCITY

$$M = mv$$

UNITS -   Kg-M/S

Use the three step process like before (Ch. 20)

EXAMPLE
- WHAT IS THE MOMENTUM OF A 500 Kg BOULDER ROLLING AT 10 M/S?

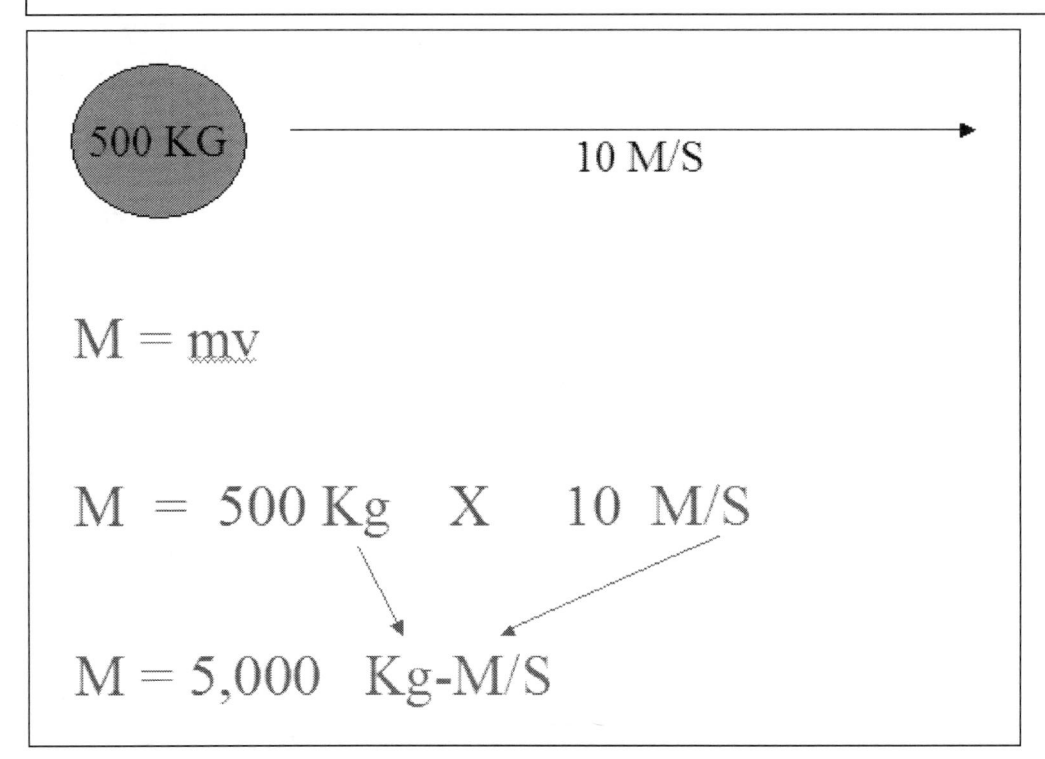

500 KG → 10 M/S

$M = mv$

$M = 500 \text{ Kg} \times 10 \text{ M/S}$

$M = 5,000 \text{ Kg-M/S}$

Notice the units for the answer comes from the set up, just like before.

Now we can compare the strength of moving objects, just by calculating the momentum.

It should be obvious now that getting hit by a fast car is much worse than being hit by a fast feather. It should also be obvious that a stationary car does not hurt at all! But wait, is getting hit by a moving car the same as hitting a parked car? No, not even close. If you get hit by a moving car, the energy hitting you is its mass (a lot) times its velocity. This results in a big momentum and you need to see a doctor. If you are running (at the same speed as the previous car) and smash into a parked car, it may hurt but the momentum is your mass (much smaller) times your velocity (the same). This is a small momentum and although you may get a boo boo, it is not that bad. It might be a bit embarrassing, so don't tell anyone you got hit by a parked car.

The Law of Conservation of Momentum

Newton explains the Law of Conservation of Momentum:
https://www.youtube.com/watch?v=4-O89u_loqY

The Law of Conservation of Momentum describes how momentum is not created or destroyed, just like the Law of Conservation of Energy, which says energy cannot be created or destroyed. So since momentum is not destroyed it must go somewhere, and all of it goes somewhere. The Law states that the momentum into a group of objects is the same as the momentum coming out. The following cartoon will explain.

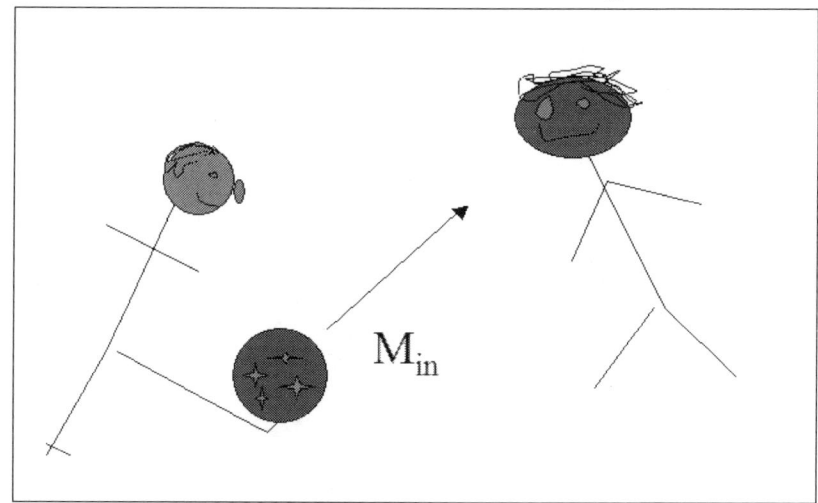

The soccer ball is the momentum *into* this system.

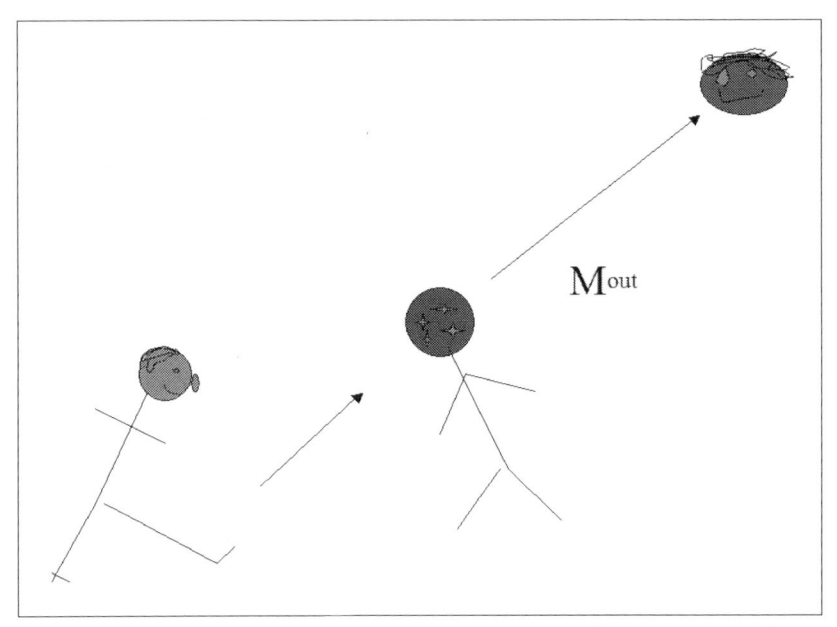

The guy's head is the momentum *out*. The ball stops since it gave all its momentum to the head.

When you play pool, you hit the cue ball at the break with a lot of momentum. When the ball hits the others (and stops) it spreads its momentum among all the other balls.

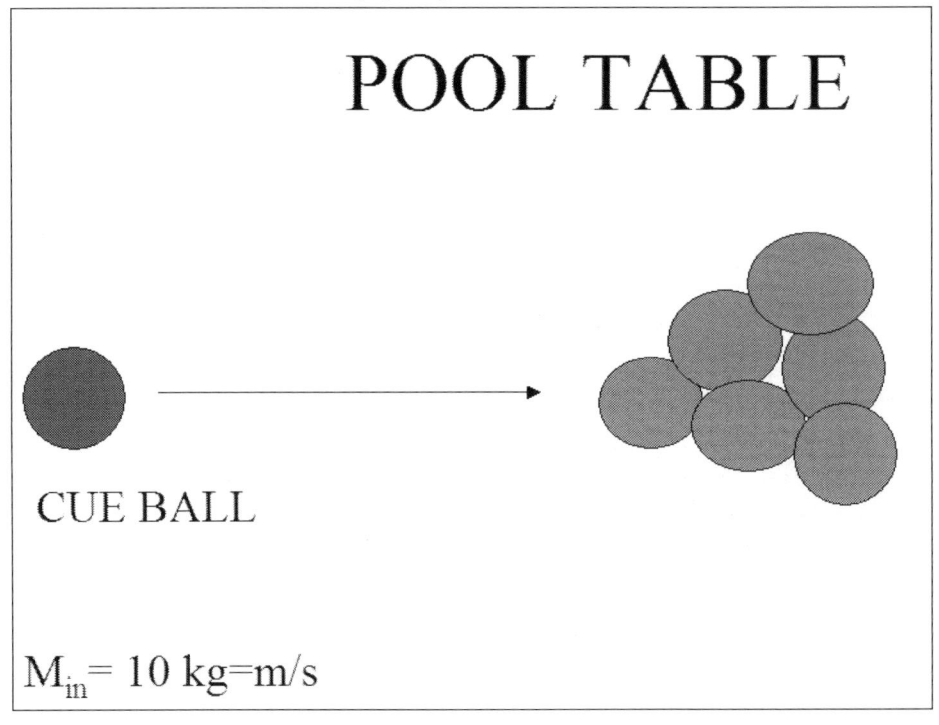

POOL TABLE

CUE BALL

M_{in} = 10 kg=m/s

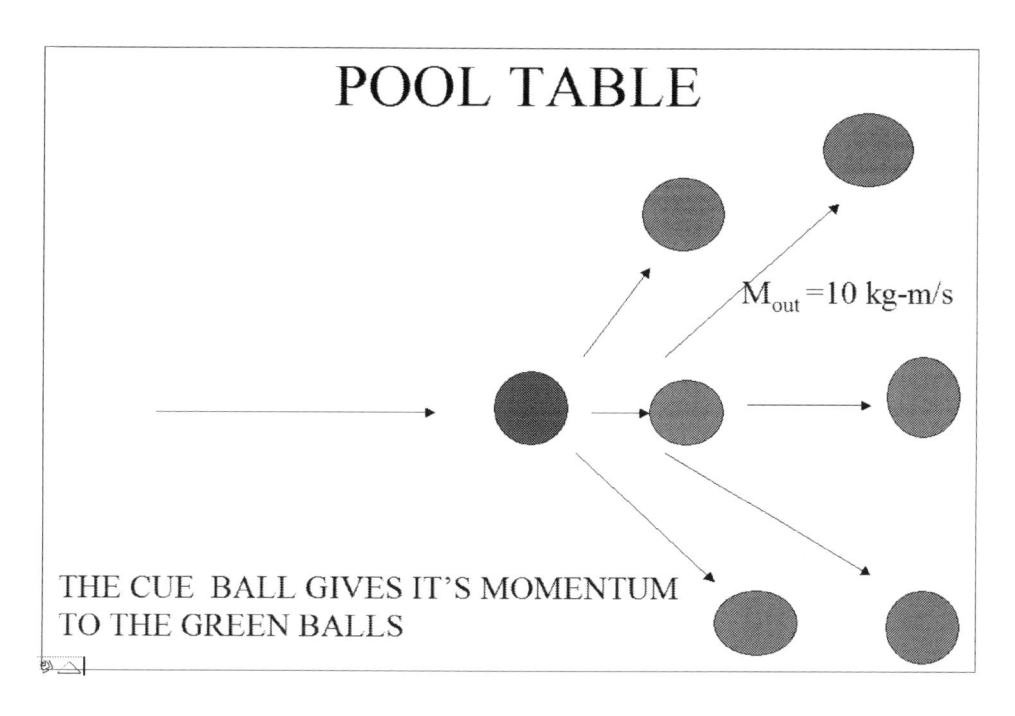

POOL TABLE

M_{out} =10 kg-m/s

THE CUE BALL GIVES IT'S MOMENTUM TO THE GREEN BALLS

Sometimes a couple of cars will be stopped at a red light when someone runs into them from behind. Most of the cars do not move while the momentum travels through them, but the one in the front gets knocked forward. One car *in* (momentum in), results in one car *out* (momentum out).

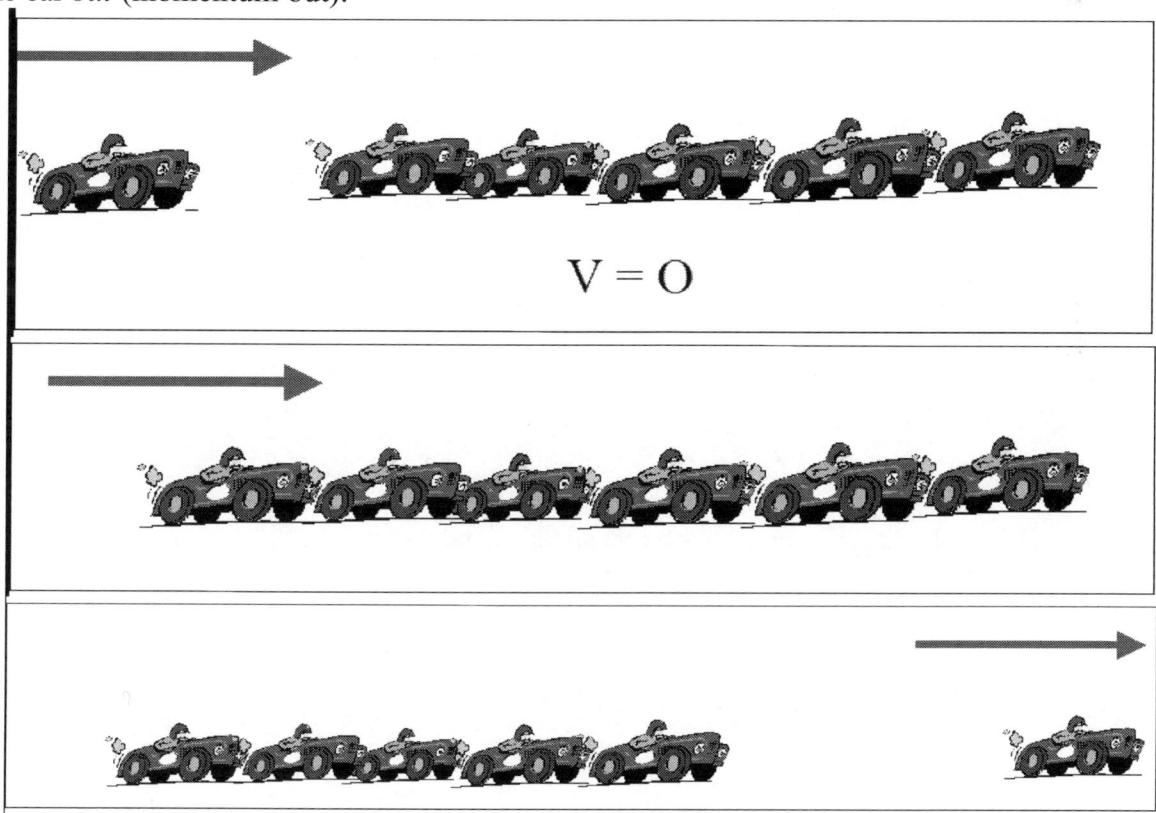

V = O

This is easy to see with a thing called Newton's cradle, a bunch of balls that smack back and forth. Check out this video of one:
https://www.youtube.com/watch?v=0LnbyjOyEQ8

A bigger and better one: https://www.youtube.com/watch?v=mFNe_pFZrsA

It is easy to understand this when all the objects have the same mass, but what happens if one object is bigger than the other? The answer is that the larger mass object goes slower, and the light one goes faster. This is why a bullet flies faster, than the gun (I mean projectile accelerator) that fired it.

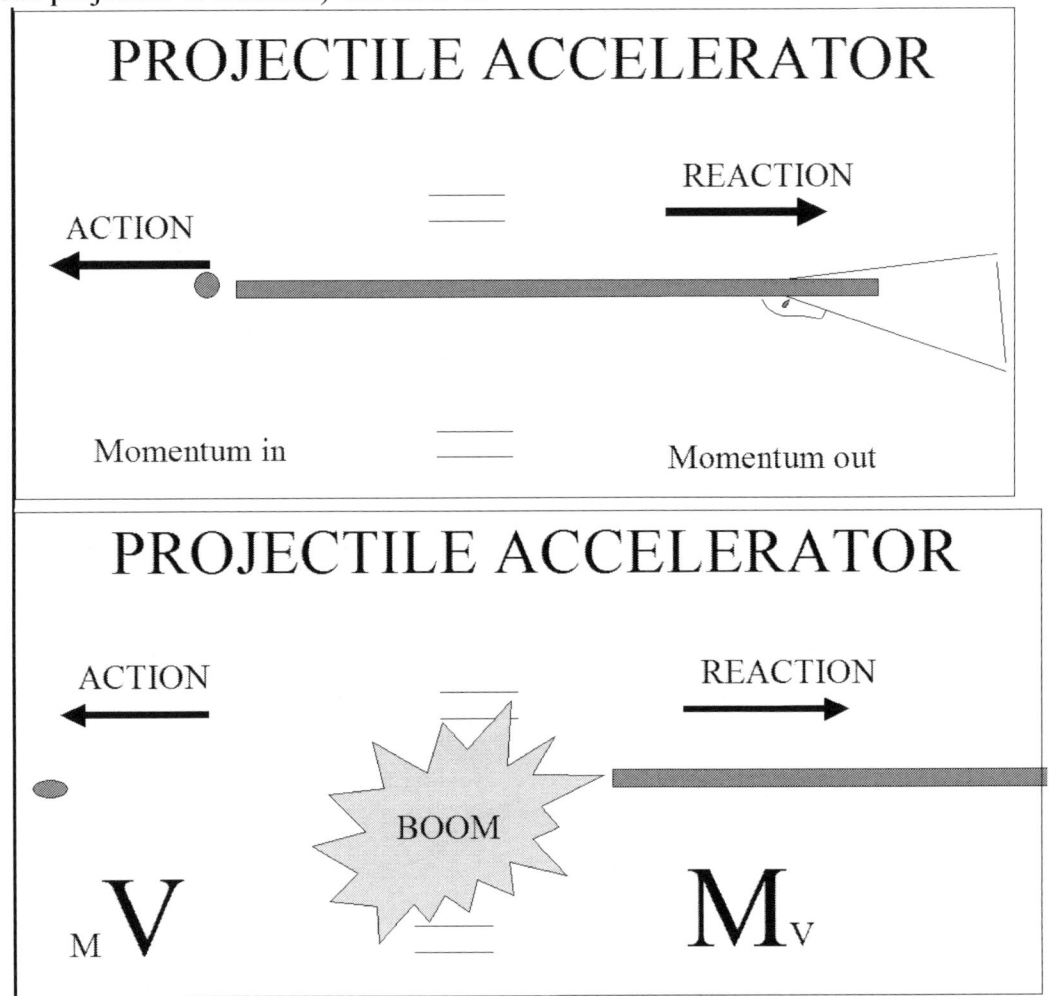

PROJECTILE ACCELERATOR

ACTION ← — REACTION →

Momentum in — Momentum out

PROJECTILE ACCELERATOR

ACTION ← REACTION →

BOOM

$_MV$ M_v

This is a cool experiment you can do with a basketball and a tennis ball. The momentum of each ball is the same but the smaller tennis ball has the greater velocity.
https://www.youtube.com/watch?v=-5QiWu8iAQo

My little rocket shows the same thing, the rocket flies fast, the launcher flies slower:
https://www.youtube.com/watch?v=jbp0rKLv5tY

Review
The Law of Conservation of Momentum states that, the momentum into a system is equal to the momentum out of the system. Momentum is not created or destroyed but

132

transferred. Momentum is an objects mass times its velocity. It is the strength of a moving object.

   Flash cards at quizlet: https://quizlet.com/130674092/momentum-chapter-10-volume-2-flash-cards/

A day in the life of Earwig Hickson III

   There was a big building fire not far from my house. It was horrible, smoke and flames destroying a structure and changing many people's lives. No one was hurt, and the firemen did a fantastic job putting out the fire. It changed my life too. I was fascinated by the firemen shooting columns of water onto the fire from far away. They were brave and looked to have an exciting job. Now, many children want to be firemen when they grow up. Saving people is something most of us want to do, but I did not want to be a fireman, I wanted the fire hose.
   One of my chores at home was to water the garden, flowerbeds, and all the trees in our yard, and we had a lot to water. We also had a big yard. Every Saturday, I would water the plants one at a time with the garden hose, walking from one to the other until the hose could not stretch any longer. At that point I had to fill a five-gallon bucket with water and carry it from plant to plant. It was a long hard job for a skinny kid, but my Dad said work built character, and that would make me a better person. I actually thought I had enough character, but he disagreed.
   When I saw the firemen with their wonderful hoses, I got an idea. What if I had a hose like that, and could water the entire yard with it, in minutes instead of hours. I set to work. There was a fireplug near my house for water and I managed to find an old fire pumper tanker truck at the local junkyard. Apparently it got a flat tire and the town had no money in budget to fix tires, so the taxpayers bought a new pumper tanker. I did not need the truck, just the big water pump on the back. I towed the giant fire truck home with my bike, and then got to work. I found big long hoses on E-bay and made the most powerful watering system in the world. But I forgot about momentum and the Third law of motion!
   When I turned on the pump water shot all over my yard, just as I planned but I shot all over the yard too. My tiny mass was no match for the powerful force coming out of the hose. I shot back with an incredible velocity, then shot up into the air, where I let go of the hose and crashed back to Earth. The hose whipped around like a snake hitting and smashing the trees and flowers in the yard. My Dad's truck had quite a dent in the side before I turned the pump off. Now I know why it takes two big firemen to hold a fire hose.
   Scientists never say they fail, they like to say, they just do not know the answer, yet. I did not fail either; I was just trying to solve the wrong problem. I had seen in a science book (this one in fact) how some guys had attached water hoses to some jet skis and used the water to fly! I had the technology to do what they did. I had the force, I had the

momentum, but as my father pointed out later, not the brains I needed for an experiment like this.

Four hoses were attached to the giant water pump; on each of my legs I had duck taped a hose nozzle pointing down, on each arm another one pointing out. I put on my Iron Man mask and bike helmet and hit the power switch. The next thing I knew I was climbing down out of a tree.

The mass of the water shot down, at an unbelievable velocity, creating a powerful momentum force. The momentum down was the same as my momentum up, and since my mass was so small, my velocity was large. I shot up like a rocket; in fact I was a rocket!

I was fired upward until I ran out of the long hoses, flipped over and shot down, whipped back and forth hitting the house and the ground in an alternating sequence like a balloon someone forgot to tie a knot on. Finally I managed to stabilize myself high in the air as the fire hoses stretched like guitar strings holding me back. The view was spectacular, especially the little fire pump tanker far below with the on-off switch and all the controls! Finally, the duck tape on one of my arms unfortunately failed and I spun out of control into the upper branches of an Oak tree, where the hoses became tangled after flailing like a giant octopus. I became stationary again. Eventually the pump ran out of gas, the water stopped and I climbed down.

Scientists never say they fail; they like to say they just do not know the answer, yet. I must correct them though, that was a fail.

Quizlet flash card review: https://quizlet.com/130674092/chapter-28-momentum-flash-cards/

Links

Napoleon learns about how the Third law is similar to momentum: https://www.youtube.com/watch?v=LoNiSfaKF64

Einstein explains momentum: https://www.youtube.com/watch?v=f6EYrILYeW8

Newton explains the Law of Conservation of Momentum: https://www.youtube.com/watch?v=4-O89u_loqY

Newton's cradle: https://www.youtube.com/watch?v=0LnbyjOyEQ8

A large Newton's cradle: https://www.youtube.com/watch?v=mFNe_pFZrsA

A 30 minute documentary of the moon landing and flight: https://www.youtube.com/watch?v=lRwKUScppvQ

Video of the Saturn V lift off: https://www.youtube.com/watch?v=FzCsDVfPQqk

Apollo 12 moon trip, after the rocket got struck by lightning: https://www.youtube.com/watch?v=FBhIDjWaByg

Fun things to Google
Apollo 11
Apollo 13
NASA moon walk
Lunar rover
Pool – trick shots

UNIT 8
THE TYPES OF FORCES

Chopping a wooden board using just air pressure to hold it down.
https://www.youtube.com/watch?v=sOe2Pk_FjEQ

Chapter 29
Newton's Universal Law of Gravitation

Video of this power point lesson: https://www.youtube.com/watch?v=9ilc9ZBeaSY

The wonders of science

There is one in the center of the Milky Way galaxy. Lucky for us the closest one is 1600 light years away. A light-year is the distance light can travel in a year or about 6 trillion miles, that is 6,000,000,000,000 miles. But the closest one to Earth is 1600 light years or 9,600, 000,000,000,000, much too far for humans to travel to, or the thing to hurt us. We are safe. The objects I am describing are called black holes.

A black hole forms when a large star runs out of its hydrogen fuel and starts to cool down. Without the heat to keep the star big in size, its gravity begins to make it shrink. This is called a stellar collapse or a gravitational collapse, since the gravity of a star crushes in on itself until it is very small in size and large in density. The gravity in a black hole is very concentrated and super strong, so strong that it causes light to bend and fall into the black hole along with anything else nearby. The reason a black hole is called a black hole is because light cannot get from the black hole to our eyes, so all we see is black. Black holes cannot be seen directly, but their gravity can be noticed. Scientists have detected black holes by the behavior of objects around them and something called gravitational lensing, which means the gravity bends light like a magnifying glass.

Surrounding the black hole at a distance smaller *than the diameter of the original star* is a boundary called the event horizon. This is the boundary, a certain distance from the black hole where light cannot get out. In fact nothing can get out of a black hole that has fallen inside this boundary; it is the point of no return.

So a black hole has a lot of gravity near it but that gravity *does not reach out any farther* than the gravity of the original star did. So if there were planets orbiting around the original star, they are probably still orbiting around it. Black holes are not going to "eat" the universe.

But what would happen if you fell into one, if there was one close enough to fall into? This is the weird part. As you got pulled into the black hole you would just feel like you were falling, you would feel weightless, that is until you got closer to it. There is a point where you would notice your feet are being pulled faster than the rest of you. You would begin to be stretched! This is called spaghettification and you would get really long like a piece of spaghetti and eventually snap in half. You may not notice when you pass the event horizon but then again you might see very bright light that may be orbiting the black hole like a satellite. Weirder yet, you might see the back of your head in every direction you look because the light you see is in orbit. Once inside the event horizon (if you were still alive), looking down you would probably see darkness but looking up you would probably see violet light (a violet sky) because the light waves are shortened as they fall in. Whatever you do see will be much distorted, like looking through super powerful glasses. You would fall in normal time and your watch would act as it always has.

But your friend, watching you, would see things different. The first thing they would notice is that the closer you get to the event horizon, the slower you *seem* to fall. Time for you is different than for your friend. This is called gravitational time dilation, and is caused by the large gravity bending the spacetime of the universe! Your friend will be watching you fall for a long, long time and never actually see you fall in. You would simply start to turn red in color and fade away until you disappear just as you reach the event horizon. Another friend watching both of you with a telescope from far away would see your watch ticking very slowly as you fell in the black hole, but your friend's would be ticking much faster.

Good thing there are no black holes anywhere near Earth.

Falling into a black hole: The singularity and spagettification
https://www.youtube.com/watch?v=OGn_w-3pjMc

Neil DeGrasse Tyson - Death By Black Hole:
https://www.youtube.com/watch?v=h1iJXOUMJpg

The strange fate of someone falling into a black hole:
http://www.bbc.com/earth/story/20150525-a-black-hole-would-clone-you

How is time changed inside a black hole:
http://www.skyandtelescope.com/astronomy-resources/time-changed-inside-a-black-hole/

Black Holes Explained – From Birth to Death (5 minute video):
https://www.youtube.com/watch?v=e-P5IFTqB98

What Would Happen if You Fell Into a Black Hole? (6 minute video):
https://www.youtube.com/watch?v=6H6CcbIMH7I

Black Holes - Mysteries Of The Universe HD Documentary (a 45 minute video):
https://www.youtube.com/watch?v=BWn29kYjOZQ

What you need to remember

As always you still need to know about force and how it makes a mass accelerate. For this chapter you should realize that gravity is a force, and falling objects accelerate. Also do not forget that your weight is a measure of gravity and your weight would be different if you were on a different planet. http://www.exploratorium.edu/ronh/weight/

Newton's Universal Law of Gravitation

It was Sir Isaac Newton's greatest discovery, and he had a lot to choose from. The old story about an apple falling on his head making him think of the Law of Gravity is not true, but he actually did watch apples fall outside his study window as he thought about his theory. What made him curious as he watched apples fall were two questions he had.

Why does the apple and everything else fall, or accelerate, toward the center of the Earth and not go in a different direction? What was it about the Earth that caused it to attract the apple in the first place?

Does it matter how high the apple is when it starts to fall? A few feet is easy to accept, but how about 100 feet, or 1000 feet, or for that matter, how about 1000 miles? The height did not seem to matter to the apple. He realized that the Earth's gravity reached all the way to the moon and beyond.

He came up with two ideas. The first thing he realized was that gravity had to be a *force* since the apple accelerated as it fell, and only a force can cause an acceleration. It did not just float down. The reason the Earth had so much gravity was because it was big and had a lot of mass. He reasoned that there was a relationship between mass and gravity. He also realized that the apple must have some gravity too, but since it had so little mass, it had very little gravity. He was sure anything with mass had gravity of its own, and the more mass, the more gravitational attraction. Gravity is between objects and exists everywhere in the universe.

The second idea had to do with why the sun, which has more mass, thus more gravity than the Earth, did not attract the apple towards it. He reasoned that it could not because it was too far away. He realized that the closer object (the Earth) affected the apple more because it was closer. Once he realized this, everything made sense. His theory explained the orbit of the moon around the Earth, why the moon causes the ocean tides, the Earth's movement around the sun and planets orbits around the sun.

Try playing this gravity simulation game (make a solar system):
http://www.testtubegames.com/gravity_flash.html

The Law
You can think of Newton's Universal Law of Gravity as having three parts.

All masses have gravity of their own. Everywhere in the universe (this is what *universal* means, everywhere). Every object is attracted to every other object too.

The more mass an object has, the more gravity it has. Double the mass and the gravity becomes double. Since gravity causes your weight, if you double your mass, you double your weight.

Gravity is proportional to the distance between two objects. As the distance between them doubles, the gravity becomes ¼ as great. This is called the inverse square law which just means gravity gets weaker at a faster rate than the distance.

2 times the distance = ¼ the gravity
3 times the distance = $1/9^{th}$ the gravity
4 times the distance = $1/16^{th}$ the gravity. And so on.

NEWTON'S LAW OF UNIVERSAL GRAVITATION

- 1. ALL OBJECTS IN THE UNIVERSE ATTRACT EACH OTHER BY THE FORCE OF GRAVITY
- 2. THE SIZE OF THE FORCE DEPENDS ON THE MASS OF THE OBJECTS
- 3. AND THE DISTANCE BETWEEN THEM

IN ENGLISH

- EVERYTHING HAS GRAVITY
- MORE MASS = MORE GRAVITY
- THE SUN HAS MORE GRAVITY THAN THE EARTH
- THE CLOSER AN OBJECT, THE STRONGER THE GRAVITY ON US
- THE EARTH EFFECTS US MORE THAN THE SUN

The first part explained – Everything has gravity of its own
Newton explains the first two parts of his Universal Law of Gravitation:
https://www.youtube.com/watch?v=kgtfRTKYIsw

This is rather simple. Every atom in the universe has a tiny, tiny, tiny bit of gravitational attraction to every other molecule. This force is super weak, but if you get enough atoms together, say in a planet or a star, it adds up to a very strong gravitational force. It takes a lot of atoms to make enough gravity to notice. In fact this is how planets and stars formed in the first place, from dust clouds billions of years ago. The atoms and molecules in the dust were attracted to each other and formed a big clump. This is why planets are round too. The average of all the gravity of all the atoms is in the center of the planet and they are all attracted there.

Now think about the Earth just floating out in space? Oddly enough, one object by itself really does not have any gravity, there must be a second object, and the gravity is between them! The force of gravity between them is called weight. To measure your weight you need to put a scale between the Earth and you. The gravity between the two of you squishes the scale. If there is no second object to squish the scale, it reads zero. This means the Earth is weightless until you or something else pulls on it.

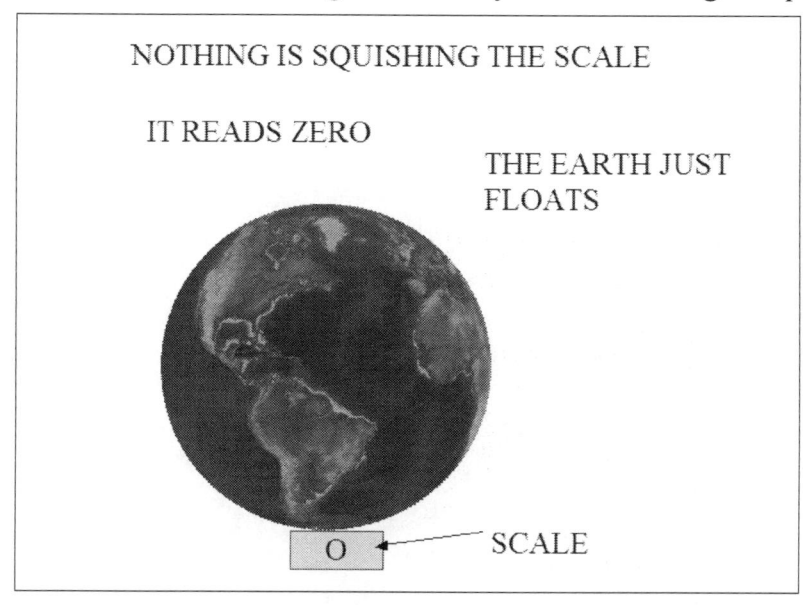

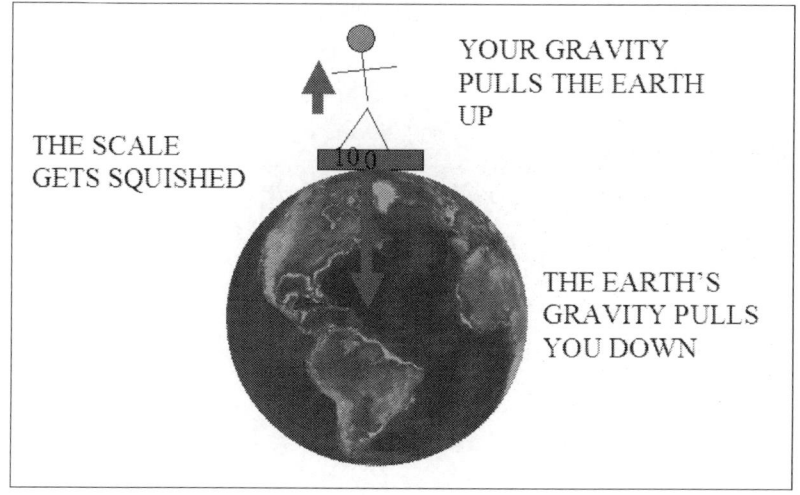

NOTHING IS SQUISHING THE SCALE

IT READS ZERO

THE EARTH JUST FLOATS

SCALE

YOUR GRAVITY PULLS THE EARTH UP

THE SCALE GETS SQUISHED

THE EARTH'S GRAVITY PULLS YOU DOWN

Imagine what this means. You are actually attracted to your pencil and your pencil is attracted to you. You are also attracted to me since we both have mass, and I am attracted to you. You are attracted to the Earth most of all though, because it has a lot more mass than me and is much closer.

The Earth is attracted to the Moon causing it to orbit the Earth, but the Earth is attracted to the Moon also. This is why we have high and low tides; the Moon's gravity pulls the water a few feet closer to it.

The second part explained – more mass means more gravity

This should be much easier now. The Earth has a lot more mass than you do so things are attracted to it more than they are attracted to you. But, remember your weight is a combination of your mass (gravity) and the Earth's mass (gravity). This may sound odd but if your weight increases, it may not be you that is getting fatter. It could be that the Earth is getting fatter because it does not matter which mass increases.

Here you are standing on a bathroom scale and the force between you and the earth is 100 pounds

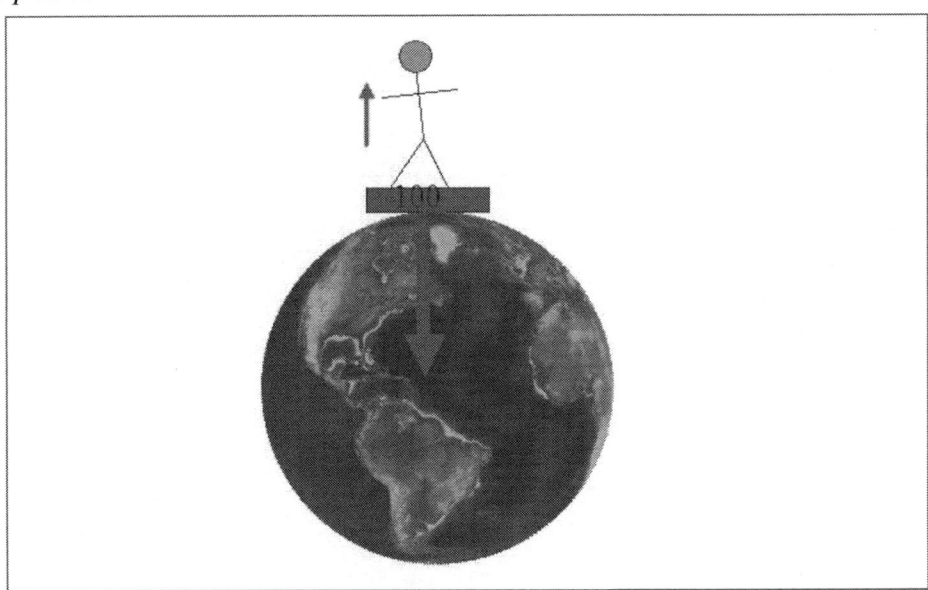

After a big breakfast your mass doubles and the scale reads 200 pounds.

142

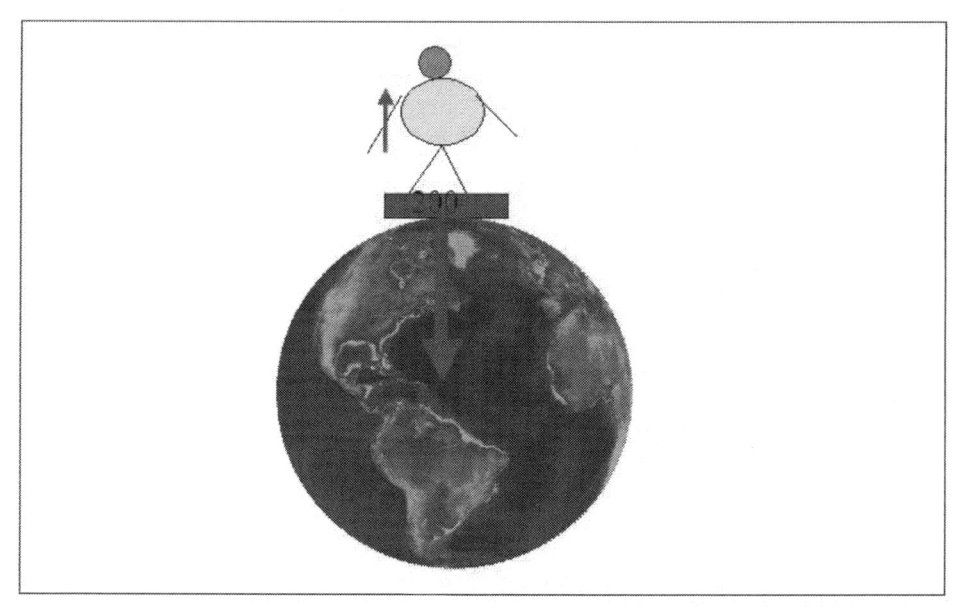

But what if your mass stayed the same but the Earth's ate the big breakfast and doubled its mass. The scale would still read 200 pounds.

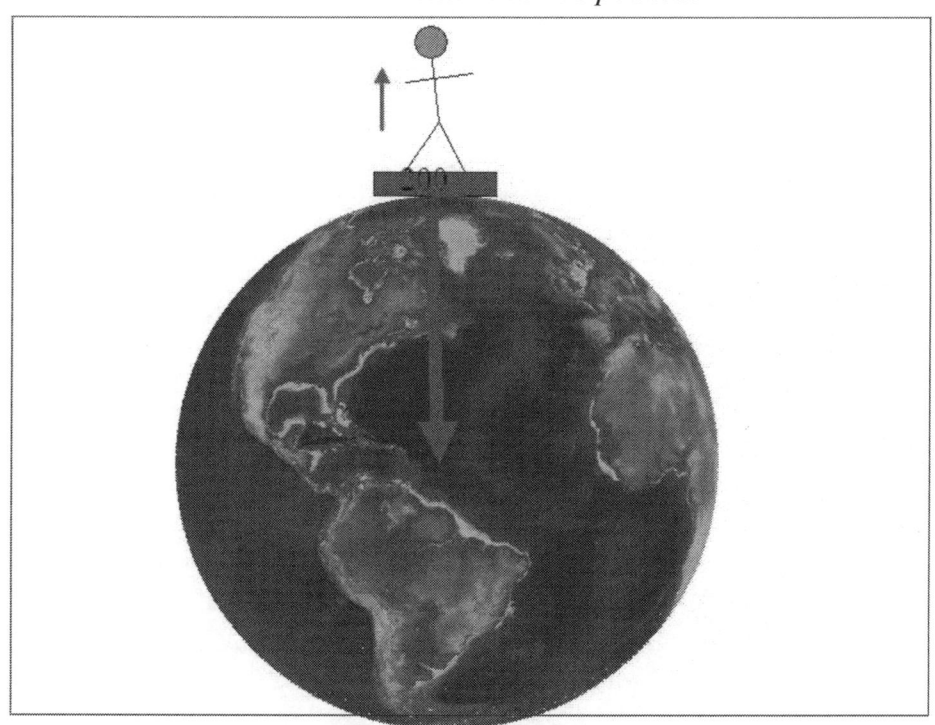

I often try and blame the Earth for my increase in weight, but it does not work.

The moon has gravity because it has mass but since it is so small (1/6 the mass of the Earth) it has less mass (1/6th as much as the Earth. The Astronauts that walked on the moon had a blast.

Apollo 17 Bunny hopping on the Moon Gene Cernan astronaut moon walk video: https://www.youtube.com/watch?v=7tFP4ha2IOQ

Apollo 16 EVAs 2 (falling down on the Moon): https://www.youtube.com/watch?v=7ciStUEZK-Y

Astronauts tripping on the surface of the Moon: https://www.youtube.com/watch?v=x2adl6LszcE

The third part explained – Closer means more gravity
Newton explains the last part of his Universal Law of Gravitation: https://www.youtube.com/watch?v=sUVUGng05iQ

Gravity goes out away from the Earth forever but it gets weaker. This may seem simple but it has some surprising results. You know that you are attracted to the Earth more than anything else because of its mass and how close it is. You are attracted to the Moon also, but not as much because it is so far away. But what about astronauts and the International Space Station, 250 miles above the Earth? Are they attracted by the Earth's gravity? It seems like they should be but aren't they floating in no gravity? The answer is they are attracted to the Earth almost as much as you are. They are NOT gravityless, but for some reason people think they are. Admit it; you thought there was no gravity pulling on them too, until I told you otherwise. The reason they *seem* to float is for a totally different reason, which I will explain in chapter 32.

For now let's see just how strong the Earth's gravity is out in space. The third part of the Law of gravity basically says that when the distance between objects is doubled the gravity between them becomes $1/4^{th}$ as strong. Now the distance to the center of the Earth is 4000 miles, and we are going to imagine you are standing on a bathroom scale in your house, and your weight is 100 pounds. To reduce your weight to $1/4^{th}$ or 25 pounds, I would have to take you 4000 miles into outer space (double the distance from the Earth's center). If I let you go, you would not float, but would instead slowly fall back to Earth accelerating the whole way until you became a shooting star, burning up in the atmosphere. The International Space station is only 250 miles high so its weight is just about the same (actually about 15% less) as it was on Earth. If I take you double the distance again from the center of the Earth to 12,000 miles, your weight would be about 6 pounds. Double it again to 24,000 miles and you would weight a bit over a pound. By the time I got you to the Moon's orbit (250,000 miles) you would only weigh about $1/250^{th}$ of a pound. But the Earth is still trying to pull you back and if I let you go, down you would go. Slowly at first but accelerating the whole way until you crashed into it.

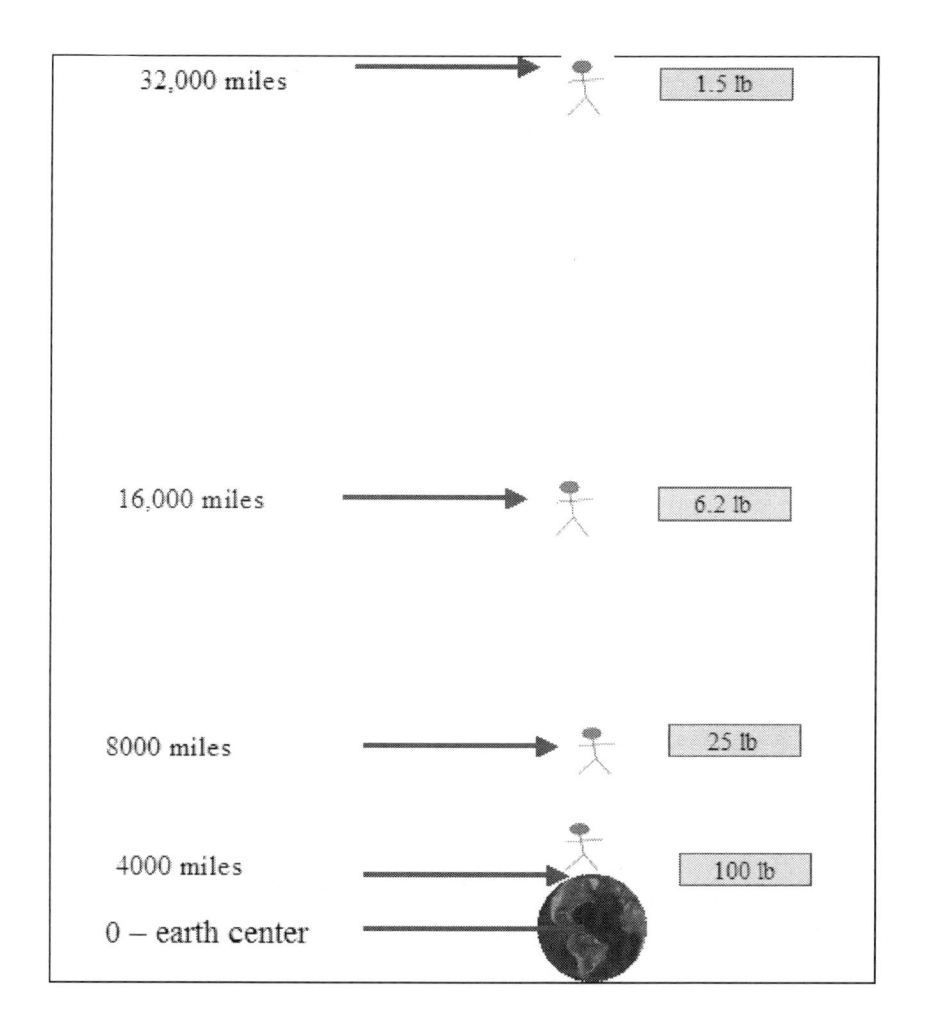

```
32,000 miles  →  🚶  [1.5 lb]

16,000 miles  →  🚶  [6.2 lb]

8000 miles    →  🚶  [25 lb]

4000 miles    →  🚶  [100 lb]
0 – earth center →
```

One last example. Imagine the moon which is being held in orbit by the Earth's gravity. How many times harder would the Earth be pulling on the moon, if it were half the distance? It would be 4 times stronger.

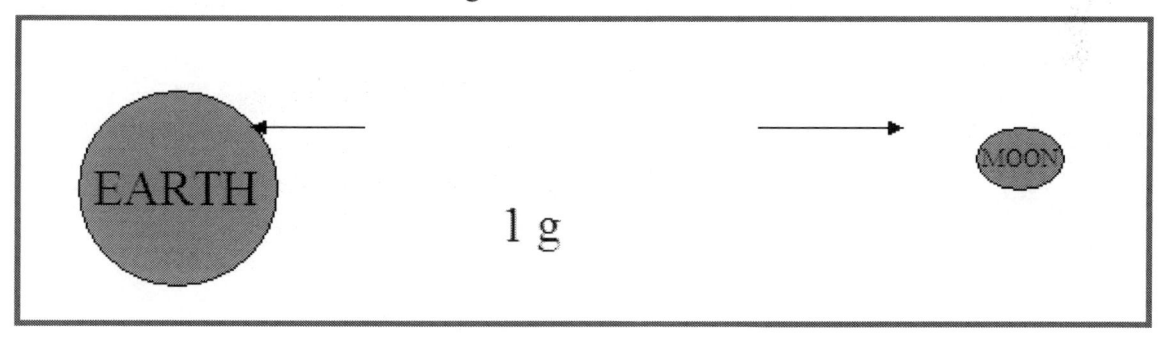

```
EARTH  ←                    →  MOON
            1 g
```

EARTH ←——→ MOON

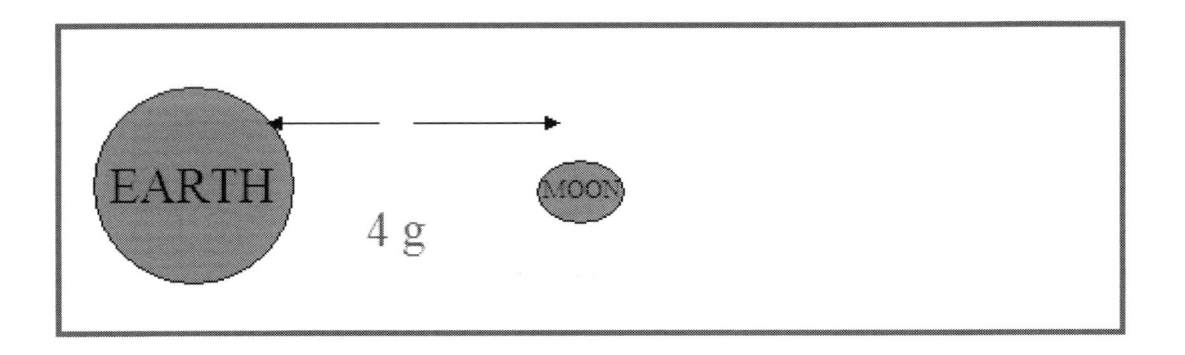

4 g

The Gravity on other planets
The Moon, Mars, Jupiter, the Sun, Pluto, and many other bodies out in space have different masses than Earth. If you could go to the moon, for instance, you would weigh 1/6 of your weight on Earth. This is because the Moon has 1/6 the mass of the Earth. I should add that your *mass does not change* (the molecules in you) only your weight, which is the Moon's gravity pulling on you. Jupiter on the other hand has 2 ½ times more mass than Earth, therefore 2 ½ times more gravity. You would weigh 2.5 times more than on Earth. Pluto is fun, it only has $1/36^{th}$ the mass of Earth, and a 100-pound person would only weigh 6.7 pounds. This is one of the factors that caused Pluto to not be a planet anymore. It is too small!

Try this simulation, where you can put your Earth weight into a formula and see your weight on other planets. It is very cool: http://www.exploratorium.edu/ronh/weight/

Gravity formula
Just to ruin all you fun I am going to show you the formula Newton came up with for calculating the gravity between objects.

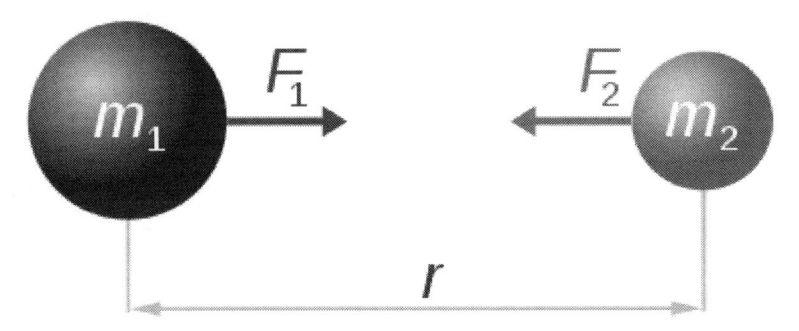

m_1 F_1 →   ← F_2 m_2

r

146

$$F = G \frac{m_1 m_2}{r^2}$$

F = the force between masses
G = gravitational constant (6/674x10-11 n)
m_1 = the first mass
m_2 = the second mass
r = the distance between masses

This is a nice formula and all, but instead of using it you can just go to the following web sites to play around with it:
http://www.ajdesigner.com/phpgravity/newtons_law_gravity_equation_force.php
Or https://www.easycalculation.com/physics/classical-physics/newtons-law.php

Review

Anything with mass has gravity of its own. The more mass an object has, the more gravity it has. The closer two objects are to each other, the stronger the gravitational attraction.

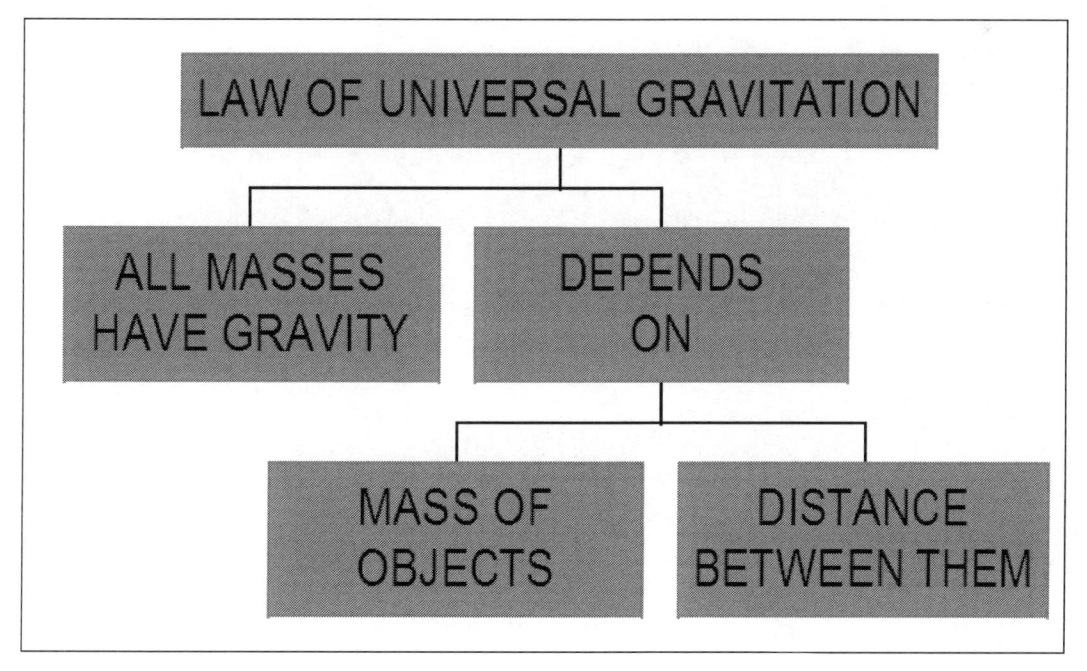

```
                    LAW OF UNIVERSAL GRAVITATION
                    /                          \
         ALL MASSES                    DEPENDS
         HAVE GRAVITY                     ON
                                        /    \
                              MASS OF          DISTANCE
                              OBJECTS          BETWEEN THEM
```

Flashcard at Quizlet: https://quizlet.com/131920256/chapter-11-vol-2-newtons-universal-law-of-gravitation-flash-cards/?new

A day in the life of Earwig Hickson III

I woke up the other day with nothing to do. I was bored and needed a project to keep me out of trouble. I decided to dig to China.

I found a shovel in the shed and got to work. I dug for hours but no China. I kept digging until I was 1 mile deep, still no china. By the time my hole was 100 miles deep, I was getting tired and still no China. When I got to 2000 miles deep, I noticed something, I felt heavier than normal. I climbed out with my very long ladder, and did two things, got the bathroom scale and looked up how far it was to China. It turns out the Earth is 8000 miles in diameter which meant China was 8000 miles away. The center of the Earth is 4000 miles away and I was only half way there. I also learned that as I got closer to the center of the Earth it should pull harder on me and my weight would get greater. I threw the bathroom scale down the hole and made the long climb down.

Now my weight at the surface was 100 pounds, but when I checked my weight at the bottom of my hole it read 400 pounds. This was interesting so I kept digging. I dug 1000 more miles (half way to the center of the Earth) and my weight went up to 1600 pounds. Half way deeper (500 more miles) and my weight went up to 6400 pounds. Every time I got halfway closer to the center of the Earth my weight increased 4 times.

Eventually I got to the center of the Earth and started digging *up* to China. By dinner I was done. I had dug the first hole to China. I climbed back up my very long ladder and decided to do an experiment. I would jump in the hole to see what would happen!

I jumped and accelerated the whole way. When I got to the Earth's center I just kept going, only now I was slowing down. When I got to China, I came to a stop, but before I could climb out, I started falling back down. I went back and forth all the next day until I stopped at the Earth's center. Then I climbed out.

That would have been the end of my experiment except that I covered the hole with a big sheet of cardboard. Unfortunately my Dad went outside that night for some reason, which he later regretted, and in the dark, stepped on the cardboard. We did not see him until the next day when he climbed out. He was not happy.

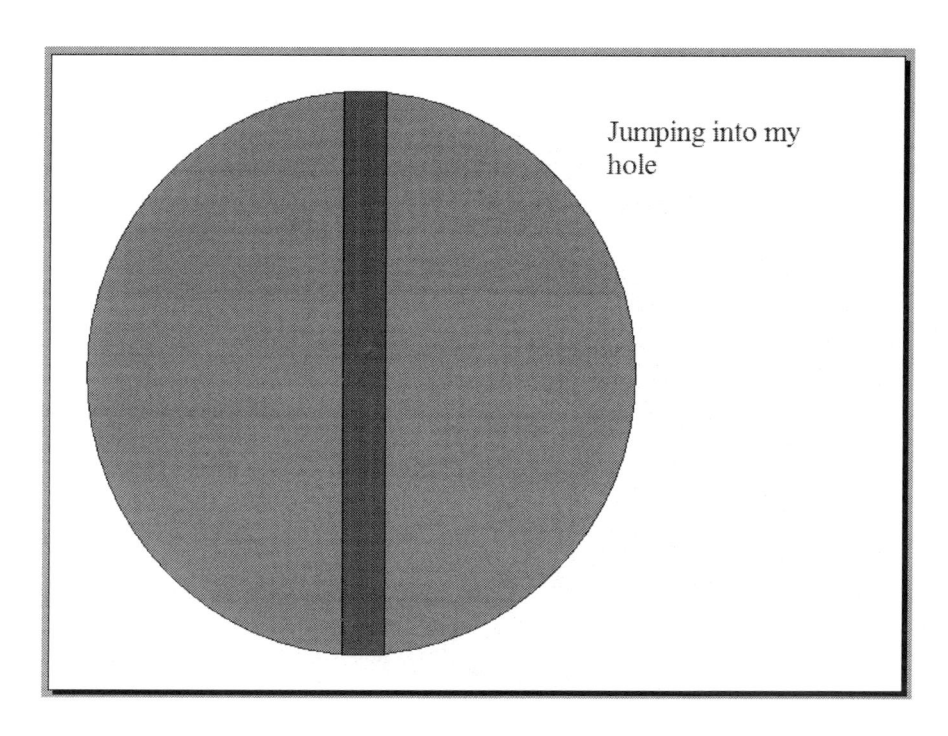

[Figure: A circle (Earth) with a vertical tunnel through its center, labeled "Jumping into my hole"]

Links
Gravity hill in Lewisberry PA: https://www.youtube.com/watch?v=bsMZkm66uSI

Newton explains the first two parts of his Universal Law of Gravitation: https://www.youtube.com/watch?v=kgtfRTKYIsw

Newton explains the last part of his Universal Law of Gravitation: https://www.youtube.com/watch?v=sUVUGng05iQ

What would happen if I drilled a tunnel through the center of the Earth and jumped into it?
http://science.howstuffworks.com/environmental/earth/geophysics/question373.htm

A video showing what would really happen if you dug a hole all the way through the Earth... and Jumped In??? Seriously!
https://www.youtube.com/watch?v=6TZVCxCiMHE

Gravity train – travel anywhere in the world in 42 minutes: https://www.youtube.com/watch?v=EapvQ3ALYJY

Find out how much you weigh on other planets: http://www.exploratorium.edu/ronh/weight/

A fun little game where you can make a solar system complete with planets and asteroids and see how gravity affects them:
http://www.testtubegames.com/gravity_flash.html

Super planet crash game: http://www.stefanom.org/spc/#

Gravity Launch. A gravity game where you launch a rocket and try to orbit the moon and hit things: http://sciencenetlinks.com/interactives/gravity.html

The strange fate of someone falling into a black hole:
http://www.bbc.com/earth/story/20150525-a-black-hole-would-clone-you

How is time changed inside a black hole:
http://www.skyandtelescope.com/astronomy-resources/time-changed-inside-a-black-hole/

Black Holes Explained – From Birth to Death (5 minute video):
https://www.youtube.com/watch?v=e-P5IFTqB98

What Would Happen if You Fell Into a Black Hole? (6 minute video):
https://www.youtube.com/watch?v=6H6CcbIMH7I

Black Holes - Mysteries Of The Universe HD Documentary (a 45 minute video):
https://www.youtube.com/watch?v=BWn29kYjOZQ

Neil DeGrasse Tyson - Death By Black Hole:
https://www.youtube.com/watch?v=h1iJXOUMJpg

How tides work: https://www.youtube.com/watch?v=NqDEaFjIXPw

Falling into a black hole: The singularity and spagettification
https://www.youtube.com/watch?v=OGn_w-3pjMc

Apollo 17 Bunny hopping on the Moon Gene Cernan astronaut moon walk video:
https://www.youtube.com/watch?v=7tFP4ha2IOQ

Apollo 16 EVAs 2 (falling down on the Moon):
https://www.youtube.com/watch?v=7ciStUEZK-Y

Astronauts tripping on the surface of the Moon:
https://www.youtube.com/watch?v=x2adl6LszcE

Fun things to Google
Gravity train
Newton's Law of Universal gravitation
Hole through the center of the Earth
Why are astronauts weightless?
Your weight on other planets
Gravity simulations
Black holes
Neutron stars
Wormholes

Gravity train

Chapter 30
Falling objects

Video of this power point lesson:
https://www.youtube.com/watch?v=gJRvMSMdcfc

The wonders of science

His name was Galileo Galilei. He was smart; He was an Italian astronomer in the 1600's. At the time, everyone believed the Earth was the center of the universe, they kind of had to, it was the law. Back then if you challenged the establishment, you were punished. This was the time of the Spanish Inquisition, which did not want anyone to upset the "current beliefs". Science was not popular. Galileo was caught in this mess, and as a scientist, got into major trouble. He liked to experiment with things he had observed; at first it was no big deal. He noticed that a swinging chandelier swung the same number of times per second no matter how far it was swinging. He experimented with pendulums and discovered it was the length of the string that mattered, not the mass or how big the swing (https://phet.colorado.edu/en/simulation/pendulum-lab). He also started doing experiments about falling objects.
https://en.wikipedia.org/wiki/Galileo's_Leaning_Tower_of_Pisa_experiment.

It is said that he dropped different size cannon balls from the leaning tower of Pisa, and they landed at the same time. The problem with this story is that the leaning tower of Pisa https://en.wikipedia.org/wiki/Galileo's_Leaning_Tower_of_Pisa_experiment was not leaning in his day; it would have been the upright tower of Pisa. What he did was measure the rate of objects rolling down an incline plane (with his pulse, since there were no clocks back then), but he came up with the correct answer anyway. All falling objects accelerate at the same rate toward the Earth. Newton explains Galileo's experiment: https://www.youtube.com/watch?v=TxCLqbNEYgg

This idea eventually evolved into Newton's Universal Law of Gravitation. All falling objects fall at the same acceleration, and in an Ultra-high vacuum would land at the same time. Unfortunately for Galileo, this went against common sense.

The final nail in Galileo's coffin was his observations about Jupiter's moons. You see, he built a telescope, and in it he saw moons orbiting the planet Jupiter. This was not to be accepted, at the time, since everything was supposed to orbit the Earth. This supported the ideas of Copernicus, and his theory of "Commentariolus", which stated that the sun was the center of the solar system, and the Earth revolved around it. This was a controversial idea, in his day, and not a popular one. Galileo began to believe this idea, that the sun was the center of the solar system, or, more importantly, that the Earth was not the center of the universe, and he paid dearly. In 1633, he was placed under house arrest for the next 300 years http://www.history.com/this-day-in-history/galileo-is-convicted-of-heresy; He was eventually cleared in 1990, 300 years after his death. Scientists had a hard life back then.

Now Galileo is considered the father of the scientific method. He has come a long way.

What you need to remember
The larger the mass an object has, the stronger its gravity.

The closer the object, the more its gravity affects you.
Newton's second law formula. F = ma

Your weight

You know things fall when you drop them, and you know gravity causes this. You probably even know that gravity causes you to have weight, so what else is left? There is actually quite a bit.

First a quick review of Newton's Second Law of motion and its formula. You may remember the formula

$$F = M \times A$$

F means force, and since weight is a force I can substitute W for F. A means acceleration and since gravity is an acceleration, I can substitute G in place of A.

$$W = M \times g$$

- (g = GRAVITY)
- W = WEIGHT (FORCE OF GRAVITY)

With this formula I can calculate my weight on any planet I wish, if I know the acceleration of gravity for that planet and my mass in kg.

> MY WEIGHT ON EARTH WHICH HAS AN ACCELERATION DUE TO GRAVITY OF ABOUT 10 M/S/S
>
> W = M x G
>
> W = 90 KG x 10 M/S/S
>
> W = 900 N
>
> - 900 N IS ABOUT THE SAME AS 200 POUNDS

> MY WEIGHT ON THE MOON WHICH HAS AN ACCELERATION DUE TO GRAVITY OF ABOUT 1.6 M/S/S
>
> W = M x G
> W = 90 KG x 1.6 M/S/S
> W = 144 N
>
> - 144 N IS ABOUT THE SAME AS 33 POUNDS

Of course why use the formula when you can just go to http://www.exploratorium.edu/ronh/weight/ and have it calculate it for you. Now you know how they do this.

Falling things

Things fall down, but what direction is down? Is it the same for everyone? Not really. Down actually means toward the center of the Earth. To us the center of the Earth is on the other side of our feet so we call that direction down. But what direction is down in China on the other side of the Earth? To a person in China, it is toward the center of the earth (and thus their feet) also, but to US, when they drop something it goes in the direction for OUR up. To us they are kind of upside down.

154

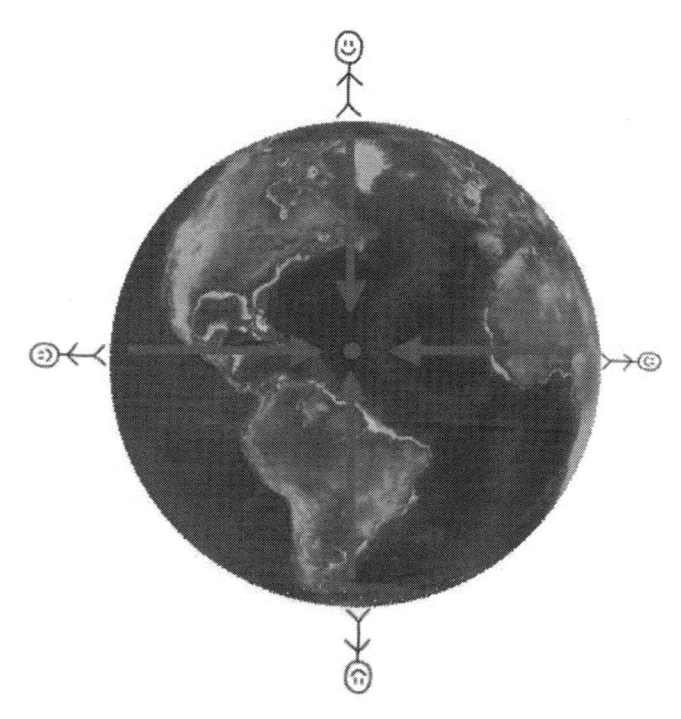

So, falling objects go down but did you know they *accelerate the whole way*? They actually accelerate 10 m/s faster each second they fall, for an acceleration due to gravity of 10m/s/s (for you that comes out to about 32 ft/s/s). You can actually estimate the depth of a deep hole by counting how many seconds it takes a rock to hit the bottom. You do not notice this acceleration much because air resistance kind of messes things up. If there was no air in the way it would happen every time and falling objects would be going incredibly fast.

This is what a hammer would do if dropped with NO AIR in the way. Notice how the hammer falls faster each second it falls.

155

0 SECONDS= 0 M/S

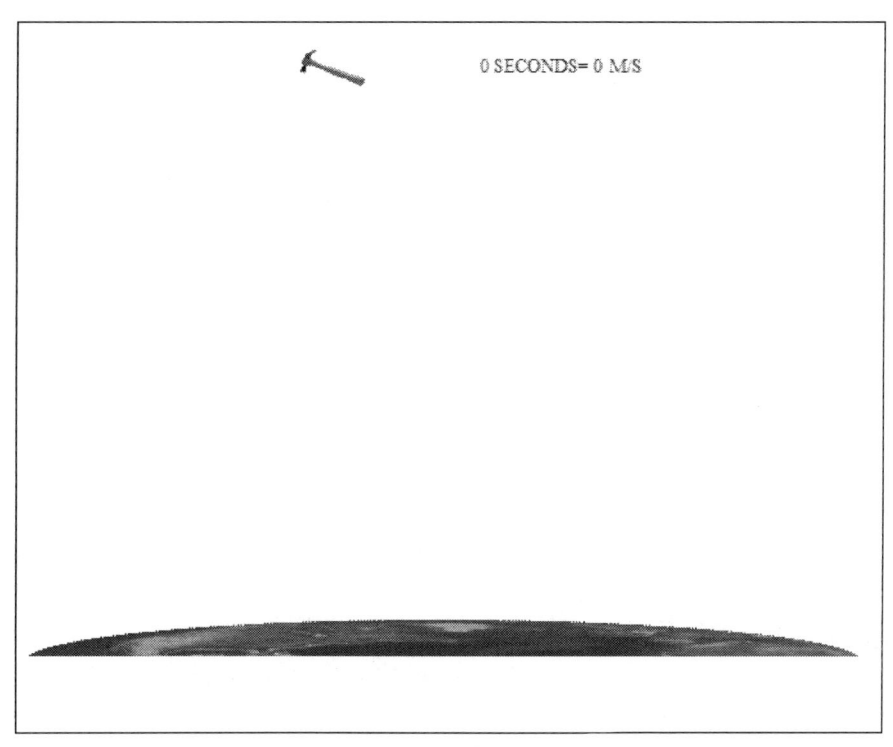

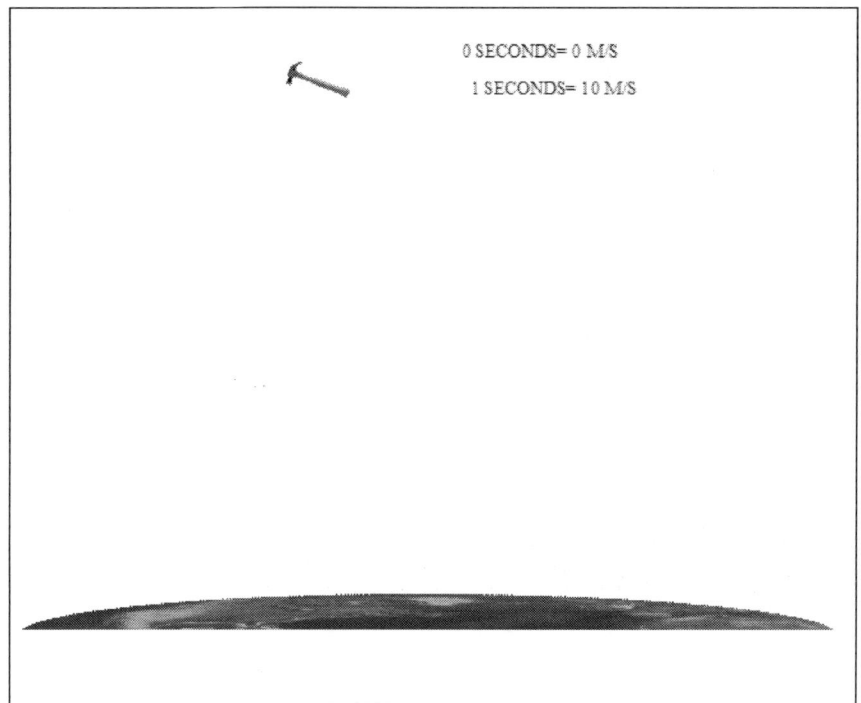

0 SECONDS= 0 M/S
1 SECONDS= 10 M/S

0 SECONDS= 0 M/S
1 SECONDS= 10 M/S
2 SECONDS= 20 M/S

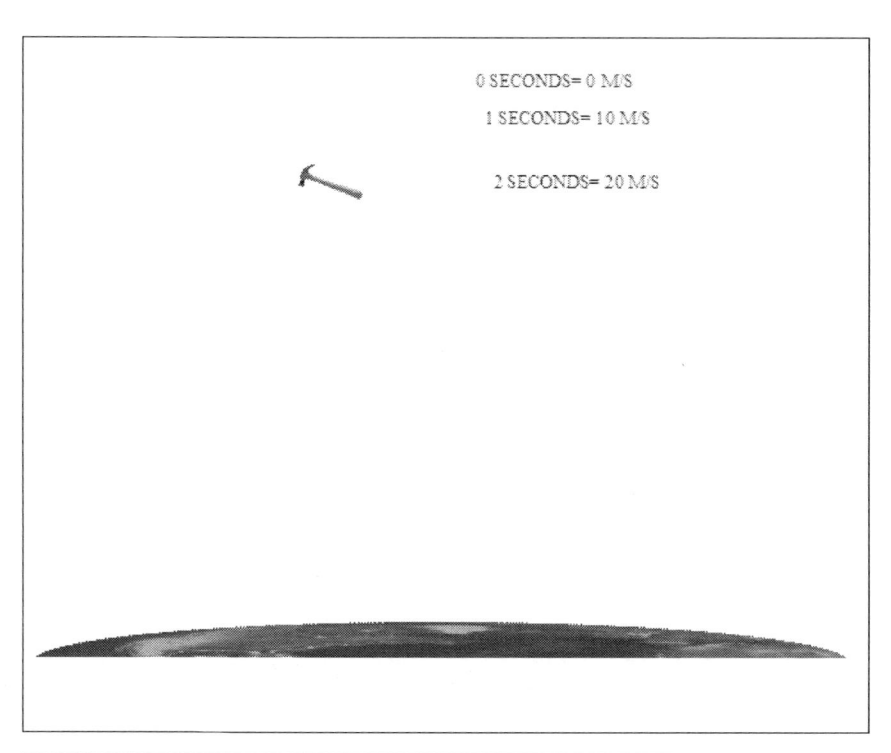

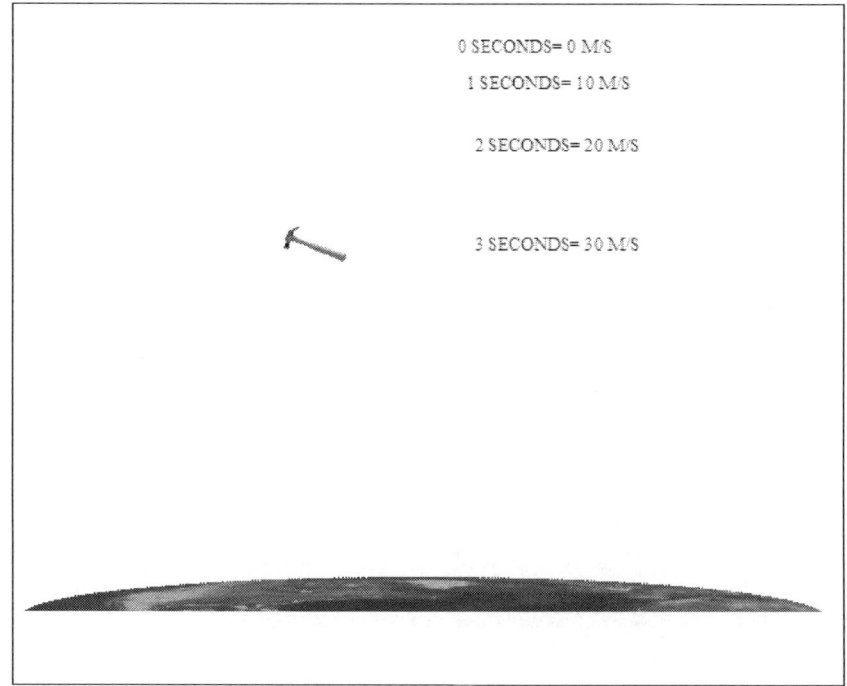

0 SECONDS= 0 M/S
1 SECONDS= 10 M/S
2 SECONDS= 20 M/S
3 SECONDS= 30 M/S

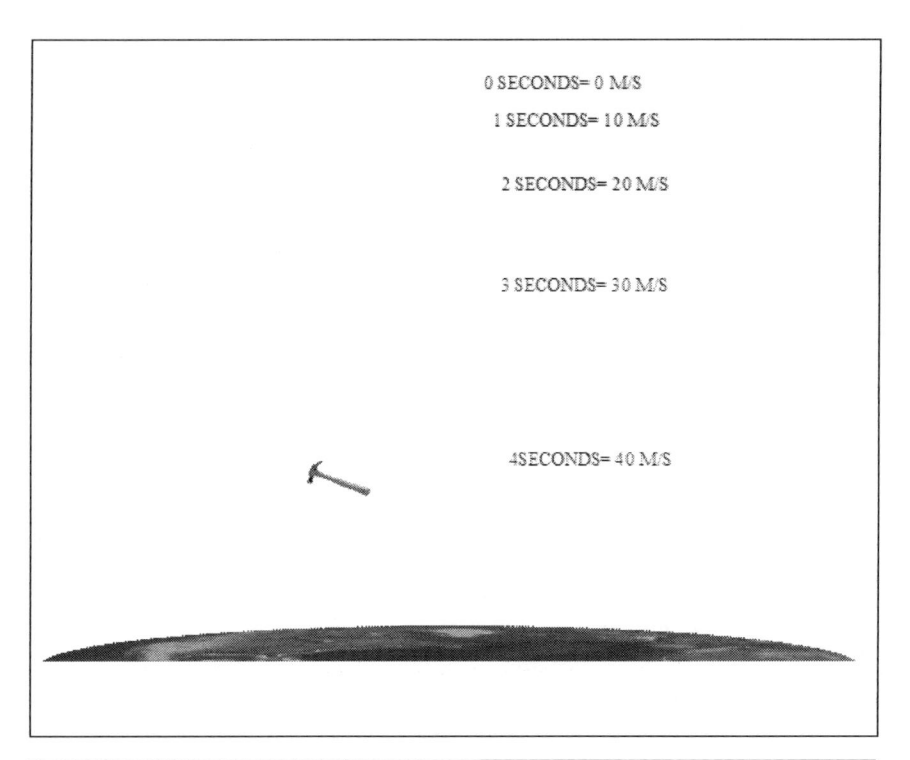

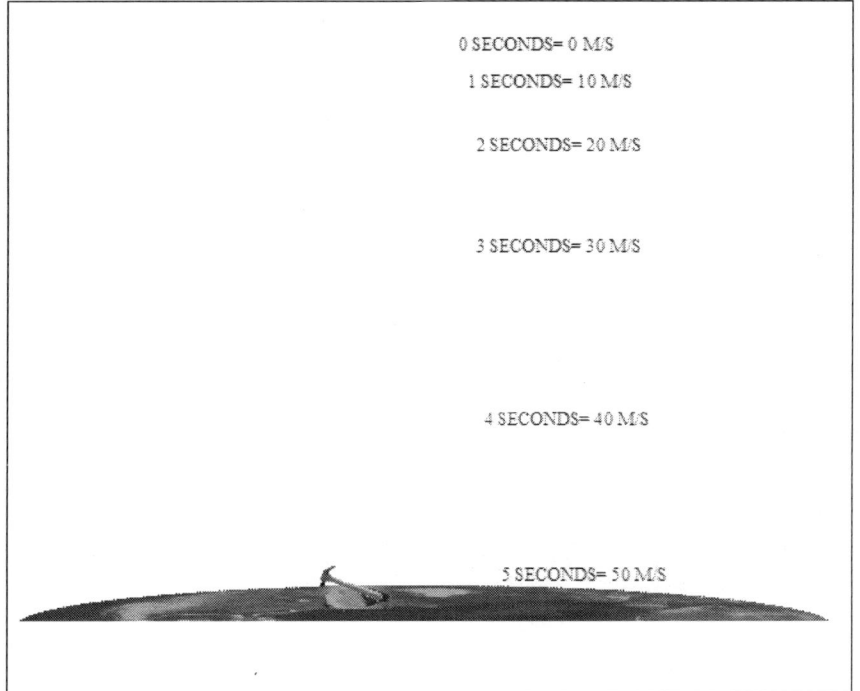

```
0 SECONDS= 0 M/S
1 SECONDS= 10 M/S
2 SECONDS= 20 M/S
3 SECONDS= 30 M/S
4 SECONDS= 40 M/S
```

```
0 SECONDS= 0 M/S
1 SECONDS= 10 M/S
2 SECONDS= 20 M/S
3 SECONDS= 30 M/S
4 SECONDS= 40 M/S
5 SECONDS= 50 M/S
```

Now for the weird part, the mass of the object does not matter. That is correct. A feather would accelerate at 10 m/s/s just like the hammer, as long as *no air is in the way*. This is what Galileo proved so long ago. This implies that if you were to drop a hammer and a feather at the same time they should hit the ground at the same time! In fact they do, but only if there is no air in the way. This experiment was actually done on the moon where there is gravity but no air. I was eight years old when I saw that live on T.V., and I was amazed. We just do not see that on Earth with its air molecules.

Hammer and feather dropped on the moon:
https://www.youtube.com/watch?v=oYEgdZ3iEKA

- ALL FALLING OBJECTS (NO MATTER THE MASS) ACCELERATE AT THE **SAME RATE**
- BECAUSE THERE IS ONLY **ONE** EARTH
- PULLING WITH ONLY **ONE** ACCELERATION

THIS MEANS

- HEAVY THINGS FALL AT THE SAME <u>ACCELERATION</u> AS LIGHT THINGS
- THEY *WANT TO* LAND AT THE SAME TIME
- AIR FRICTION (DRAG) MESSES IT UP

This is what is happening on the moon.

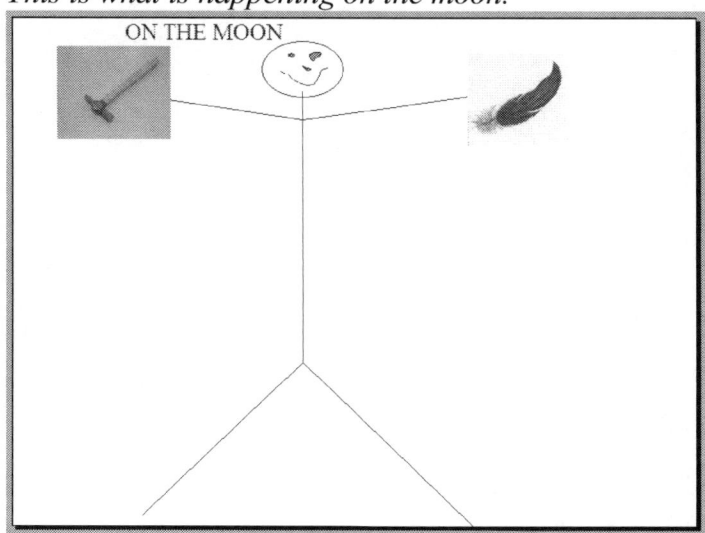

ON THE MOON

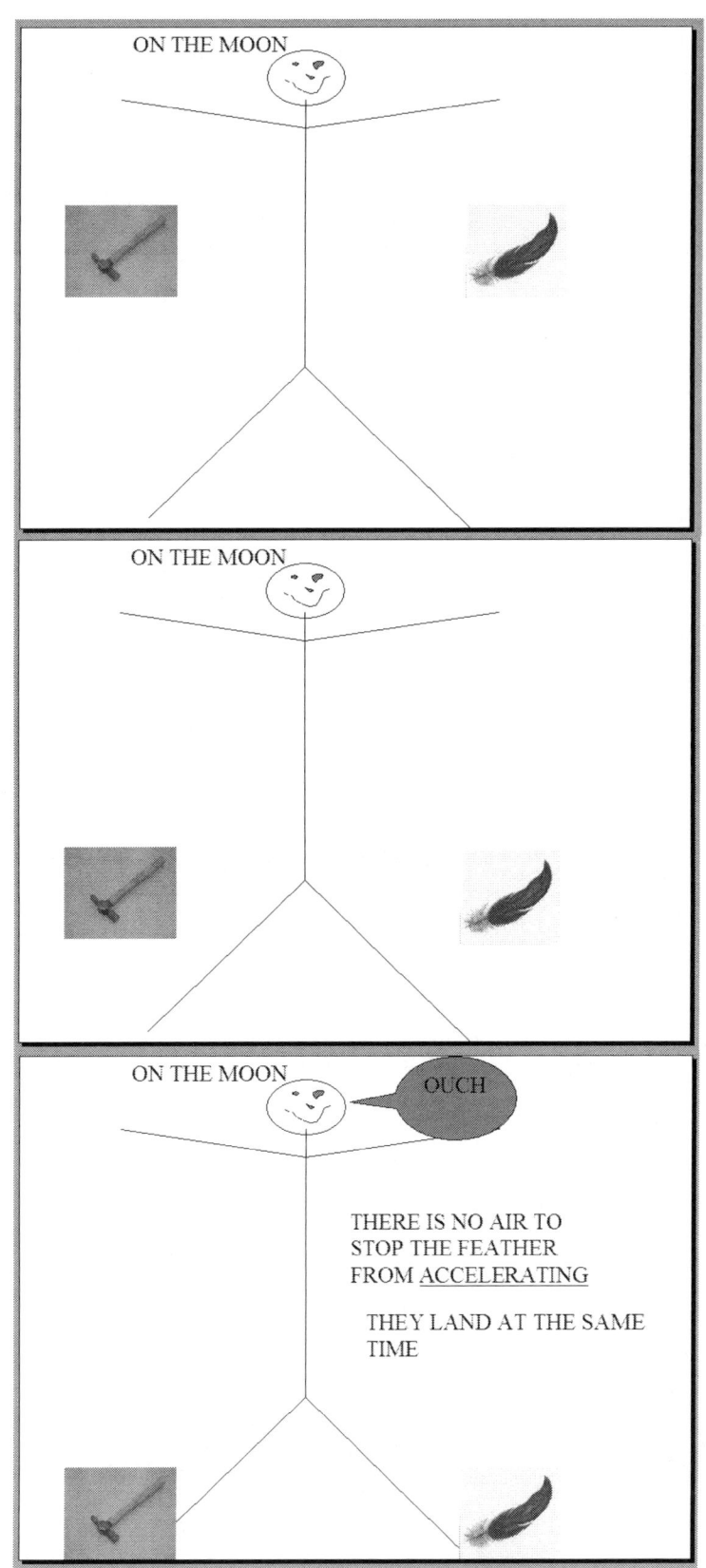

But on Earth where there is an atmosphere and a lot of air, friction stops the feather from accelerating, so it lags behind, but the hammer just plows through.

ON EARTH THERE IS AIR IN THE WAY

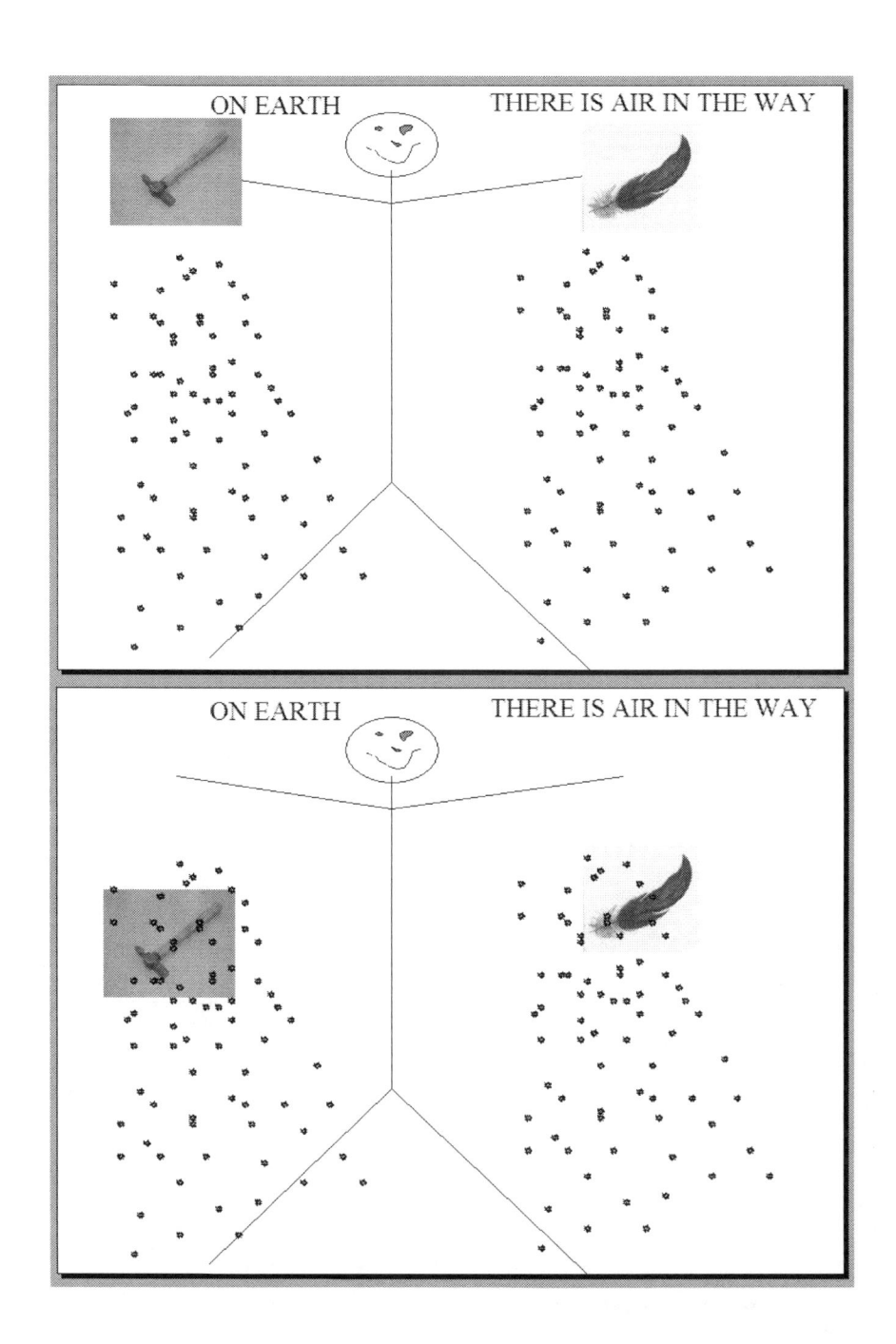

ON EARTH THERE IS AIR IN THE WAY

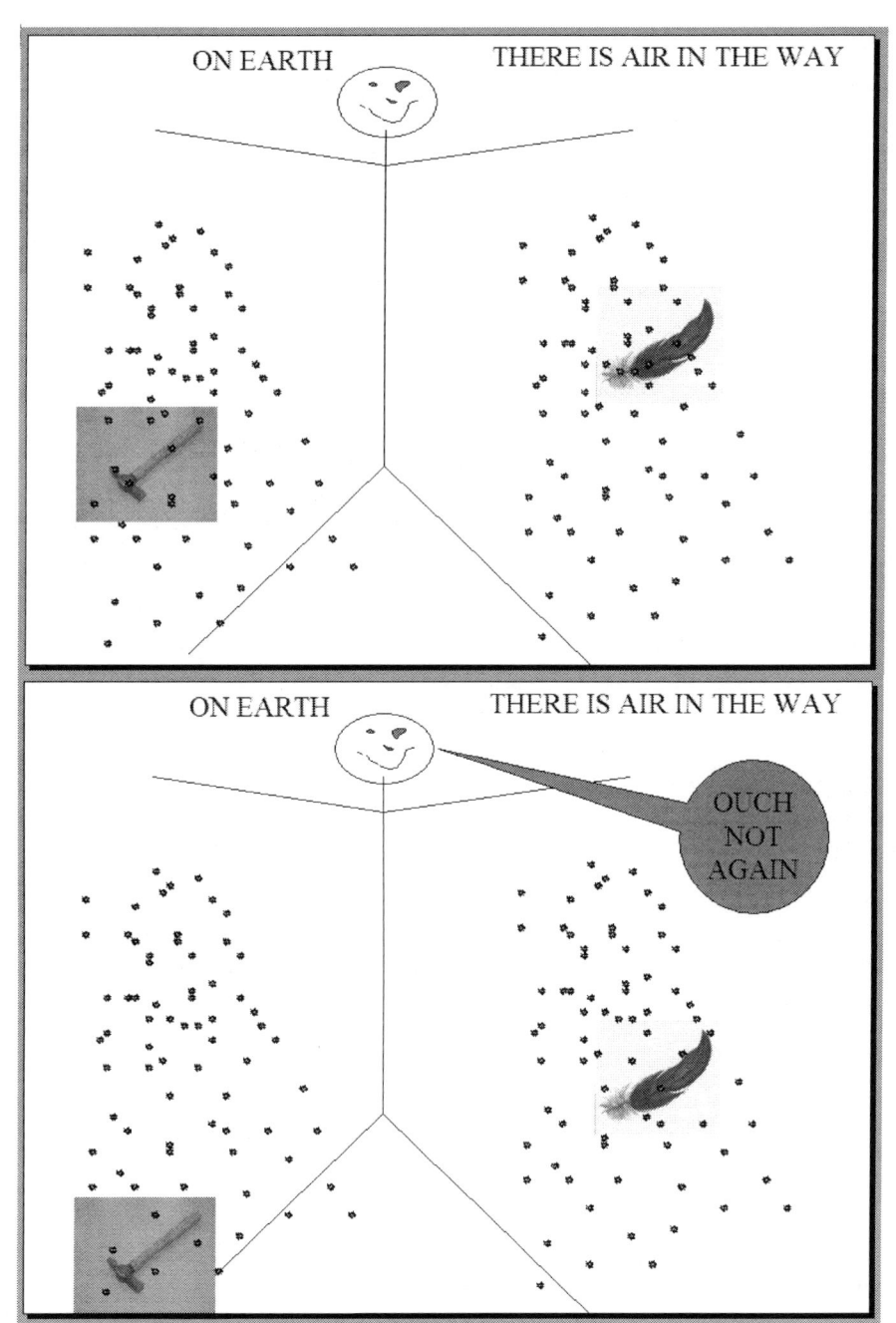

For a really cool video of how this works check out this link where a bowling ball and some feathers are dropped on a total vacuum: https://www.youtube.com/watch?v=E43-CfukEgs

Acceleration due to gravity on other planets

Since every planet has a different mass, and gravity comes from mass, they each have a different acceleration due to gravity.

PLANETS

- THE EARTH'S ACCELERATION OF GRAVITY IS 10 M/S/S (9.8)
- EACH PLANET HAS A **DIFFERENT** ACCELERATION OF GRAVITY
- ON OTHER PLANETS
- YOU WOULD WEIGH DIFFERENT
- BUT YOUR **MASS** IS THE **SAME**

Here is the web site for calculating you weight on other planets again, since I think it is cool: http://www.exploratorium.edu/ronh/weight/

Terminal Velocity

Now that we know falling objects want to fall at the same rate, but in air we know they do not, let's look at this closer.

Depending on the size (not the mass) or more properly the surface area compared to the mass or the density, air resistance affects different things differently. A large feather has a large surface area and a small mass, meaning a small density (feathers are fluffy). Feathers have a lot of surface area to hit, so the acceleration stops quickly. (It does not slow it down, just stops it from going faster). When Galileo did his gravity experiments he used iron cannon balls of different sizes, but the same density.

AIR RESISTANCE

- AIR RESISTANCE STOPS FALLING OBJECTS FROM ACCELERATING
- IT DOES NOT STOP THEM
- IT DOES NOT SLOW THEM DOWN
- JUST STOPS THEIR ACCELERATION

The terminal velocity for a typical skydiver is about 120 miles per hour (baggy clothes would reduce this though). This means a skydiver will accelerate for about 11 seconds and then fall at a constant velocity for the rest of the way. When he opens his parachute he slows down to about 10 miles per hour for a softer landing, since the parachute has more surface area for the air to hit. By the way do you think a parachute will help an astronaut on the moon? Since there is no air on the moon astronauts and parachutes land at the same time, no help at all. Terminal velocity does not exist in a vacuum (area of no air).

> - **AIR RESISTANCE DOES NOT EXIST IN A VACUUM**
> - **(VACUUM IS AN AREA OF NO AIR)**
> - **THE MOON**

Review
All falling objects accelerate at the same rate. In a vacuum all falling objects would land at the same time. Terminal velocity is the maximum speed a falling object can reach due to air friction.

Flash cards on Quizlet for this chapter: https://quizlet.com/143903012/chapter-30-falling-objects-flash-cards/?new

A day in the life of Earwig Hickson III

It was my 8th birthday and for my present, I wanted to go skydiving. I had seen it on T.V. and it looked fun! My Mom did not think so and refused my request. For my birthday she said we were going to visit the Grand Canyon. My Dad got me new underwear.

So off we went on our trip to the Grand Canyon. It was cool, I will admit, but it sounded more like a present for my Mom and not me.

We were staying at a fancy hotel where your sheets were actually changed everyday and even the towels were washed. It was a nice place. I used to wander the halls while my family was still asleep and that is when it happened. I saw the dirty sheets hamper sitting in the hall, with no one watching it. I did not see dirty sheets. I saw a parachute.

I started sewing. Soon I had a giant sheet of cloth, just right for a parachute. My family was still asleep. I tied some rope to the parachute and then to me. I snuck out of the hotel to the edge of the canyon. I was getting my nerve up when just as I was ready to jump, a Park Ranger saw me. He forbid me from jumping. He said it was illegal and a very dumb thing to do. It was at that time I realized just how stupid an idea it was. The Grand Canyon is a mile deep with sharp rocks all the way down. I realized I was afraid of heights. I made up my mind to abandon my experiment and go back to the hotel, but then I tripped.

Off the edge I went, falling by the sharp rocks faster and faster, my parachute fluttering behind. Suddenly the parachute caught some air and inflated, I felt yanked upward, as my velocity decreased. I was floating down very gently. I took some pictures as I drifted to the bottom of the canyon.

I landed gently near the Colorado River. I was far from my hotel. I had to run back up. When I returned to the hotel, I was exhausted. I fell into bed for a nice long sleep. Ten minutes later my Mom woke me up with a surprise. "Happy Birthday, Earwig, we are going to walk to the bottom of the Grand Canyon today. What a story that will make to tell your friends!"

Links
A basketball and a 5 pound medicine ball dropped at the same time:
https://www.youtube.com/watch?v=aRhkQTQxm4w

Galileo's experiment explained by Newton:
https://www.youtube.com/watch?v=TxCLqbNEYgg

Misconceptions About Falling Objects: https://www.youtube.com/watch?v=_mCC-68LyZM

How Far Away is the Moon? (The Scale of the Universe):
https://www.youtube.com/watch?v=Bz9D6xba9Og&nohtml5=False

Pendulum lab – Phet - (https://phet.colorado.edu/en/simulation/pendulum-lab

The problem with this story is that the leaning tower of pizza
https://en.wikipedia.org/wiki/Galileo's_Leaning_Tower_of_Pisa_experiment

Hammer and feather dropped on the moon:
https://www.youtube.com/watch?v=t4JeussHgk4

A really cool video of how this works. Check out this link where a bowling ball and some feathers are dropped on a total vacuum: https://www.youtube.com/watch?v=E43-CfukEgs

Here is the web site for calculating you weight on other planets again, since I think it is cool: http://www.exploratorium.edu/ronh/weight/

Fun things to Google
Skydiving
Galileo
Falling objects in a vacuum
Hammer and feather on the moon
Terminal velocity
Weight on other planets
Acceleration due to gravity

Chapter 31
Gravity in the solar system

Video of this power point lesson: https://www.youtube.com/watch?v=TZOhcfcdpXk

The wonders of science

The Earth is very small, almost insignificant when compared to the other objects in the universe.

The Solar system is big, very big. Not as big as the Milky Way galaxy, or even the space between stars, or the universe, but it is big enough. The closest object to the Earth is the moon and that is 240,000 miles away; it takes 3 days for a spacecraft to get there. Mars at its closest approach to Earth is 34 million miles away; so far that at its closest it would take seven months to get there. Jupiter at its closest is 365 million miles away, at its farthest 600 million. It took the New Horizon space probe 9 ½ years to travel 3 billion miles to Pluto.

The solar system is huge, and the planets are always moving so sometimes they are closer to the Earth than at other times. Humans are exploring the solar systems with unmanned space probes.

Notice the planets are not just lined up like in a model; some are on each side of the sun.

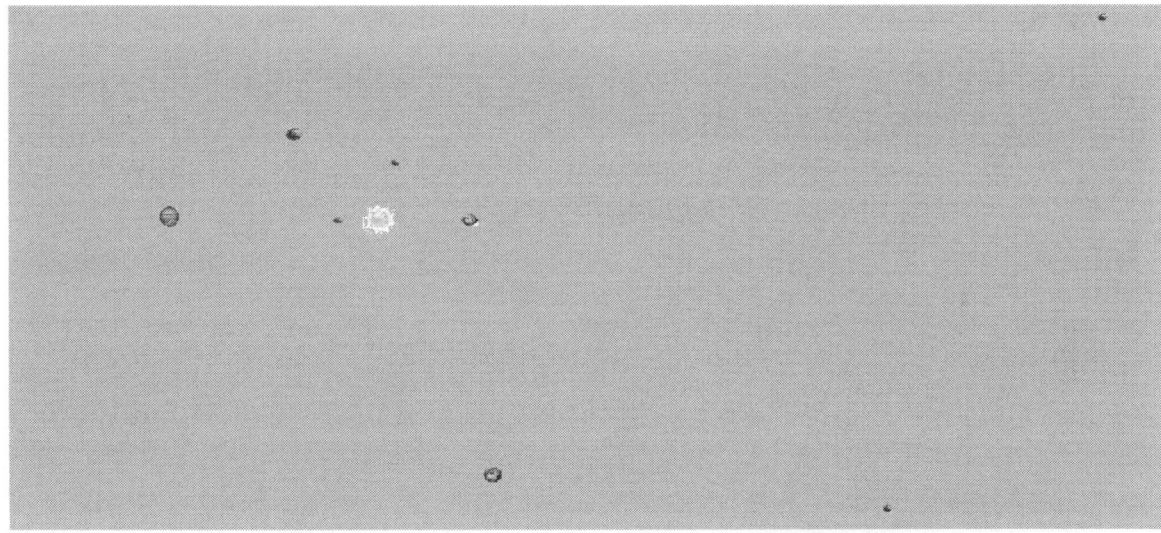

Traveling long distances like this is one thing, waiting for the target planet to get close enough another. The real problem is fuel. There are no gas stations on the way. A spacecraft must carry enough fuel to escape the Earth's gravity and make a billion mile trip. How does NASA do this? They cheat. They use a trick called gravity assist or planetary slingshot. What they do is increase the velocity of the space probe without using any fuel; instead they use the *momentum* of the planet (remember mass x velocity). There is a lot of momentum in a moving planet. The spaceship flies close to the planet and steals some of the planet's momentum. The planet slows down a tiny bit and the probe increases velocity a lot. Using this method NASA has been able to build smaller, cheaper rockets that go farther and faster. There is a lot to explore out there.

A video explaining gravity assist: https://www.youtube.com/watch?v=0iAGrdITIiE

What you need to remember
The universal law of gravity states:
Anything with mass has gravity of its own.
Larger masses have more gravity than small masses.
The closer together two objects are, the more their gravity effects each other.

Gravity in our solar system
The planets
All planets have gravity of their own. The sun has gravity of its own, and every asteroid, and comet has gravity of its own. All these objects are attracted to every other object in the solar system, but the sun is the boss, since it has the most mass. The sun is at the center of the solar system and all the other objects orbit around it, but not in a circle, in an ellipse or flattened circle. That means the Earth (which also orbits in an ellipse) is sometimes closer to the sun than at other times.

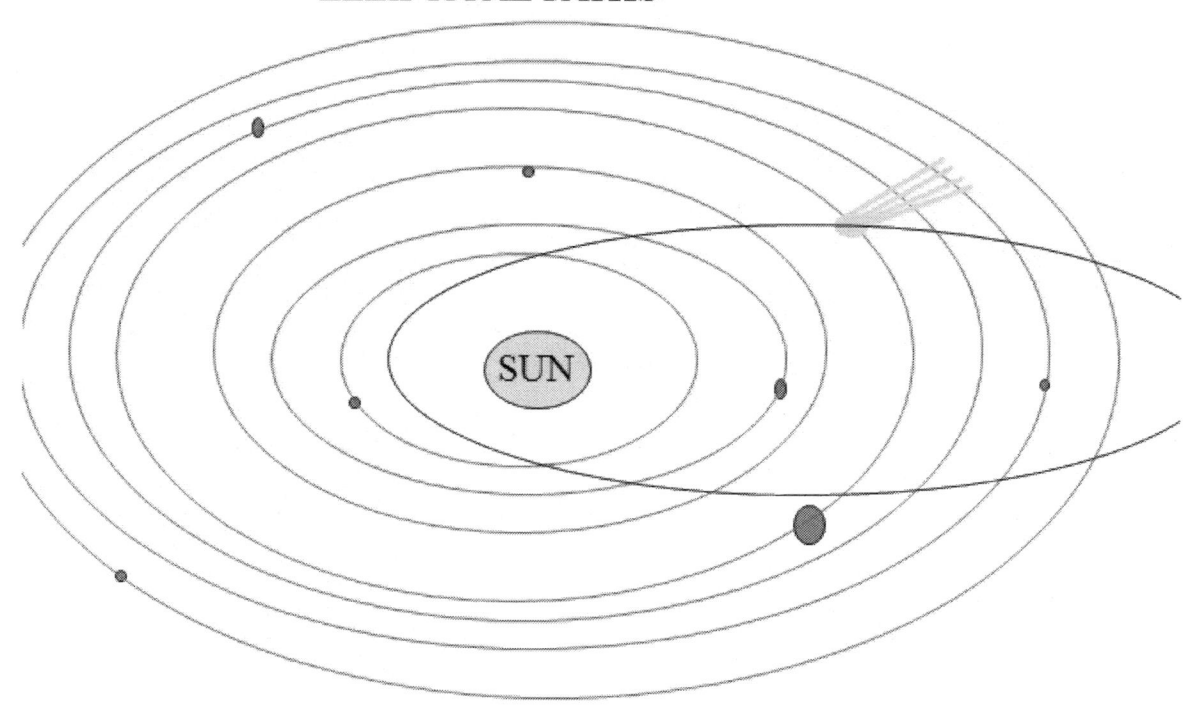

ELLIPTICAL PATHS

Another neat thing about the orbit of the planets is that the closer to the sun they are the faster they move. Mercury, the closest goes around the sun in 88 days; the Earth 365 days, Jupiter, 12 Earth years. Closer means faster. Some planets have rings, Saturn has the biggest but other planets have them too (Jupiter, Uranus, and Neptune). No one is sure what causes them but they are made of dust and ice particles and their formation probably has something to do with the planets orbiting moons and the dust being too far from the planet for the gravity to pull it in. Each planet has a different acceleration due to

gravity, Jupiter has the most. Find your weight on other planets:
http://www.exploratorium.edu/ronh/weight/

Eclipse
Every now and then the Earth, Moon, and the Sun line up in such a way that a shadow is cast on either the Earth or the moon. A lunar eclipse is the most common to see, and is when the Earth casts a shadow on the moon. The moon can sometimes turn red during a lunar eclipse. A lunar eclipse always occurs at night (lunar – moon = night) and the Earth is always between the sun and the moon. Find the next lunar eclipse in your area: https://www.google.com/#q=when+is+the+next+lunar+eclipse+in+the+united+states
Video of a lunar eclipse: https://www.youtube.com/watch?v=lcRp1jKJmJU

• IN A LUNAR ECLIPSE
• SUN – EARTH - MOON

A solar eclipse is when the moon casts a shadow on the Earth. These are the least common to see because the shadow is usually cast on another part of the Earth. Solar eclipses are the coolest. The size of the shadow is not very big so only a small part of the world can see a total solar eclipse when they occur. I have seen three and none were a total eclipse. A solar eclipse can only occur in the daytime (solar = sun = day) and the moon is always between the Earth and the Sun. As I write this the next solar eclipse to be seen in the United States is on August 26, 2017. The total Eclipse will be seen 400 miles from my house, and I will have to work that day. Find the next solar eclipse in your area: https://www.google.com/#q=when+is+the+next+solar+eclipse+in+the+united+states

If you are lucky enough to see a solar eclipse make sure you do not look directly at it without a special filter, it can blind you. You can look directly at a lunar eclipse if you want. How to view a solar eclipse safely: http://www.timeanddate.com/eclipse/eclipse-tips-safety.html

Video of a solar eclipse: https://www.youtube.com/watch?v=3_qo2CdcyC0

ORDER OF A SOLAR ECLIPSE

- SUN – MOON – EARTH

SOLAR ECLIPSE

The tides

When I go to the beach I love to build a sand castle at low tide and try to save it as the tide comes in. I always fail and realize there is something wrong with me but that is what I like to do. In the Bay of Fundy in Canada the difference between low tide and high tide can be 45 - 50 feet! At the beach I go to, it is only a few feet. Fishermen watch the tides since the fish follow them, so do boaters. Tides are interesting but what causes them?

Video of the tide at the Bay of Fundy:
https://www.youtube.com/watch?v=OP0cpXpw8yk

If you did not guess already the tides are caused by the gravitational pull of the Moon (and a lesser existent the Sun). Just as the Earth pulls on the Moon, the Moon pulls on the earth. Since water is more flexible than land, the oceans actually slosh around the Earth making one part deeper and one part shallower. A weird part, that I am not going to get into is, there are actually two high tides and two low tides per day, so the tides change about every six hours. The second high tide is on the opposite side of the Earth from the moon, caused by the Earth being pulled toward the Moon, leaving the water behind and

deeper. Now the scene in the movie *Despicable Me* should make sense:
https://www.youtube.com/watch?v=gAmo3FcaovM

What causes the tides video: https://www.youtube.com/watch?v=5ohDG7RqQ9I

A tidal bore, where the high tide actually comes up a river like a wave:
https://www.youtube.com/watch?v=LWumonz87rA

The seasons

Due to the elliptical path the Earth's orbit around the sun, it is a different distance from the sun at different times of the year. I know what you are thinking, in the summer it must be closer and in the winter it must be farther away, but NOOOOOOOO! This is not how it works. In fact the Earth is closer to the Sun in the winter (northern hemisphere) and farther away in the summer. The distance of the Earth to the sun DOES NOT CAUSE THE SEASONS; the seasons are caused by a totally different reason.

The Earth is *tilted* relative to the sun at about 23 degrees; this is what causes the seasons. In the summer the Earth is *tilted toward* the sun and in the winter it is *tilted away*. When sunlight hits the atmosphere it can either go right in (to the surface), or skip off (into space). It is kind of like throwing a rock into a pond. If you throw the rock up into the air, it goes straight to the bottom (like sunlight in summer), but if you "skip" a flat rock parallel to the water, it skips across (like sunlight in winter). The amount of sunlight that reaches the Earth's surface is what causes the seasons.

Bill Nye explains what causes the seasons:
https://www.youtube.com/watch?v=KUU7IyfR34o

What causes the seasons: https://www.youtube.com/watch?v=q4_-R1vnJyw

Notice the Earth's axis (axis of spinning) is tilted relative to the Sun.

EARTH ORBIT IS AN ELLIPSE

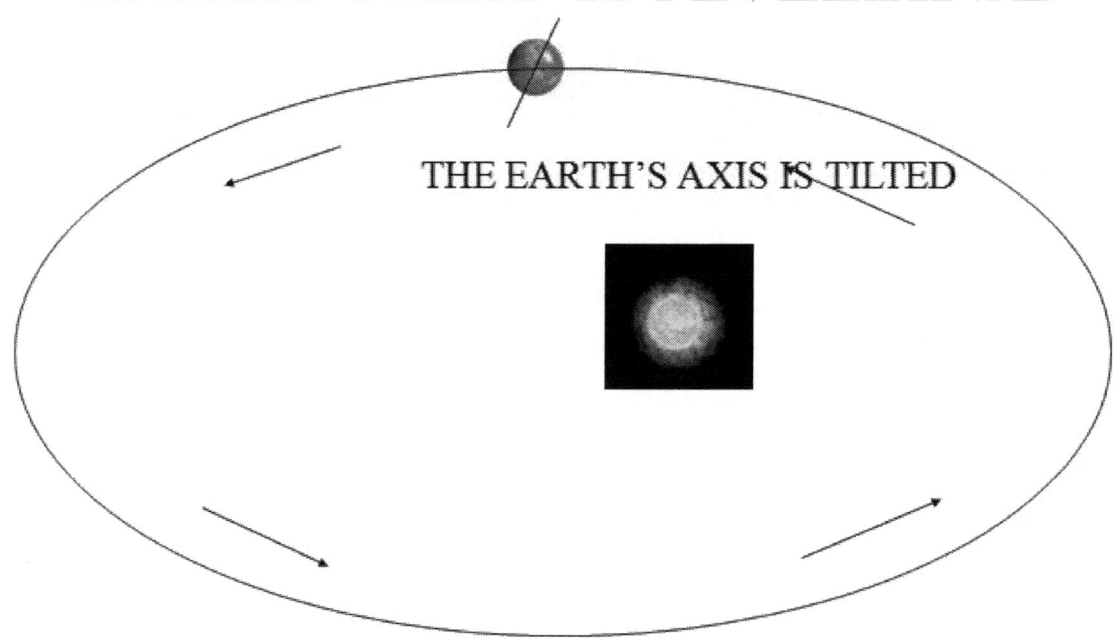

THE EARTH'S AXIS IS TILTED

EARTH ORBIT IS AN ELLIPSE

SUMMER

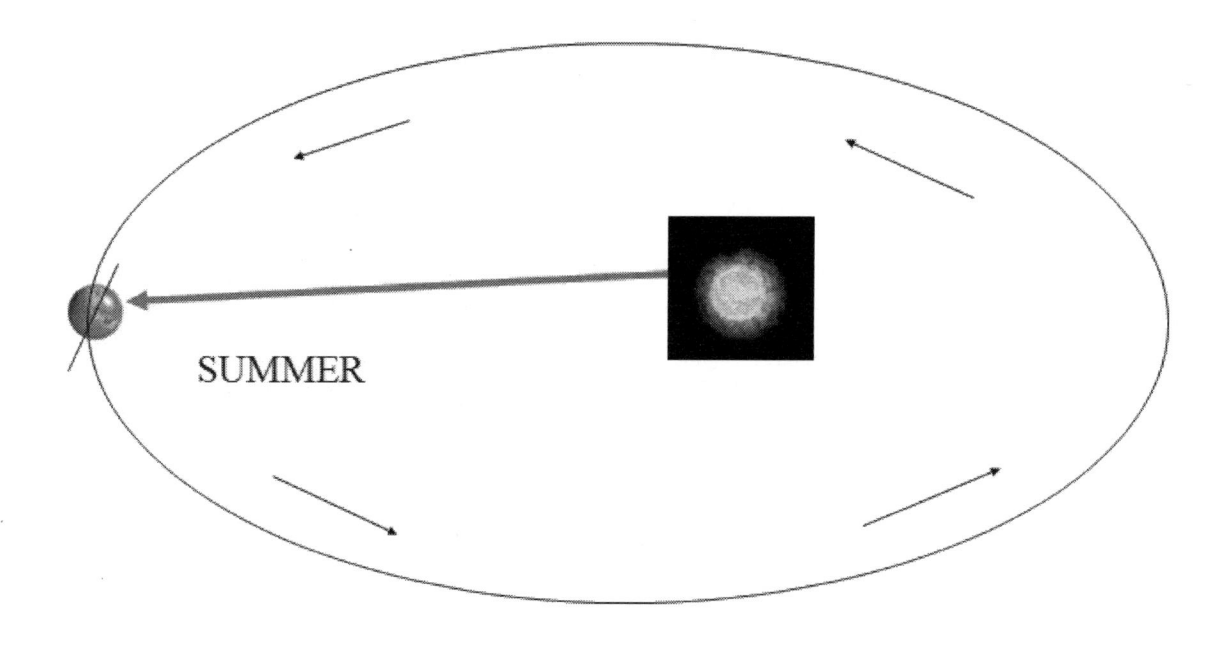

EARTH ORBIT IS AN ELLIPSE

FALL

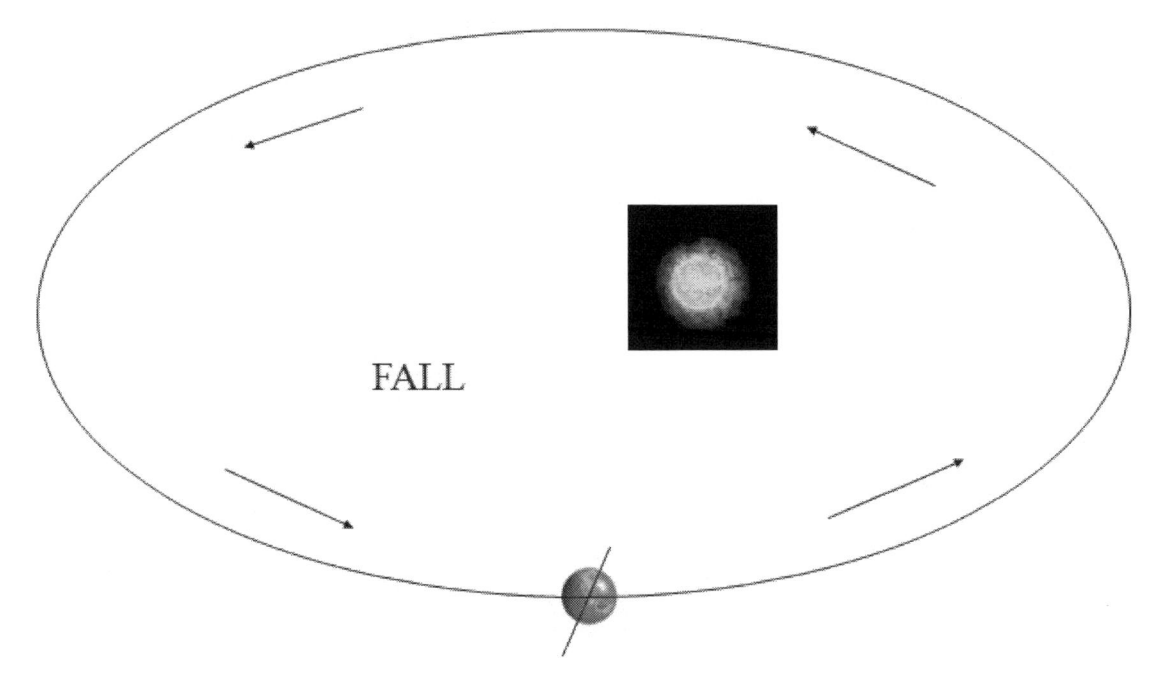

Notice the light is skipping off the atmosphere.

EARTH ORBIT IS AN ELLIPSE

WINTER

173

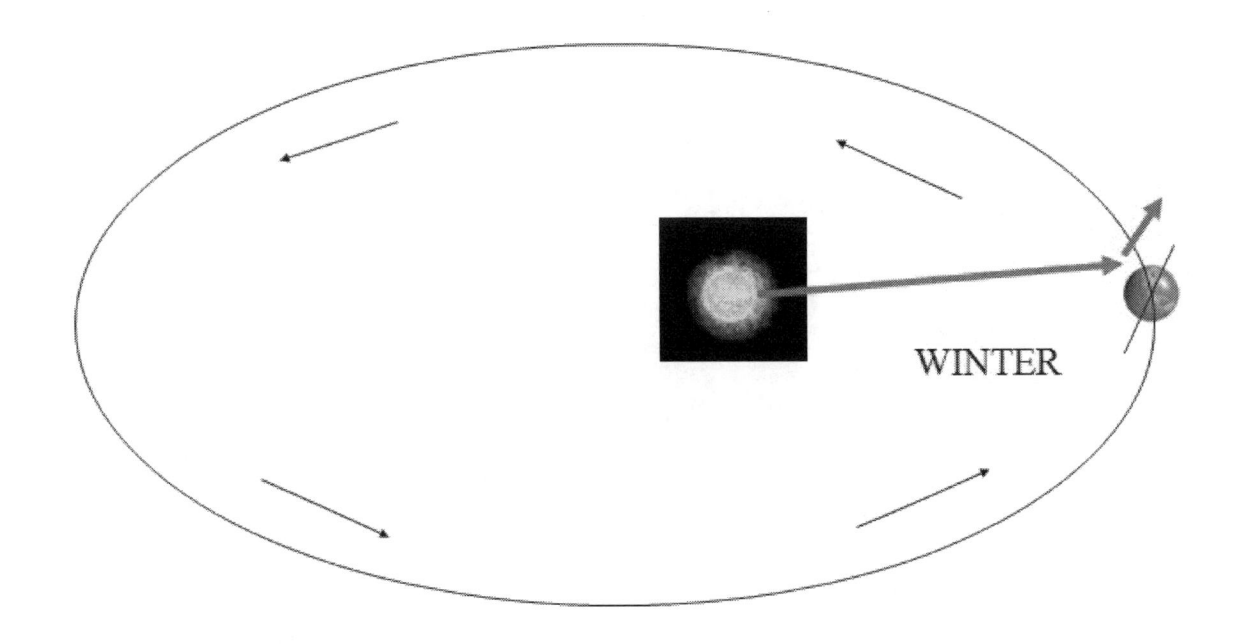

EARTH ORBIT IS AN ELLIPSE

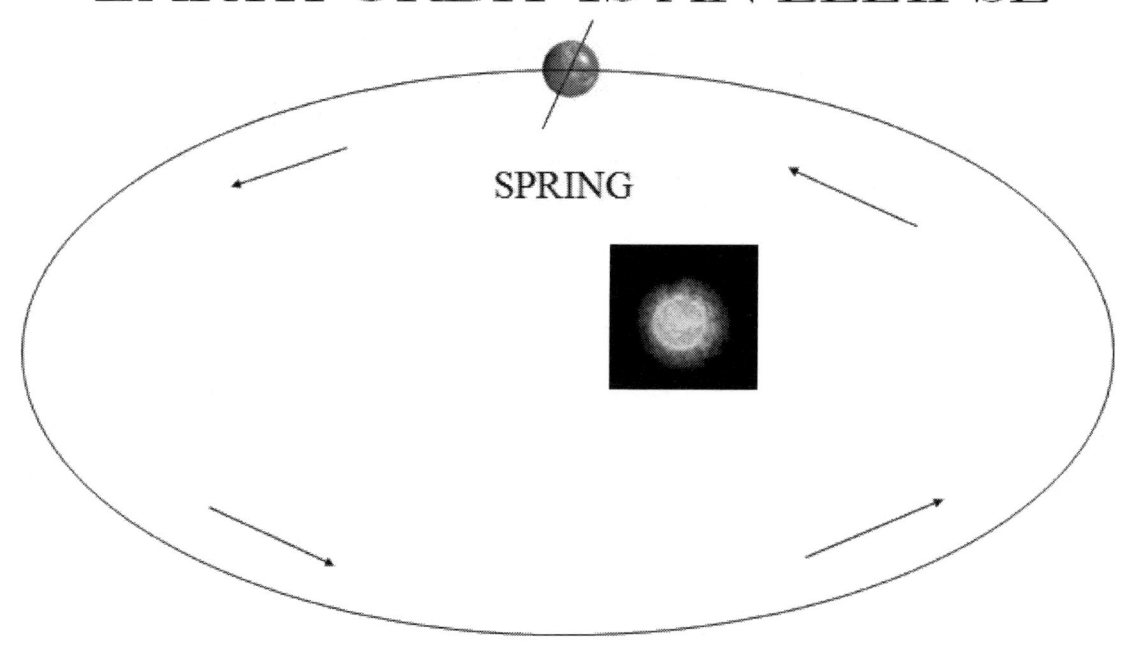

SPRING

In the summer the sunlight shines straight down to the Earth's surface.

IN SUMMER THE EARTH

- IS FARTHEST FROM THE SUN
- BUT WE ARE TILTED TOWARD IT

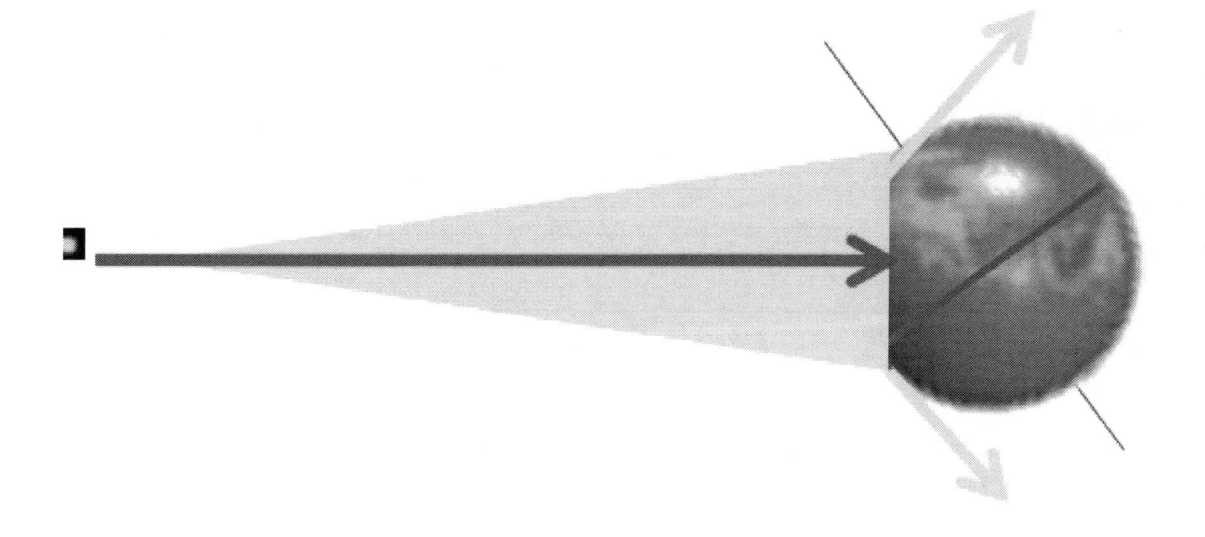

Review

Planets follow an ellipse shaped orbit. A lunar eclipse occurs at night and the Earth is between the Sun and Moon. A solar eclipse occurs in the daytime and the Moon is in the middle. The tides are caused by the gravity of the Moon. The seasons are caused by the tilting of the Earth axis, not its distance from the sun.

Quizlet flashcard review: https://quizlet.com/144279036/chapter-31-gravity-in-the-solar-system-flash-cards/

A day in the life of Earwig Hickson III

I was looking for seashells. It was low tide (the best time) and I was having a fine time. I found starfish in a tidal pool and a sand dollar on the beach. I almost caught a small lobster. As I wondered along some cliffs I found a very small cave opening. It was only about 2 feet high and one foot wide, just big enough for my small body to wiggle through.

My family was on vacation. We were touring Canada. We were now at the Bay of Fundy, a really cool place where we dug up clams, saw a tidal bore, some lighthouses, and a boring flower garden. What I liked though were the tidal pools filled with trapped creatures. I found sea anemones and stuck a feather I found into two that were nearby; they sucked in the feather until they kissed. I caught small fish and crabs. I found starfish and snails, it was really cool.

My greatest discovery was the cave. It was only exposed at low tide and I absolutely knew, without a doubt, that it was used by pirates to hide their treasure. What better place? I crawled in. My flashlight showed a wonderful scene. Starfish, and barnacles were pasted to the ceiling (my first hint). Limpets and snails were everywhere; it was a biologist's paradise. The cavern I was in was very large; the little opening went into a cavern with a tunnel going to another large room. I searched every nook and cranny but found no treasure chest, but the animals I found made up for it all. I was where few (if anyone) had ever been.

Later I found out that the tides at the Bay of Fundy rise very quickly, 45 feet in six hours that comes out to over seven feet per hour, or about 1 ½ inches per minute. I wish someone had mentioned that to me because I was in there for about an hour. When I crawled back to the entrance, I found water was uncomfortably deep and rushing in the entrance. The little cave opening was three feet under water. Suddenly I remembered, the high tide at the Bay of Fundy was 45 feet higher than low tide. I looked at the starfish on the ceiling of the cave and saw my fate. I was not going to have to do any homework ever again! The choice was death or homework, I thought about it but decided homework was the better option. I took a breath and dived.

I was bruised and cut from the waves but managed to make it back to the campground. My father was cooking supper over the campfire. Mom was boiling water for some kind of noodles; my sister was looking at her smart phone. Since the police or hospital did not call, they had no worries. I went to bed.

Links

A video explaining gravity assist: https://www.youtube.com/watch?v=0iAGrdITIiE

Find your weight on other planets: http://www.exploratorium.edu/ronh/weight/

Find the next lunar eclipse in your area:
https://www.google.com/#q=when+is+the+next+lunar+eclipse+in+the+united+states

Video of a lunar eclipse: https://www.youtube.com/watch?v=lcRp1jKJmJU

Find the next solar eclipse in your area:
https://www.google.com/#q=when+is+the+next+solar+eclipse+in+the+united+states

Video of a solar eclipse: https://www.youtube.com/watch?v=3_qo2CdcyC0

How to view a solar eclipse safely: http://www.timeanddate.com/eclipse/eclipse-tips-safety.html

Video of the tide at the Bay of Fundy:
https://www.youtube.com/watch?v=OP0cpXpw8yk

Now the scene in *Despicable Me* should make sense:
https://www.youtube.com/watch?v=gAmo3FcaovM

What causes the tides video: https://www.youtube.com/watch?v=5ohDG7RqQ9I

A tidal bore, where the high tide actually comes up a river like a wave: https://www.youtube.com/watch?v=LWumonz87rA

Bill Nye explains what causes the seasons: https://www.youtube.com/watch?v=KUU7IyfR34o

What causes the seasons: https://www.youtube.com/watch?v=q4_-R1vnJyw

Fun things to Google
Gravity assist
NASA probes
Planets we have landed on
Tidal bore
The seasons
The Bay of Fundy
Tides
Lunar eclipse
Solar eclipse

Chapter 32
Space

Video of this chapter's power point lesson:
https://www.youtube.com/watch?v=QOA3hvE1bUY

The wonders of science
There is a plane. It is a big plane and it has no seats in the back for passengers. The walls are padded. It does not fly to different airports in the world and it is not a warplane. It is a space plane, well not exactly a space plane, but one where outer space is kind of in the back. It is a Boeing 727 jet with all the seats removed forming a big empty flying box in the back. In this box people go, astronauts train back there, movies (like Apollo 13) have been filmed back there, and regular people can get a ticket do it. It is called the vomit comet a reduced-gravity aircraft, for practicing and researching weightlessness. The plane flies upward at an angle of 45 degrees, then back down again at an angle of 45 degrees, going up and down like a yoyo, but as it flies in this big curve or parabola, everyone in the back begins to float, since they are falling at the same rate as the plane. At different dive angles people can experience the gravity on the moon or mars. For 30 – 40 seconds everyone is weightless, just like an astronaut. Most people get nausea at this point and many get sick. When the passengers are floating, so does what they threw up, but when the plane has to pull up for another dive, it experiences 2g's or double normal gravity. This is when the throwup flies across the plane, thus the plane's nickname of the "vomit comet".

While weightless, astronauts and scientists can practice or do experiments without the cost of going to space. The Apollo 13 (film) took place in space and many parts of the movie were filmed in the vomit comet.

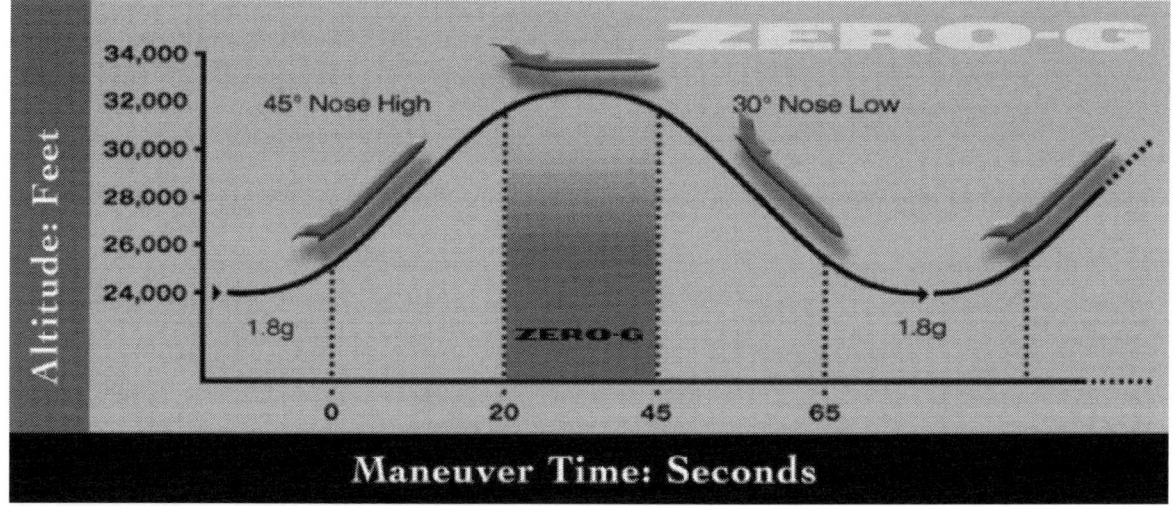

A short video of reduced-gravity aircraft and what happens:
https://www.youtube.com/watch?v=NkXrpbOEWC4&nohtml5=False
Zero Gravity Flight – Weightlessness:
https://www.youtube.com/watch?v=HQbAwE83phk&nohtml5=False

What you need to remember

Centripetal force is the inward force that causes an object to travel in a circle. The object wants to tangent to a circle. All objects fall to the earth at the same rate, and if air is not in the way, will land at the same time.

Microgravity

Astronauts appear to float around like there is not any gravity, but this is not really true. There is just about as much gravity in the International Space Station or on the Space Shuttle as there is on Earth. The Earth is trying to pull astronauts down, like anything else. The word microgravity actually means "tiny" "gravity" but this is not accurate either. What the astronauts, the Space Shuttle, and the International Space Station are doing is falling, so a better term for this is free fall. The astronauts and their spaceship are falling together at the same rate, in space where there is not any air, but they are not falling into the Earth, they are falling *around* it. This is called an orbit around the earth. So astronauts are in free fall and appear to float around since the spaceship is in free fall too.

SPACESHIPS FALL AROUND THE EARTH

ASTRONAUTS FLOAT INSIDE BECAUSE THEY ARE FALLING AT THE SAME RATE AS THE SPACESHIP

An easy way to imagine this is in a broken elevator, one on top of a tall building. If the cable breaks a person inside would float around while the elevator accelerated downward. If you get into a working elevator and stand on a bathroom scale, your weight will be less, going down, and more going up.

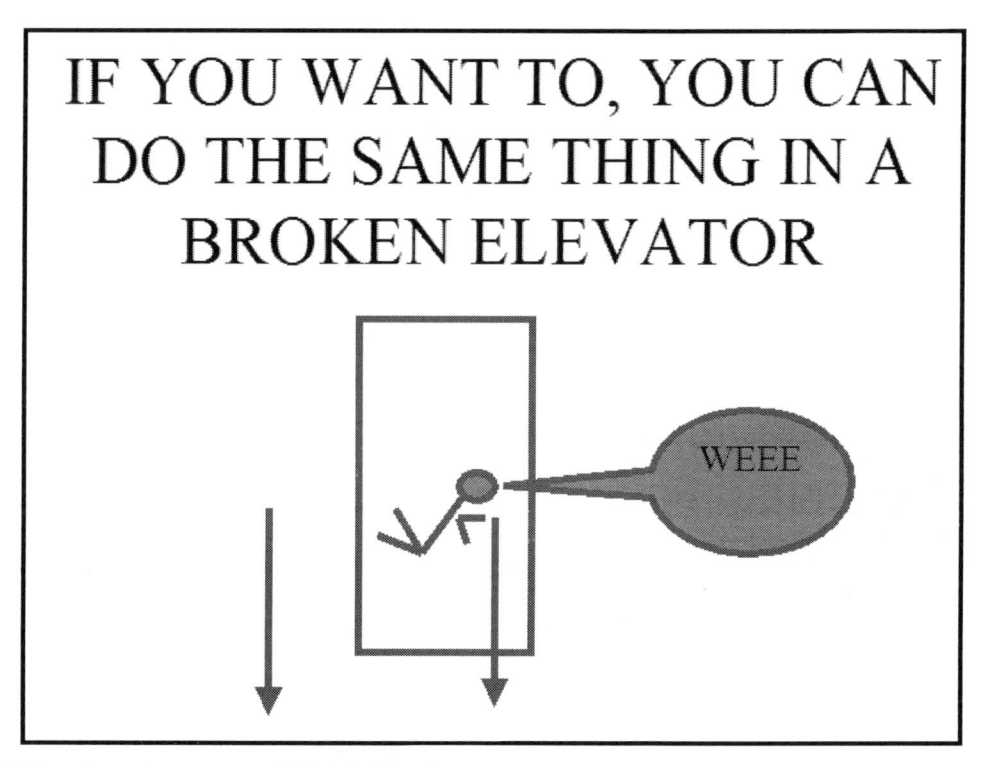

IF YOU WANT TO, YOU CAN DO THE SAME THING IN A BROKEN ELEVATOR

WEEE

Why Are Astronauts Weightless?
https://www.youtube.com/watch?v=iQOHRKKNNLQ

The reason spaceships do not hit the Earth is because they are going so fast, tangent to the Earth's center that they keep missing the planet. A spaceship needs to move at 7 miles/second to stay in orbit, it does fall toward the center of the Earth, but the Earth curves away at the same rate.

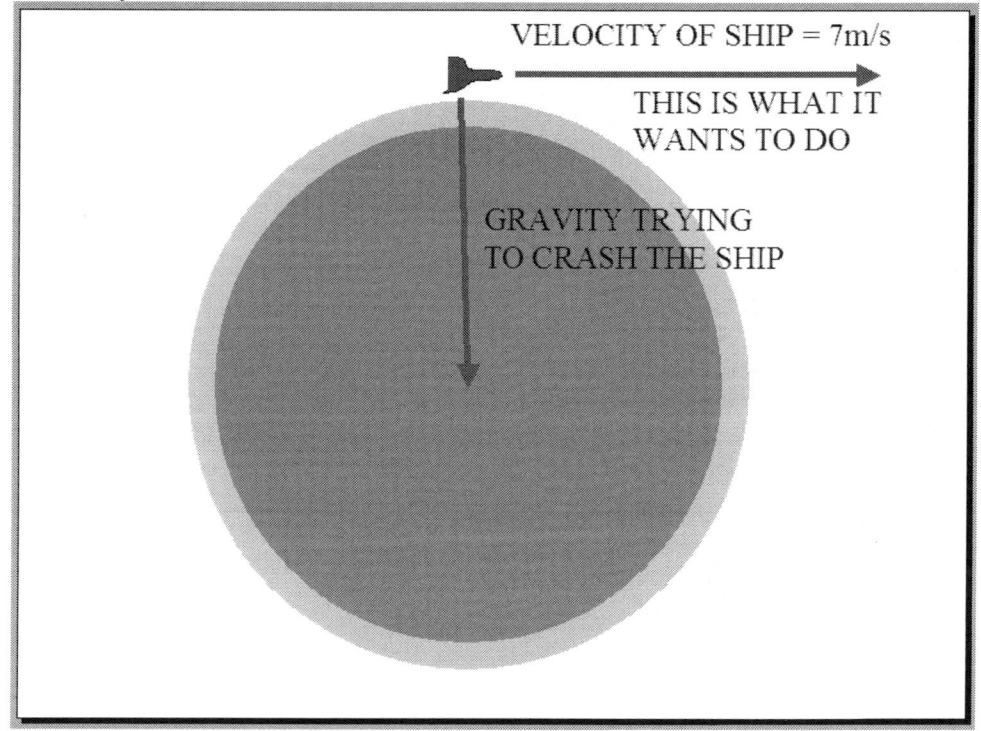

VELOCITY OF SHIP = 7m/s
THIS IS WHAT IT WANTS TO DO
GRAVITY TRYING TO CRASH THE SHIP

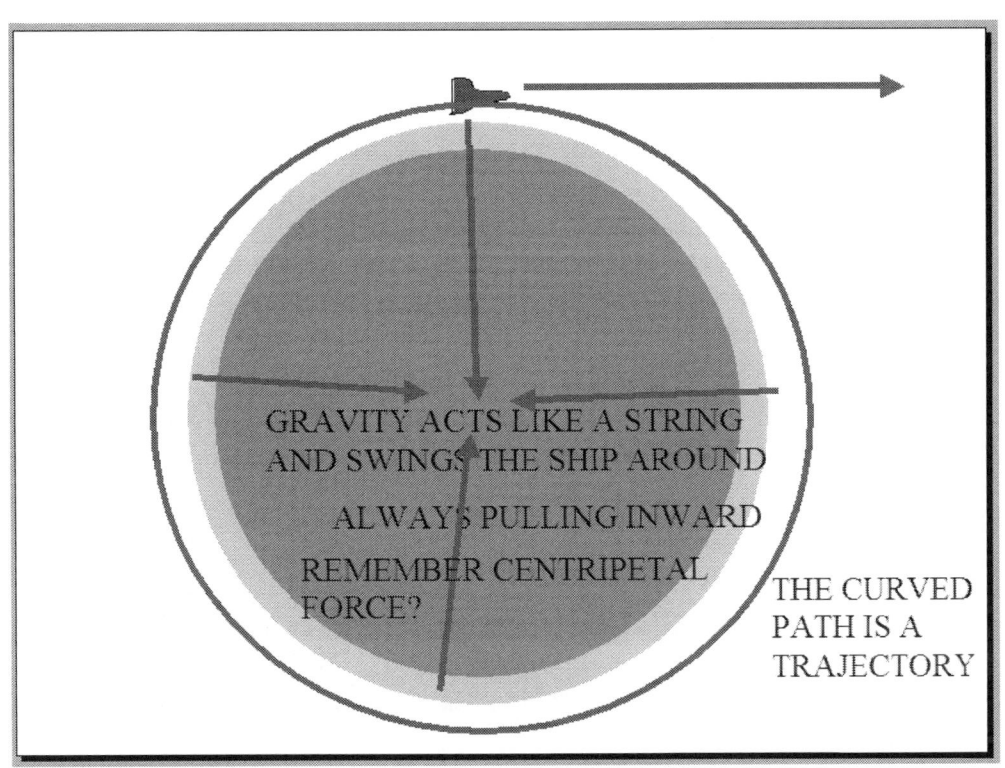

An orbit looks like this.

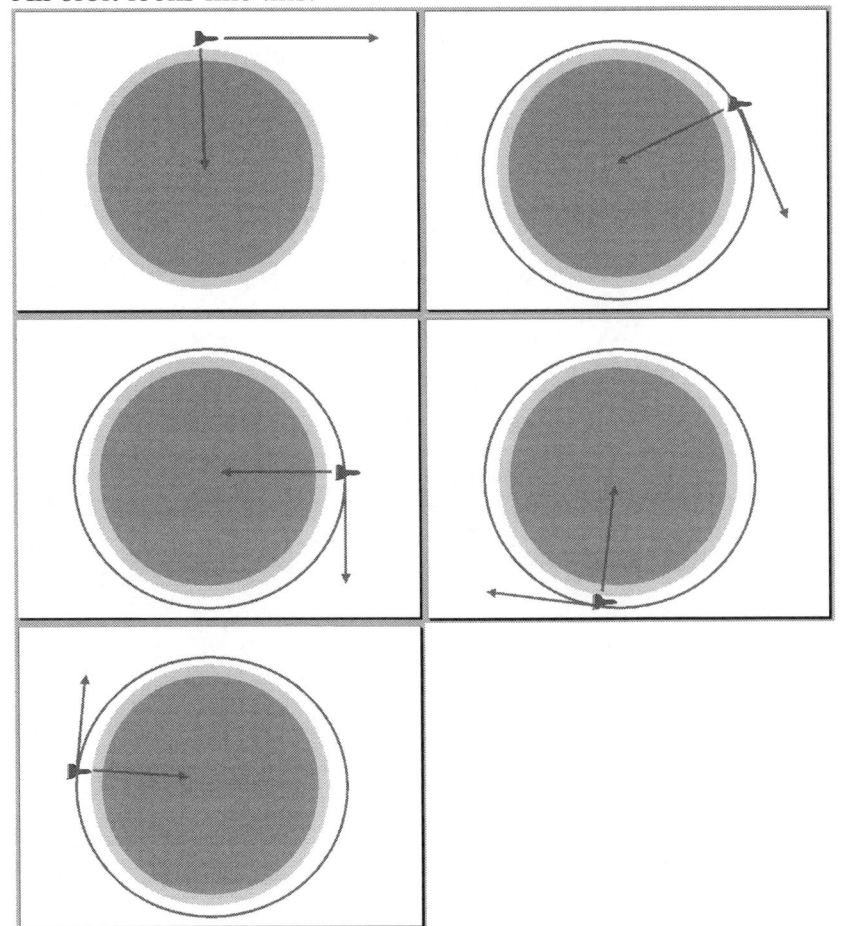

181

How to get into orbit

Let us pretend that you get a brand new spaceship for your Birthday. You immediately run outside to test it out. You would probably point it straight up, get inside and launch straight up to space, but if you did, it would be your last birthday. What goes straight up, comes straight back down. You need to go tangent to the orbit you want. So to get into orbit you go up for a short while to get above the Earth's atmosphere but then immediately turn level with the Earth's surface and hit the gas, going faster and faster, until you reach 7 miles per second. Then you turn off the gas and coast around the Earth until you are ready to come home.

I bet you can't guess who figured this out, Sir Isaac Newton (again). He imagined a mountain, and on top of the mountain, a cannon shot cannon balls (sideways – parallel to the Earth's surface) at a faster and faster velocity. He reasoned that a velocity could be archived where the cannon balls curved as they fell, matching the curvature of the Earth, and kept going around it. This is exactly what NASA does, except they use rockets instead of cannon balls.

How Do Satellites Get & Stay in Orbit?
https://www.youtube.com/watch?v=IC1JQu9xGHQ

Space Basics – a 20 minute NASA video that explains how to get into orbit, stay in orbit, weightlessness and more: https://www.youtube.com/watch?v=1GxsvKP9szs

So what good is the space program?

At about 2% of the United States budget, NASA has done some amazing things; some you never heard of, some you use everyday and do not know NASA was involved, some so incredible, people still do not believe it happened (the moon landings). Here is a short list of some of the things you may never have realized came out of NASA research:

GPS
Cell phones
2 space stations
Probes landing on Mars, Venus and orbiting Pluto, Saturn, and other planets
Weather satellites
TV satellites – satellite TV
Mylar insulation
12 men walking on the moon
A car on the moon – lunar rover
Space telescopes like Hubble and many others
Exploring Mars and the moon so other planets for signs of life
SETI – Search for intelligent life
Asteroid search to find the big ones that might hit us
Faster higher-flying airplanes
Google Earth
Memory foam
Lightweight scratch resistance eyeglass lenses (all because early astronauts wanted a window)
Insulation
Cordless tools
Water filters

Invisible braces
Freeze dried food
Work out machines
Insulation pump
Infrared ear thermometers
CT Scan
Aircraft de-icing systems
Cochlear ear implants
Land mine removal
Solar panels
Fuel cells
Smoke detectors
High quality artificial limbs
Velcro
Teflon
Global climate change research
Spy satellites
LEDs – light emitting diodes
Firefighter gear
Home blood pressure kit

Review
Spaceships fall around the Earth. Astronauts are in freefall, not zero gravity. To get into orbit a spacecraft must fly parallel to the Earth's surface very fast.

Flashcards for this chapter at Quizlet: https://quizlet.com/144309271/chapter-32-space-flash-cards/

A day in the life of Earwig Hickson III
I knew I was too young to be an astronaut, and I knew that the best way to experience weightlessness on Earth was in a diving airplane, and I knew I was not going to be getting an airplane anytime soon. But I still wanted to float around like an astronaut. How could a little kid pull this off? Otis was back and we were bored. He thought we should play catch. I thought we should play astronaut, and I had a plan.

I knew weightlessness was out of the question, but there are other ways to float. I had seen my neighbor blowing leaves off his lawn (into ours) and thought that was a place to begin. I had once tried to make a jet back with one of those leaf blowers but it was not strong enough. Perhaps I could get a bunch of them and strap them to my back! I would not be floating but the force from the air should balance my weight and perhaps lift me off the ground. I had also seen an experiment our science teacher showed us about Bernoulli's Principle where he floated a beach ball on a column of air. I knew my plan would work.

Otis and I went to every house we could find to borrow a leaf blower. We told them my Dad needed it to clear the leaves in our yard. I was hoping to get four. We got 156. A bit of overkill, but, oh well. I did not want to be short on power again.

Looking at the pile of mini windstorms an idea began to form. A terrible idea it turned out, an absolutely tremendously horrible idea. The idea would prove to triple my parents' homeowners and medical insurance. I let Otis go first. I was going to build a vertical wind tunnel like they use for indoor skydiving!

We duct taped 156 leaf blowers and my shop vacuum, from a previous experiment together, all pointing upward. We hid it in the yard behind my house and plugged them into a whole lot of power strips. I hit the switches. A blast of air shot upward. Otis in his Superman pajamas waited at the side. He dove onto the column of air and floated! It worked! He was suspended in midair, floating! His arms were swinging back and forth like he was swimming. He giggled and smiled. I was ready for my turn. I pulled Otis off the wind tunnel and jumped on. I should have thought about it first though. You see Otis is a big boy and he weighs about 170 pounds. I, on the other hand, topped out at 65 pounds. The wind tunnel was calibrated for Otis.

I shot up like a rocket. I could see Otis looking up. I screamed at him to reduce power. Otis ran away. It began to rain.

The neighbors could see me floating above my house. No one was worried for my safety; they were worried about their safety! The movie Ghost Busters had just come out and that was what was on their minds. A flying kid, in the rain with lightning flashing and thunder booming was not accepted in my neighborhood. When the fuse blew in the electric box, I fell down and ran into my room.

I am not a demon child, and I do not need to see any doctors. I am a scientist.

Links

The following links are of Canadian Chris Hadfield explaining all sorts of interesting things while in the International Space Station. Many astronauts, thought incredible people, are not entertaining to listen to, this guy is different, and He is fun.

Chris Hadfield's Space Kitchen: https://www.youtube.com/watch?v=AZx0RIV0wss
Sleeping in Space: https://www.youtube.com/watch?v=UyFYgeE32f0&nohtml5=False

How To Barf, Puke, Vomit In Space | Video: https://www.youtube.com/watch?v=HDev0cCeyF4&nohtml5=False

Wet Washcloth In Space - What Happens When You Wring It? | Video: https://www.youtube.com/watch?v=lMtXfwk7PXg&nohtml5=False

Astronauts Drink Urine and Other Waste Water | Video: https://www.youtube.com/watch?v=ZQ2T9OJY1lg&nohtml5=False

Astronaut Chris Hadfield Ejected Dirty Underwear Into Space: https://www.youtube.com/watch?v=C1j6KLP492E

An Astronaut's Guide to the Space Toilet: https://www.youtube.com/watch?v=5JPuaRBTMKs

Proof That You Can't Cry In Space: https://www.youtube.com/watch?v=4BbuOn--ERI&nohtml5=False

Chris Hadfield - Nail Clipping in Space: https://www.youtube.com/watch?v=xICkLB3vAeU&nohtml5=False

Chris Hadfield Brushes his Teeth in Space: https://www.youtube.com/watch?v=3bCoGC532p8&nohtml5=False

How astronauts use the bathroom in space: https://www.youtube.com/watch?v=UUonYZTGBjM

In Space Everyone Can Hear You Poop | Video: https://www.youtube.com/watch?v=MgMYqxdVAlA&nohtml5=False

A short video of reduced-gravity aircraft and what happens: https://www.youtube.com/watch?v=NkXrpbOEWC4&nohtml5=False

Zero Gravity Flight – Weightlessness: https://www.youtube.com/watch?v=HQbAwE83phk&nohtml5=False

Why Are Astronauts Weightless? https://www.youtube.com/watch?v=iQOHRKKNNLQ

How Do Satellites Get & Stay in Orbit?
https://www.youtube.com/watch?v=IC1JQu9xGHQ

Space Basics – a 20 minute NASA video that explains how to get into orbit, stay in orbit, weightlessness and more: https://www.youtube.com/watch?v=1GxsvKP9szs

The Scoop on Space Poop: How Astronauts Go Potty: http://www.space.com/22597-space-poop-astronaut-toilet-explained.html
http://www.madsci.org/posts/archives/2006-01/1138467020.As.r.html

Cats in weightlessness: https://www.youtube.com/watch?v=O9XtK6R1QAk

Pigeons in weightlessness: https://www.youtube.com/watch?v=w4sZ3qe6PiI

Fun things to Google
NASA
Space shuttle
Chris Hadfield
Sky lab
International space station
Indoor skydiving

Chapter 33
Friction

Power point of this lesson: https://www.youtube.com/watch?v=ijUsxUDj5Yk

The wonders of science
Primitive humans (Homo erectus) started using fire as far back as 1 million years ago. It was not until much later 125,000 years ago or so, when fire was controlled. This means a person could make their own fire rather than wait for a lightning strike or something. Controlling fire changed human culture. By the time Neanderthal man (Homo neanderthalensis) came around, fire could easily be made and controlled, resulting in a better diet from cooking food, protection from predators, and warmth. This allowed primitive man to expand into cooler climates and begin to control the environment. The fire making methods they used relied on friction. The first method was probably the hand drill; a simple method probably discovered accidentally when someone was trying to drill a hole in a piece of wood. The fire plow was probably next and was actually invented specifically to make heat and fire. The bow drill was a vast improvement since the drill could be spun much faster. The pump drill was the most advanced fire making method until it was replaced by flint and steel. Clever people these Stone Age humans.

What you need to remember
Friction slows down moving objects. Friction is a force. Friction produces heat.

Friction
Newton and Napoleon introduce friction:
https://www.youtube.com/watch?v=llZ1Zl57r1Q

Friction messes up our nice perfect laws of motion. You see it every day, but do not notice, or appreciate it. It is everywhere, slowing down moving objects. We are so used to friction slowing things down; we do not realize that moving things want to keep moving. It was Newton who showed us that they *do* want to keep moving. Friction is almost invisible, but affects us every day. When you fall off your moving bike, it is friction that saves you from sliding forever. When you walk, it is friction that holds your feet to the ground, to push you forward. Without friction you would not be able to walk, you would stay where you are forever. As nice as friction is, it robs our society of energy and money. Oil must be added to a car engine to reduce friction (and heat) so you get better gas mileage. Factories must keep their machines lubricated to keep them running. Friction is expensive; it stops moving machines, which our society needs.

Friction is a force that acts *against* motion, always against motion. It forces moving objects to slow down and eventually stop. It creates heat and makes things wear out. Friction is evil, well not always; it also gives *traction* to our shoes as we walk and car tires as they grip the road. We depend on friction and curse it too. The force of friction comes from the microscopic bumps and scratches in a material, which grab each other kind of like Velcro.

FRICTION

- THE FORCE THAT ACTS <u>AGAINST</u> MOTION
- FRICTION WILL SLOW DOWN AND STOP MOVING OBJECTS

Static friction

Static means "not moving or stationary," so static friction is between two objects that are not moving, but kind of stuck to each other. Holding your hands together and pushing them so they slide, but *do not let them slide* is static friction. Static friction is what holds stationary objects together, like you to your chair when you are sitting, or a car tire on the road, or a book on a table. Without static friction, all these objects would slide off. You would not be able to walk, since static friction makes your shoe stick to the floor, so you can push away (action = reaction). Static friction just holds two objects together, it does not generate heat or wear things out, they just stay where they are. Static friction is a pretty strong force, thanks to those little bumps and scratches. It makes it hard to get objects moving when you push them. Once you get them moving, you will find it is easy to keep them moving.

STATIC FRICTION

- 2 STATIONARY OBJECTS ARE HELD TOGETHER
- STRONGEST FRICTION
- DOES NOT CAUSE HEAT
- OR WEAR

Sliding friction

Sliding friction is when objects are moving but slow down. Sliding friction is a force that makes moving objects slow down and stop. If you are on your bike, and moving rather fast, and fall off, you know about sliding friction, it makes you slow down and stop. It also causes road rash (a rather painful result of sliding over a road). Of course, this sliding friction saves you from sliding forever or until you hit a tree, but hurts nonetheless.

Sliding friction causes heat between the two objects and causes them to wear out. Car engines wear out after a while because of sliding friction. Your shoes wear out for the same reason. If you rub your hands together really fast, they will get hot, and when you look at your palms you may notices little bits of dead skin that wore away. I once had a catfish in my fishbowl that never stopped swimming into the glass. Eventually his face wore away and he died, all because of those little bumps and scratches, poor guy.

Sliding friction also occurs between moving objects and air. This type of sliding friction is called _drag_ or air friction. It makes cars and planes slow down and use more fuel to compensate. It causes meteorites to glow brightly as they enter the Earth's atmosphere. It causes a lot of heat and a lot of wear. Friction caused the Space Shuttle Columbia to be destroyed as it re-entered the Earth's atmosphere in 2003 after it lost some insulating tiles on its wing.

Sliding friction causes a lot of heat and wear because it grabs the touching molecules and rips the material apart. Sliding friction always slows down and eventually stops moving objects. Sliding friction also causes road rash, which for us motorcycle riders (or crashers) is a pain.

SLIDING FRICTION

- 2 SURFACES SLIDE AGAINST EACH OTHER
- CAUSE THE MOST HEAT AND WEAR
- INCLUDES DRAG – FRICTION IN LIQUID OR GAS

Rolling friction

Friction was a problem, even for the ancient civilizations who built the pyramids and other amazing things. It was a problem for the war machines of the time. Sliding a sled behind you was a lot of work to move valuable cargo. Eventually some unknown primitive person invented the wheel and that solved the problem. Wheels reduce friction. Note that a wheel reduces friction but does not increase force, like a _wheel and axle_, but it is totally different, it ONLY reduces friction. Sliding friction is strong and causes things to wear out. Put wheels on your wagon, and it is easier to pull and does wear away. Without wheels a wagon is not fun to pull. Rolling friction is a weak force and does not cause much wear on the object. Wheels were a wonderful invention and made it easy to move materials from one place to another.

ROLLING FRICTION

- FRICTION BETWEEN A WHEEL AND OTHER SURFACE
- WEAKEST FRICTION
- IT STILL CAUSES HEAT AND WEAR

Rolling friction is the friction between a wheel and another object (like a road. It is very weak and does not cause things to wear out very quickly but it still causes then to wear out. Tires wear out, <u>bearings</u> go bad. It is a force to deal with and costs our society money.

If you rub your hands together this time, put a pencil or pen between them, and let it roll as you move your hands back and forth. The friction is much less and it is not nearly as hot. Wheels reduce friction, wheels are wonderful.

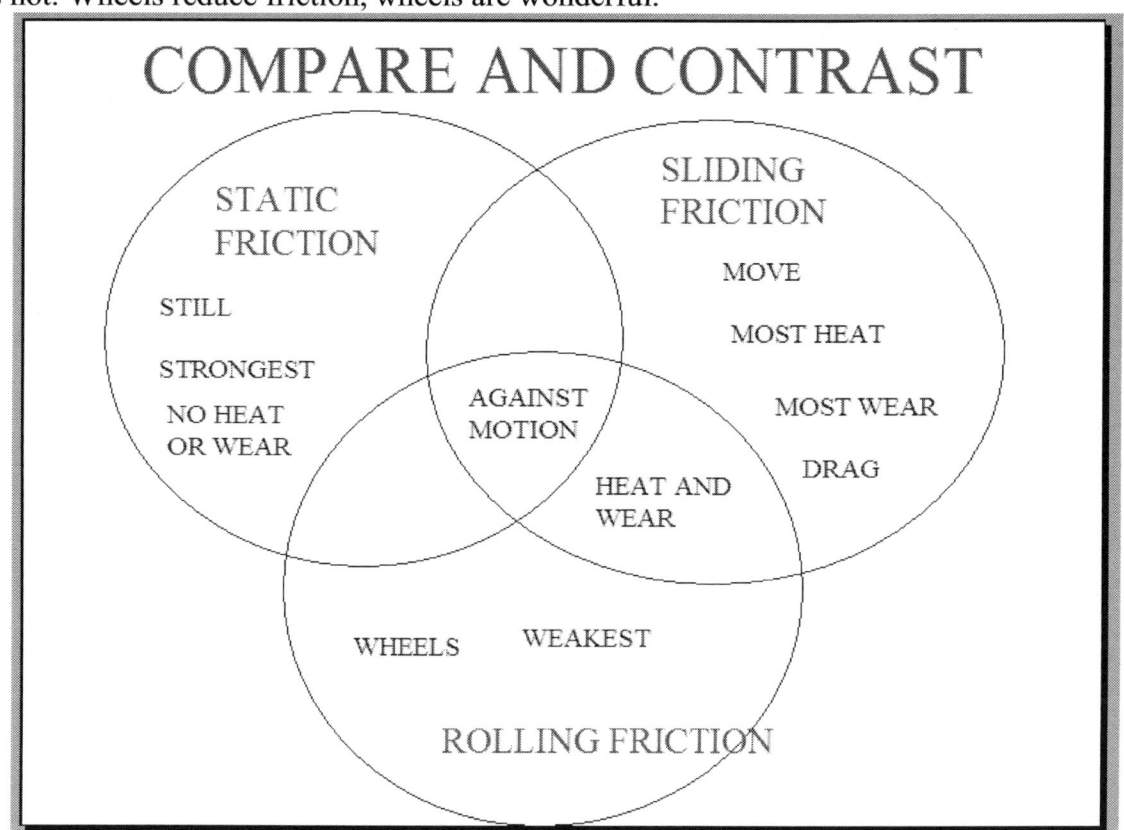

COMPARE AND CONTRAST

STATIC FRICTION
- STILL
- STRONGEST
- NO HEAT OR WEAR

SLIDING FRICTION
- MOVE
- MOST HEAT
- MOST WEAR
- DRAG

ROLLING FRICTION
- WHEELS
- WEAKEST

Intersections: AGAINST MOTION; HEAT AND WEAR

Reducing friction

Tons of money is spent to reducing friction. Friction needs to be reduced in our car engines, in factories, and between your feet and your shoes (where blisters are formed). There are three ways to reduce friction, and thus cost, if you are a machine or feet. You can polish the surface of the two parts sliding by each other. Smoother surfaces have fewer bumps and thus do not want to stick as together as much. This is why sand paper does a good job turning wood into saw dust. The grit grabs the wood. It is why old bald tires on a car or smooth soles on your shoes are slippery on ice. Or why ice is slippery in the first place. Maybe I should glue sand paper on my shoes the next time I walk on ice. You can also replace sliding friction with rolling friction (wheels on a wagon or ball bearings in a motor). Imagine trying to walk on a bunch of marbles on the floor, not much friction.

The last thing you could do to reduce friction is to use a lubricant, such as oil or silicon. Oil or silicon fills in the scratches and bumps in a material making the sliding easier. Try rubbing your hands together; when you put a bunch of butter between your hands. They will slide easily. A fresh-waxed floor is slippery too, since the wax fills the scratches and is very slippery when there is a bit of water on it.

Reducing friction on a train can save fuel and allow the train to move faster. Some people have made trains that actually float on a magnetic track, called Maglev trains. These trains can go up to 270 miles per hour. Check out the video of a maglov: https://www.youtube.com/watch?v=aIwbrZ4knpg

---

REDUCING FRICTION

- **POLISH SURFACES** – SMOOTH SURFACES HAVE LESS BUMPS
- **REPLACE SLIDING WITH ROLLING FRICTION** – WHEEL OR BALL BEARINGS
- **LUBRICANTS** – OIL, SILICON, TEFLON, WAX

Traction

Sometimes you want to increase friction. This is called traction or grip. Car tires are supposed to grip the road, because you do not want to slide when you are driving. A lot of effort and cost is put into the quality of car tires. Tire companies do not want their product to fail and cause accidents. They want them to *grip* the road, and keep you safe. Traction is *extra friction* to help grip the surface you are on. Shoes are the same way. Some shoes are designed to help your feet grip a basketball court floor, or a tennis court, the road, or the dusty trail. The soles of these shoes are designed to increase the friction between your feet and the surface you are playing on. I used to call them sneakers, but now there are tennis shoes, volleyball shoes, running shoes, basketball shoes, and any other "sport" shoe you wish. Each have a differently designed tread to help you grip the floor material.

TRACTION

- GOOD FRICTION
- CAR TIRES
- SHOES

Review
Friction is a force that goes against motion. Static friction is the strongest but does not cause things to wear away. Sliding friction causes the most heat and wear. Rolling friction is the weakest and causes the least heat and wear. Lubricants reduce friction.

Quizlet flashcard review: https://quizlet.com/144263974/chapter-33-friction-flash-cards/

A day in the life of Earwig Hickson III
My sister got a new car. I thought it was ugly but she did not think so. She kept it spotless. The inside was clean with no candy wrappers on the floor. The outside was shiny, the wheels were clean. She made sure it got washed and waxed every day. It was perfect.

The problem was she was not the one washing and waxing her car. She made me do it. Every morning she got me out of bed to wash and wax her car for her daily drive to show it off. It was getting old.

One day she was going to drive her friends up to a lake to picnic and go swimming. She demanded an extra good job cleaning her car. I decided a "good" cleaning was just what it needed. I had found a new product to try. It was a silicon-based liquid that was supposed to make things extra shiny and new looking. I put it on the car seats, the steering wheel, the gear shift, the gas and brake pedals, the door handles, everything. In fact I put 5 coats on the seats and steering wheel. I polished them too.

When she saw the car she was actually happy, it never looked better. But she was not happy when she got home from the trip. Evidently the silicon spray really makes things smooth, it reduced friction a lot. There was hardly any friction holding the passengers to the car seats. Hardly any friction on the steering wheel, and the road to the lake was very windy.

Now the First Law of Motion states that objects in motion wish to remain in motion unless a force acts on them. Friction is the force that prevents this motion. Without friction there was nothing holding my sister and her friends to their seats. They slid back and forth as the car made its way down the windy road.

The Third law of motion is about action and reaction forces. Every time they slid into the sides of the car, it hit them back. Back and forth they went, the inside of the car hitting them with each curve. The bumpy parts of the road did not help. When the doors popped open the passengers slid out. A song played on the radio called Dead Man's Curve. It was quite a ride.

I did not have to clean my sister's car anymore. I also sent a nice letter to the company that made the silicon stuff.

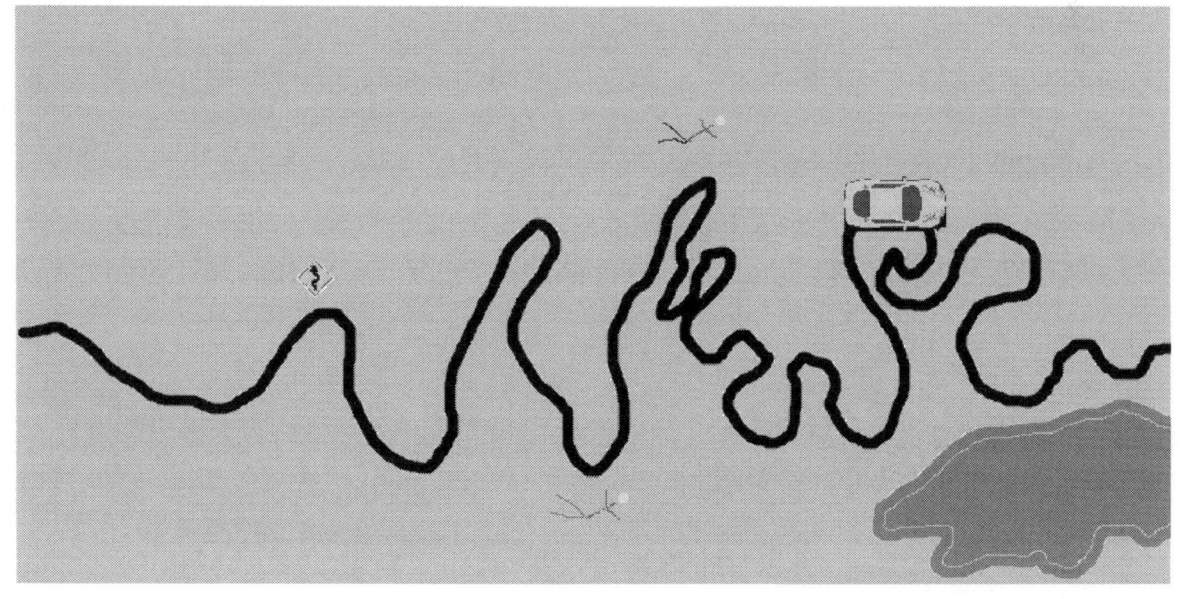

193

Links
Newton and Napoleon introduce friction:
https://www.youtube.com/watch?v=llZ1Zl57r1Q

Check out the video of a maglov: https://www.youtube.com/watch?v=aIwbrZ4knpg

Hover board like Back to the Future:
https://www.youtube.com/watch?v=ZeklZmwbjWo

Magnetic levitation: https://www.youtube.com/watch?v=vA0_5BoVpDg

Magnetic levitation of a spinning top:
https://www.youtube.com/watch?v=9zv7mBQaXPg

Producing smoke with a fire bow:
https://www.youtube.com/watch?v=APCgA9uxMGI

Making fire with friction: https://www.youtube.com/watch?v=ZEl-Y1NvBVI

Fire with a pump drill: https://www.youtube.com/watch?v=u8BXhxHWyF4

Space Shuttle Columbia disaster video:
https://www.youtube.com/watch?v=1oBTzbKx0jo

Space Shuttle Columbia disaster documentary (NOVA):
https://www.youtube.com/watch?v=nJUAGc9yH6A

A world without friction – MIT: https://www.youtube.com/watch?v=VUfqjSeeZng

A silly cartoon about friction: https://www.youtube.com/watch?v=C7NPD9W0kro

Mythbusters phone book friction: https://www.youtube.com/watch?v=AX_lCOjLCTo

Mythbusters phone book friction part 2 (with Tanks):
https://www.youtube.com/watch?v=QMW_uYWwHWQ

Nose basher bowling ball: https://www.youtube.com/watch?v=DGFfBX86ukw

Fun things to Google
Primitive fire making
Lubricants
Oil
Reducing friction
Hover board
Maglev trains

Chapter 34
Pressure

Power point lesson on chapter 34:
https://www.youtube.com/watch?v=nG3H4Q1eVAU

The wonders of science
Scuba diving (self-contained underwater breathing apparatus), invented by Jacques-Yves Cousteau can be fun but it can be dangerous. There are many things to worry about: sharks, big sting rays, equipment failure, drowning, getting lost, even someone stealing your boat and having nowhere to go when you surface far out in the ocean. Shallow water diving is rather safe as long as you are properly trained. The surface is not that far away and visibility is good. In recreational diving, the limit for a novice is 60 feet deep; an experienced recreational diver can go up to 140 feet deep. Deeper than this is called technical diving, and the risks multiply. Deep diving has many unexpected problems; people are not designed to go that deep, under that much pressure. Past 200 feet deep a diver can actually suffer from oxygen toxicity, where oxygen under high pressure can cause a sudden seizure resulting in unconsciousness and a near certain death by drowning. At 400 feet deep a diver must breathe a special gas called hydreliox, a mixture of Oxygen, Helium, and Hydrogen. The reason for this is because normal air contains 78% Nitrogen, which when under high pressure can kill a diver from a problem called decompression sickness or the Bends.

Decompression sickness is caused when a diver breathes air at a deep depth and a lot of pressure. The Nitrogen molecules he breathes are compressed and dissolve in his blood stream. Everything is fine until he comes to the surface and those tiny bubbles expand and become big bubbles in his blood vessels. Air bubbles in the blood stream is bad, resulting in extreme pain and possible death. Think of the Bends as what happens to warm soda when you pour it on ice, lots of foam. To prevent this a diver must surface very slowly as his blood gets rid of compressed Nitrogen and replaces it with less compressed Nitrogen.

Decompression sickness video: https://www.youtube.com/watch?v=R037jVYOwOU
Decompression sickness documentary:
https://www.youtube.com/watch?v=QP9MaVr5hLk
What happens in decompression sickness:
https://www.youtube.com/watch?v=p_Vmr-QJ7yM

What you need to remember
Force is a push or pull on matter. Gravity is a force.

Pressure
A *force* is a push or pull on matter. When a book or something falls on your foot it hurts, and force made it hurt, but it could have been a lot worse. What if the book landed corner first? That would be worse, because the force from the falling book would be *concentrated* all at one spot. This is why kids' darts have big wide suction cups on the front but adult darts have sharp points. Sharp points concentrate the force, and little kids

can't be trusted with sharp objects. This is why they give you dull useless scissors in school too.

Pressure is how much force is acting on a certain size area or how concentrated a force is. A small force can create a giant pressure and a giant force can make a small tiny pressure. It all depends on the area.

PRESSURE

- A MEASURE OF HOW MUCH FORCE IS ACTING ON A CERTAIN AREA.
- HOW CONCENTRATED A FORCE IS
- GRAVITY CAUSES PRESSURE (FORCE)

PRESSURE

- A FORCE SPREAD OUT OVER A LARGE AREA IS A LOW PRESSURE
- THE SAME FORCE OVER A SMALL AREA IS A LARGE PRESSURE.

A cool experiment is to make a small bed of nails and try and pop a balloon with it. One nail pushed into an inflated balloon will pop it quickly, but if you push 100 nails into the balloon (at the same time) the force is spread out and it is very hard to pop the balloon.

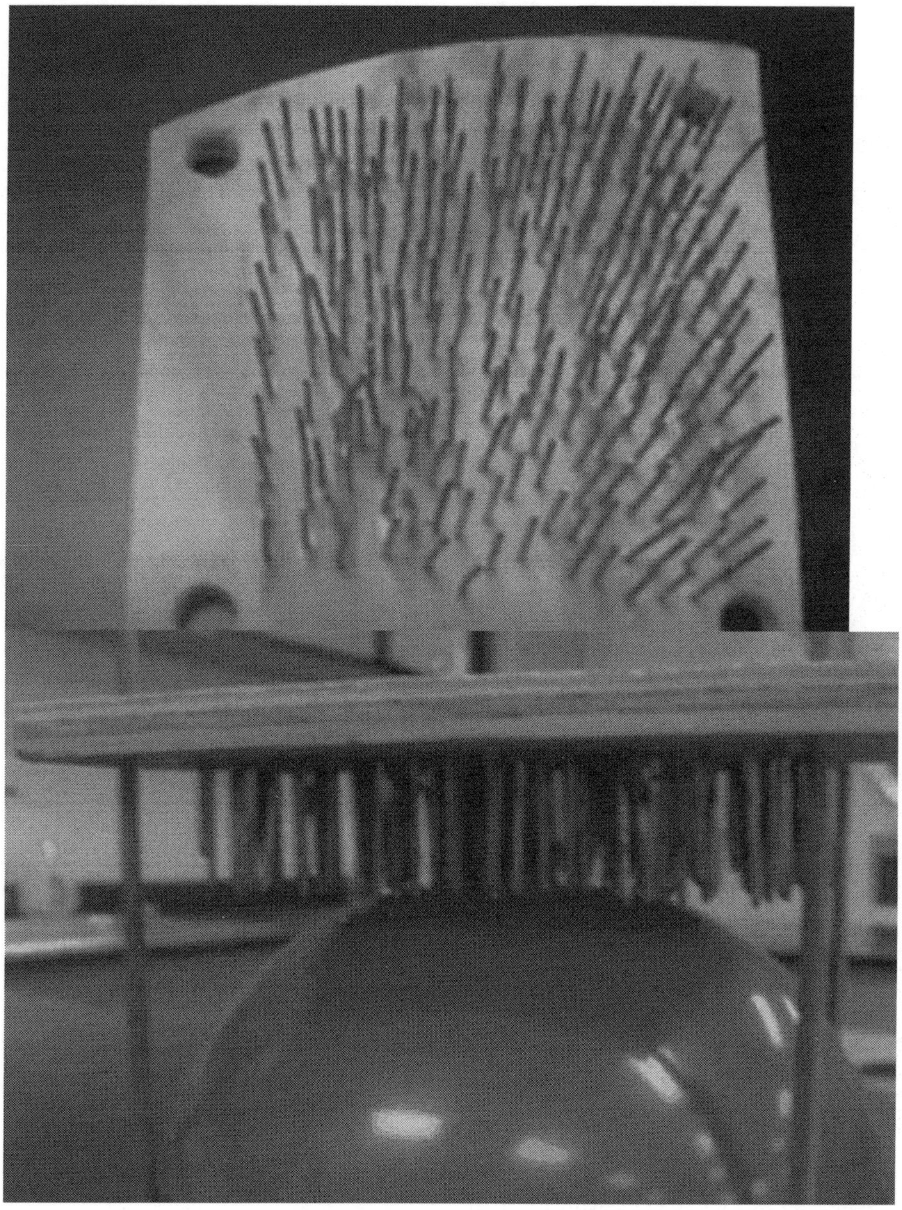

Check out this video of my small bed of nails popping a balloon: https://www.youtube.com/watch?v=4jFki8kEm7w

The reason for this is the force of my hand pushing the bed of nails down is spread out among 100 nails, so if I push with 100 pounds of force each nail only has one pound pushing into the balloon, not enough to pop it.

You hear the word *pressure* a lot in everyday language. You check the pressure of a bike or car tire, *not* the force of all the air inside, force does not matter in this case. You want to know how hard the air inside the tire is pushing outward on *each square inch*. If the pressure is too great, the tire can explode since the material is too weak to stop the pressure. My doctor is always checking my blood pressure. It seems to be her favorite hobby. She wants it to be at a certain level so my blood vessels do not explode. Blood vessels are rather fragile and if my blood pressure gets too high, it could push so hard against my blood vessels that they could pop and get a leak. That would be bad.

I have a big bed of nails too. One I can lay down on. It has 1000's of nails. It is actually comfortable to lie on since each nail is exactly the same height. More nails look kind of scary but if I were to lie down on a bed with only one nail, it may not look as dangerous but it would poke right through me. All my weight would be concentrated at a high pressure on that nail.

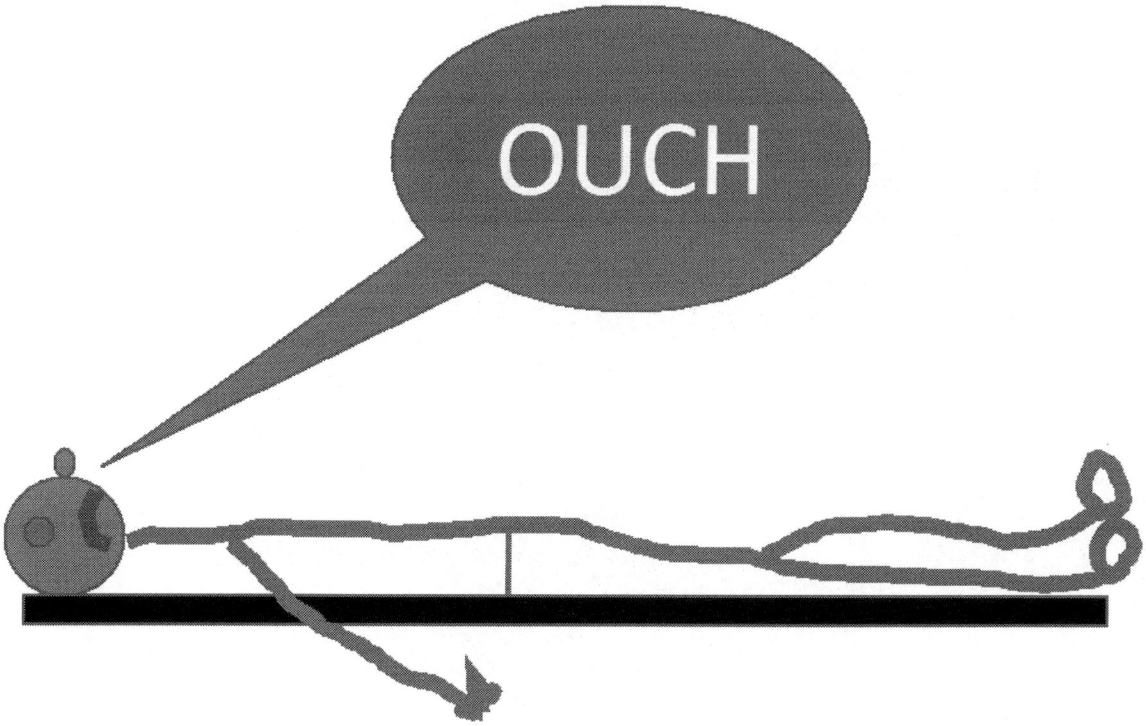
OUCH

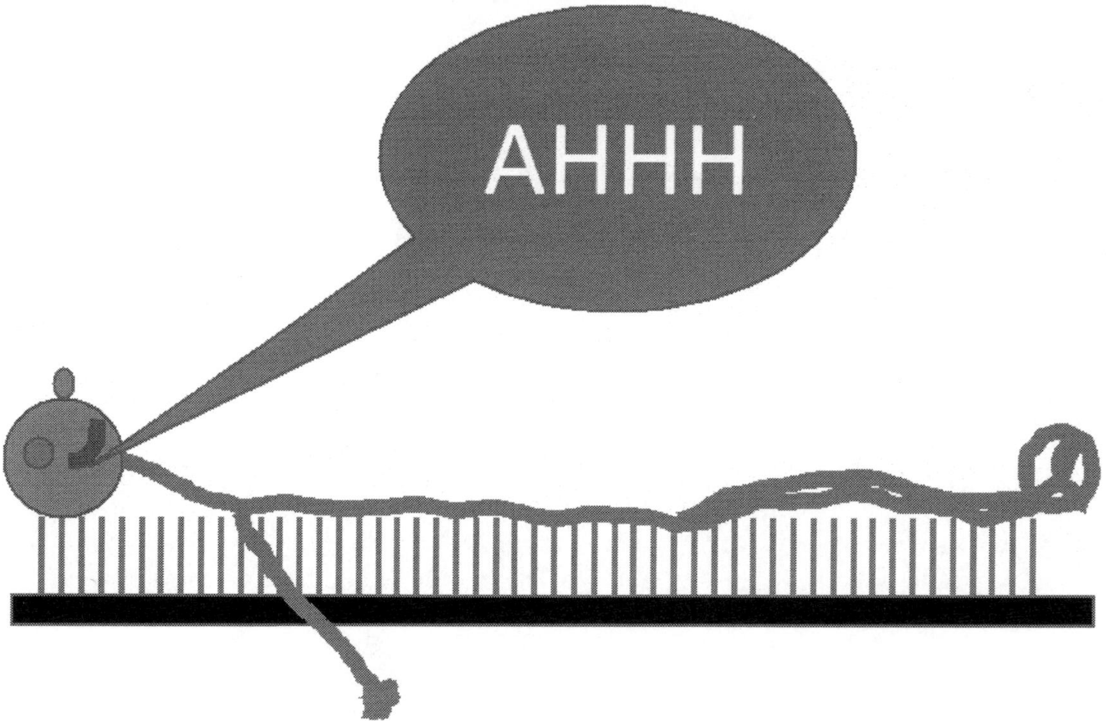
AHHH

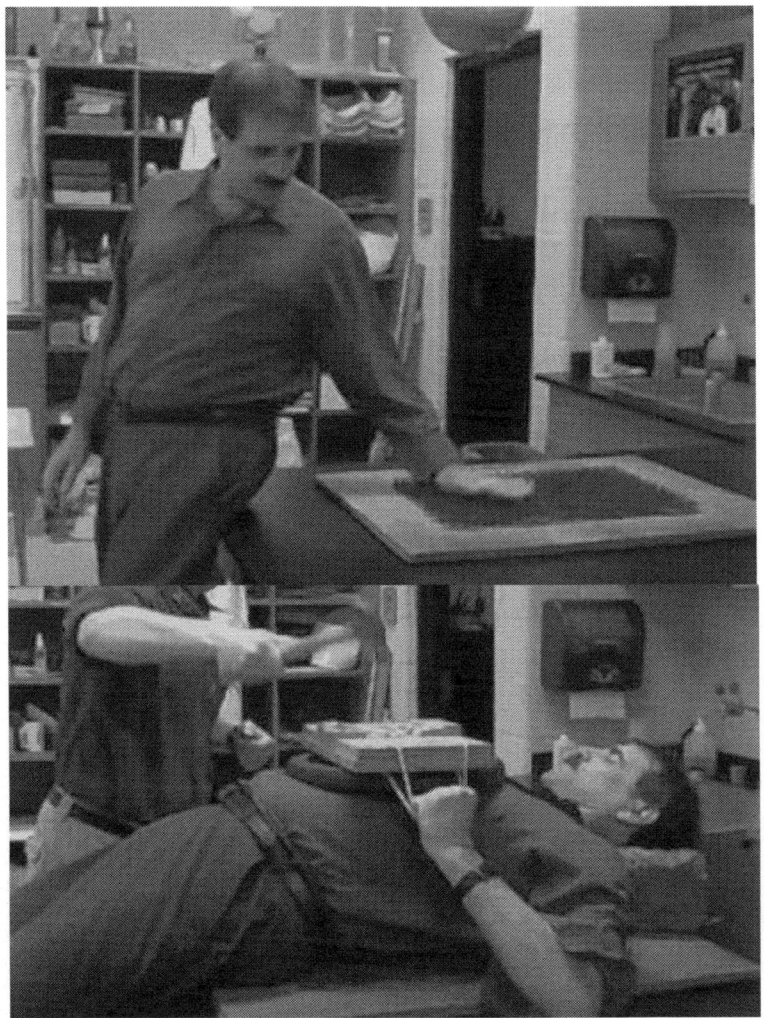

Video of me playing with my bed of nails as I get hit with hammer:
https://www.youtube.com/watch?v=HrUFHn0Ygec

Pressure formula

It is now time for another super easy formula. Since pressure is a relationship between force and area, so is the formula. You simply divide how much force is pushing on how much area.

Pressure = Force / Area

Or

P = F/A

The units are n/cm^2. Once you calculate the pressure you can compare it to other pressures or the strength of a material.

EXAMPLE

- WHAT IS THE PRESSURE OF A 200 N FORCE PUSHING ON AN AREA OF 10 CM²?

$$P = \frac{F}{A}$$

$$P = \frac{200 \text{ N}}{10 \text{ CM}^2}$$

$$P = 20 \text{ N/CM}^2$$

PRESSURE FORMULA

- PRESSURE = FORCE/ AREA

$$P = \frac{F}{A}$$

UNITS = N/CM² OR PACAL

    Pressure can be increased two ways, by increasing the force or decreasing the area. Think of different kinds of shoes. A 100-pound lady stepping on your foot with high heel shoes could smash your foot pretty good, a lot more than a 200-pound person wearing normal shoes. The very small surface area concentrates all the lady's force (weight) at one small spot; the bigger shoes spread it out.

THIS IS NOT GOOD

THIS IS MUCH BETTER!

Snowshoes are big in area and distribute your weight over that big area. If you were to go into deep snow with regular boots, you would sink to the bottom. With snowshoes you only sink a few inches, and it is much easier to walk. Imagine if you went into very deep snow with a snowshoe on one foot and a high heel shoe on the other, what would happen?

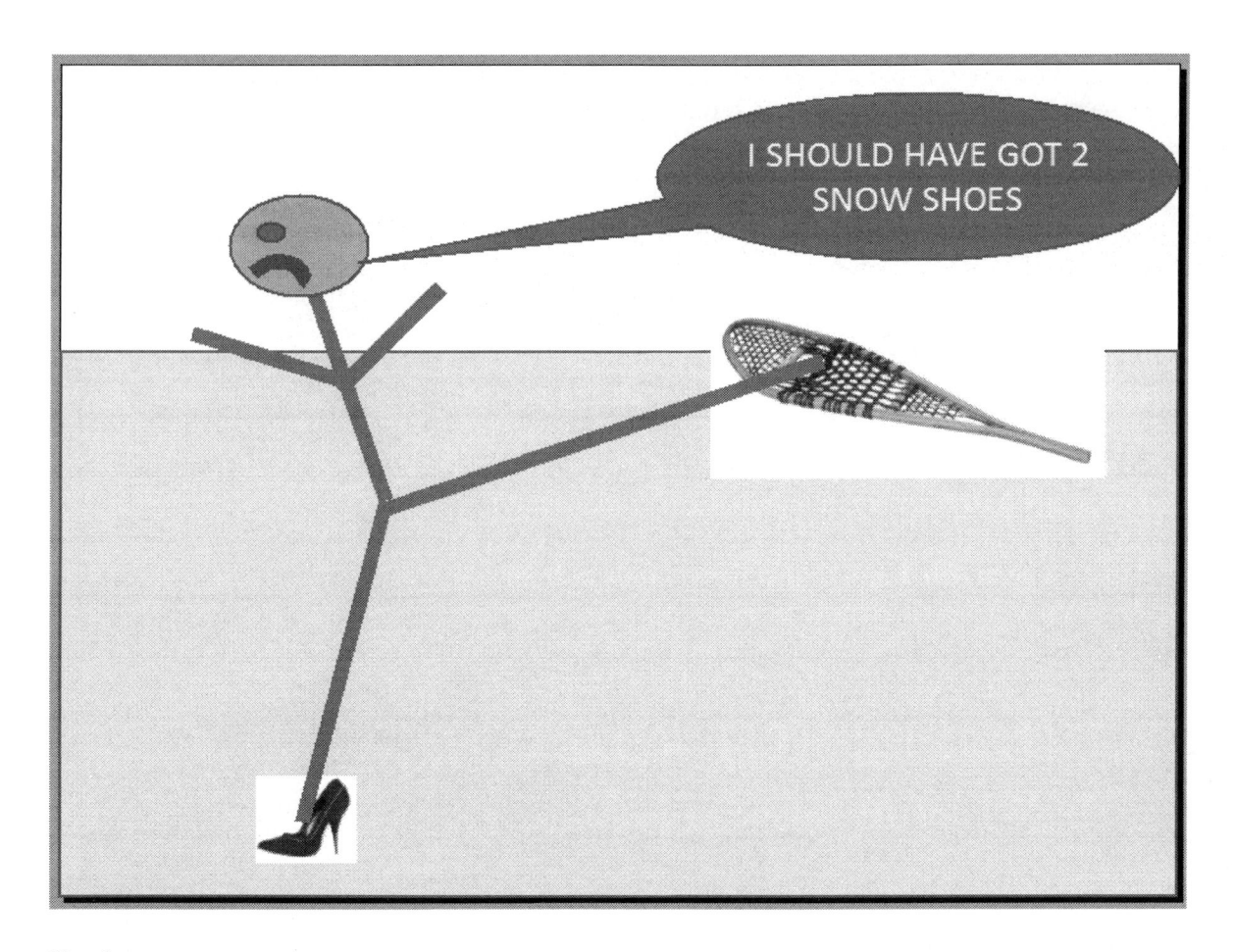

Review

Pressure is how much force is acting over how much area. Pressure = force/area. Pressure is how concentrated a force is.

Flashcard review on Quizlet: https://quizlet.com/144473286/chapter-34-pressure-flash-cards/

A day in the life of Earwig Hickson III

My neighborhood had been quiet, too quiet. It had been peaceful for a long time, too long. It almost seemed like the neighbors were getting used to me. It was uncomfortable. Otis and I were bored. It had snowed for two days and when it was done it was four feet deep, so deep that my Dad actually shoveled our driveway himself because I was too little. That never happened before.

Otis and I tried to play in the snow but it was too deep and we could not walk through it, we could not do anything. We ended up in the garage trying to make snowshoes. It turns out "real" snowshoes are hard to make. Tennis rackets just do not work. Our construction skills were not the best so we went with big wooden boards duct taped to our feet. It was dark when we finished and I got an idea. I shaped the snowshoes to look like big bare feet. We took our Bigfoot feet out into the fluffy soft snow. We ran all over the neighborhood. We walked up to people's doors, mailboxes, backyards, cars, and even did a lap around the police station. We left 20-inch long footprints in the schoolyard and the town square. They were discovered the next day. The news people came. The Game Commission people came. Sasquatch hunters from cable T.V. came. The "experts" studied the tracks, followed the tracks, and were making a T.V. show. Everyone in the town hall had a story. Everyone wanted interviewed. Everyone wanted to be on T.V. Otis and I wanted to be on T.V. too. We heard stories from people describing how Sasquatch ate their cows, scared their chickens, ran out in front of their cars, looked in their windows at night, and knocked over their outhouses. We heard about strange noises in the night and how Sasquatch stole someone's duck. Everyone was excited.

At the big wrap up of the show when the actors told their conclusions and how Bigfoot was near-by, and they might have a chance to catch it, I raised my hand. I had a simple question. "Why don't the tracks sink deeper into the snow?" I asked. "According to my calculations the force that pushed those big tracks down was only about 65 pounds, wouldn't a Sasquatch be heavier and have deeper prints? How do you know it wasn't just some kids wearing big fake feet pulling a joke?"

There was silence. The Mayor was the first to leave. The T.V. lights went out. The crowd began to thin. The fun was over. The experts went home.

When the show aired Otis and I were excited. We stayed up late to watch. We were not on the show. The producers cut us out! That was the end of my acting career.

Fun things to Google
Bigfoot
Sasquatch
The bends

Deep sea diving
Diving suit
Snowshoes

Links
Decompression sickness video: https://www.youtube.com/watch?v=R037jVYOwOU

Decompression sickness documentary: https://www.youtube.com/watch?v=QP9MaVr5hLk

What happens in decompression sickness: https://www.youtube.com/watch?v=p_Vmr-QJ7yM

Check out this video of my small bed of nails popping a balloon: https://www.youtube.com/watch?v=4jFki8kEm7w

Video of me playing with my bed of nails as I get hit with hammer: https://www.youtube.com/watch?v=HrUFHn0Ygec

https://en.wikipedia.org/wiki/Jacques_Cousteau Jacques Cousteau a deep-sea explorer.

Chapter 35
Fluids

Power point lesson of chapter 35:
https://www.youtube.com/watch?v=CKkfXHXmvYI

The wonders of science
It was supposed to be artificial rubber. It was a fail, but sometimes scientists do not fail, they just do not realize they had succeeded. It was during World War II. The Japanese Army had invaded and controlled all the major rubber producing countries in the Pacific Rim. Rubber was made from tree sap, latex, that only grew in a few areas of the world. America, who was rather angry with the Empire of Japan after the sneak attack on the American fleet at Pearl Harbor, was building up its military. And military things need rubber; tires, rafts, boats, airplanes, gas masks, and other things all have rubber parts. America needed rubber and Japan had it all.

The great chemical companies of the USA were asked to come up with a Synthetic rubber, which they did, but there were failures also. Scientists such as Earl Warrick and James Wright independently came up with a rubber substitute that was not quite right. It was a failure as rubber but it had some unique properties. It sometimes acted as a liquid, and sometimes like a solid. If you hit it with a hammer, it shattered. If you pulled it gently, it stretched and was gooey. It bounced and it was fun. Five years later someone put it in a plastic egg and sold it for a dollar, as an Easter toy. It is one of the best selling toys in history and you know it as Silly Putty.

Silly Putty is a non-Newtonian fluid, which means it acts as a solid and a liquid, and does the opposite of what you expect it to do. The Apollo astronauts even took it to space (to stick tools to walls). There are other non-Newtonian fluids such as Quicksand, Slime (toy), and Oobleck (cornstarch and water).

Try making some: http://www.pbs.org/parents/crafts-for-kids/homemade-silly-putty/
Five ways to make silly putty: http://www.wikihow.com/Make-Silly-Putty
How to make silly putty (video): https://www.youtube.com/watch?v=7KYv5s_yVm0
Three ways to make silly putty: http://www.instructables.com/id/Three-Ways-to-Make-DIY-Silly-Putty/
How to make Oobleck: http://www.wikihow.com/Make-Oobleck
Oobleck on a speaker video: https://www.youtube.com/watch?v=3zoTKXXNQIU
Oobleck: https://www.youtube.com/watch?v=FXkwv_zWL84
How to make slime: http://www.hometrainingtools.com/a/slime-recipes-project
More slime recipes: http://www.stevespanglerscience.com/lab/experiments/glue-borax-gak/

What you need to remember
Liquids have a definite volume but no definite shape. They take the shape of the container they are in and cannot be compressed.

Fluids

A fluid is any material that flows. Everybody knows a liquid is a fluid, but gases flow too, so gases are fluids. If it can change shape it is a fluid.

FLUID

- ANYTHING THAT FLOWS
- INCLUDES GASES
- AND LIQUIDS
- NOT JUST WATER OR LIQUIDS
- GASES MOVE AROUND TOO

Properties of liquids

Even though fluids include gases, for this chapter I am going to concentrate on liquids. Liquids have three properties that make them special; surface tension, cohesion, and adhesion.

Surface tension

Molecules in a liquid like water are pulled in every direction by other molecules in the liquid, except for the molecules at the surface. The poor molecules at the top of the liquid are only pulled down and sideways. They are pulled sideways to each other hard enough that they act like a skin or a piece of elastic, like the skin of a balloon. The force of surface tension makes the surface of a liquid act like a stretched sheet of elastic.

1. SURFACE TENSION

- MAKES THE SURFACE OF A LIQUID ACT AS IF IT WERE A STRETCHED SHEET OF ELASTIC
- SOME INSECTS CAN WALK ON IT
- SOAP REDUCES SURFACE TENSION

The surface tension acts like a barrier, strong enough that some insects such as water striders can walk on it. One of the jobs of soap is to destroy surface tension so water can clean something easier. One of the reasons soap is wonderful.

HAPPY WATER BUG

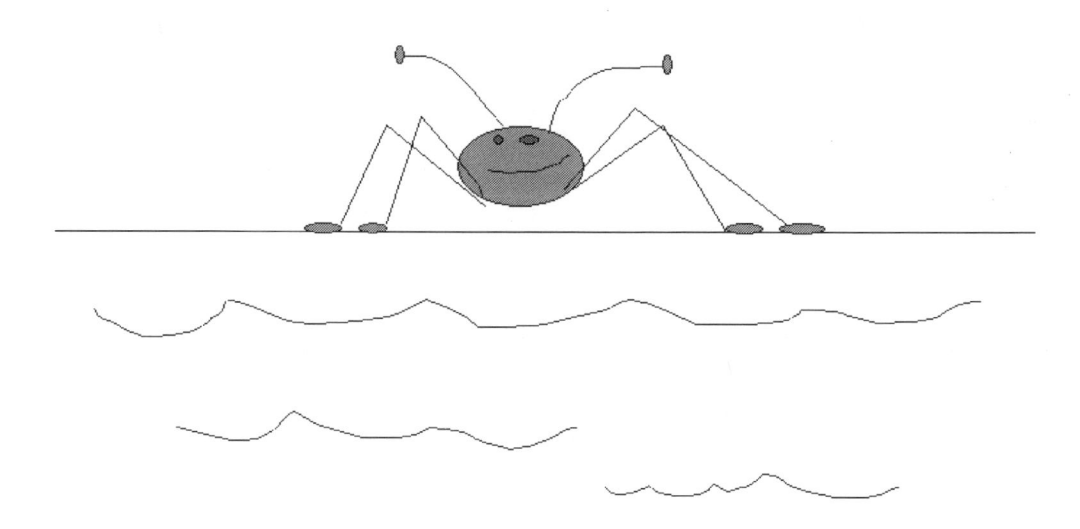

SOAP

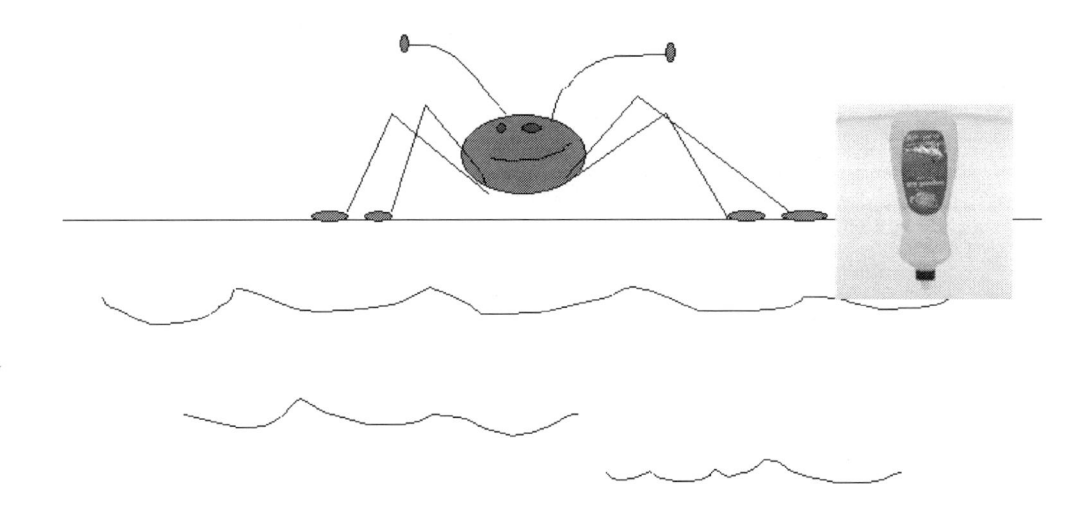

NOT A HAPPY WATER BUG

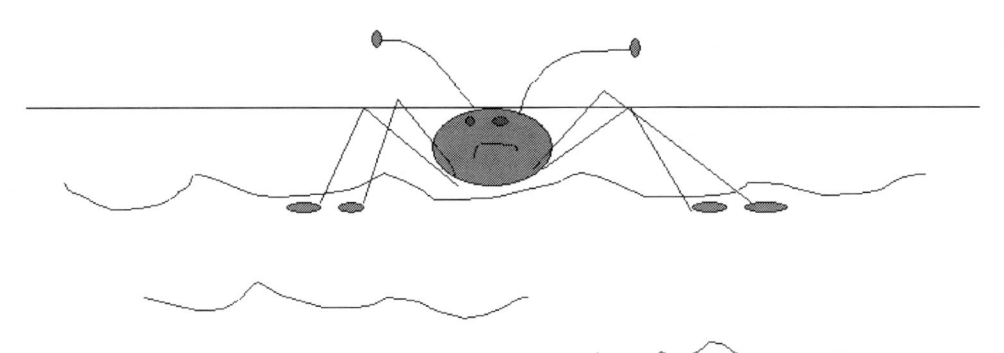

Bill Nye explains surface tension: https://www.youtube.com/watch?v=Hm52rkh68JA
Some cool and colorful surface tension experiments:
https://www.youtube.com/watch?v=WsksFbFZeeU
Pepper and surface tension experiment:
https://www.youtube.com/watch?v=vmilnEWwFPk
Water strider walking on water: https://www.youtube.com/watch?v=V-cXzZt2iVk

This has nothing to do with surface tension but is just too cool not to include. Lizards walking on water: https://www.youtube.com/watch?v=45yabrnryXk

Adhesion

Adhesion is when two different materials stick to each other. The word <u>adhesive</u> (glue) comes from the word adhesion. When you use glue to glue paper, the paper does not stick to paper, it sticks to the glue between the papers. Liquids have a tendency to stick to things (called getting wet or <u>wetting</u>). You may notice that water droplets stick to window glass when it rains. Wax will reduce adhesion. This is why people wax their cars, so the water does not stick to the car and helps it last longer. There are even products to reduce water from sticking to a car's windshield to help in visibility when it rains.

2. ADHESION

- THE FORCE THAT ATTRACTS PARTICLES OF ONE SUBSTANCE TO PARTICLES OF ANOTHER SUBSTANCE
- GLUE IS AN ADHESIVE
- WATER STICKING TO GLASS
- WAX REDUCES ADHESION

WALKING IS SUCH GOOD EXERSIZE

211

A cool little experiment using adhesion to pour water down a string: https://www.youtube.com/watch?v=kUL_317rptE
Another version: https://www.youtube.com/watch?v=iJBHckmq_o0
Certain lizards can walk on walls and glass: https://www.youtube.com/watch?v=hoIk8G7ykDI
How do geckos defy gravity? By TED-Ed: https://www.youtube.com/watch?v=YeSuQm7KfaE

Cohesion

Cohesion is when a material sticks to itself. It causes liquids like water to bead up in a glob. It causes two little globs to join up into a big glob. When I was little I used to play with Mercury, which has very strong cohesion forces. It was fun but do not do it. Mercury is very poisonous.

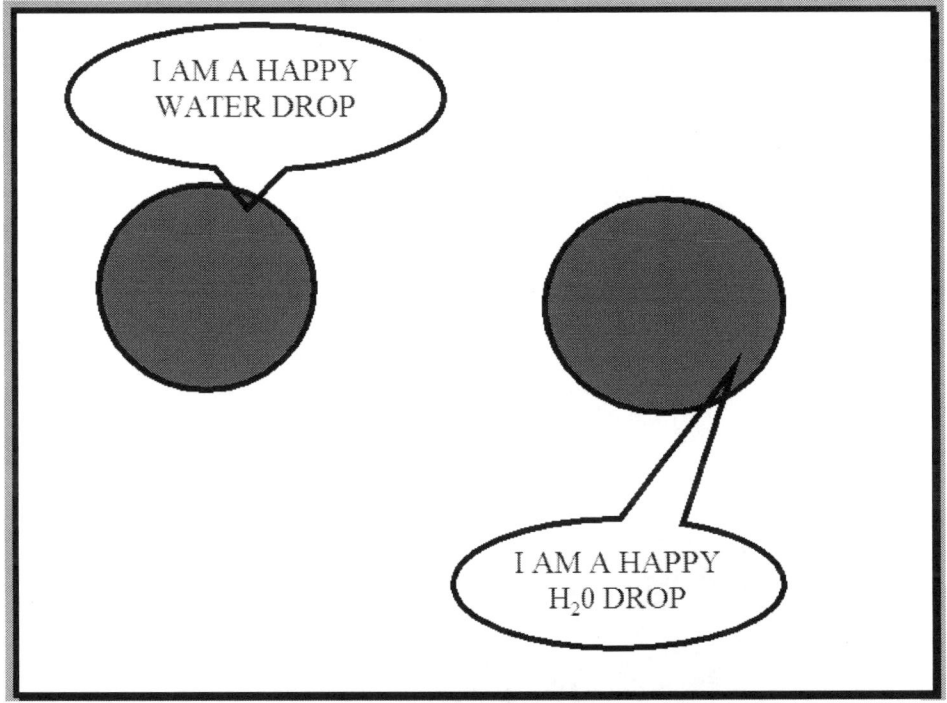

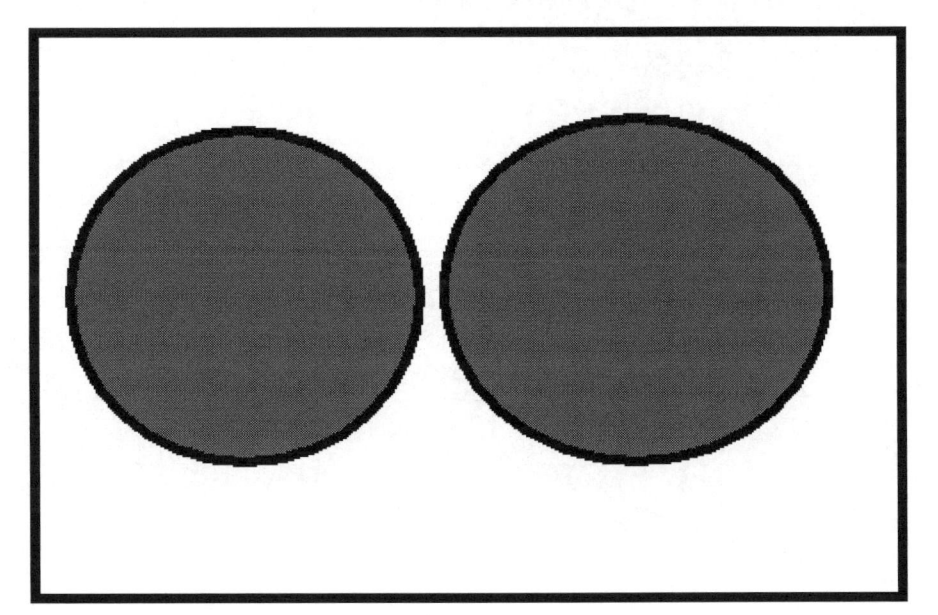

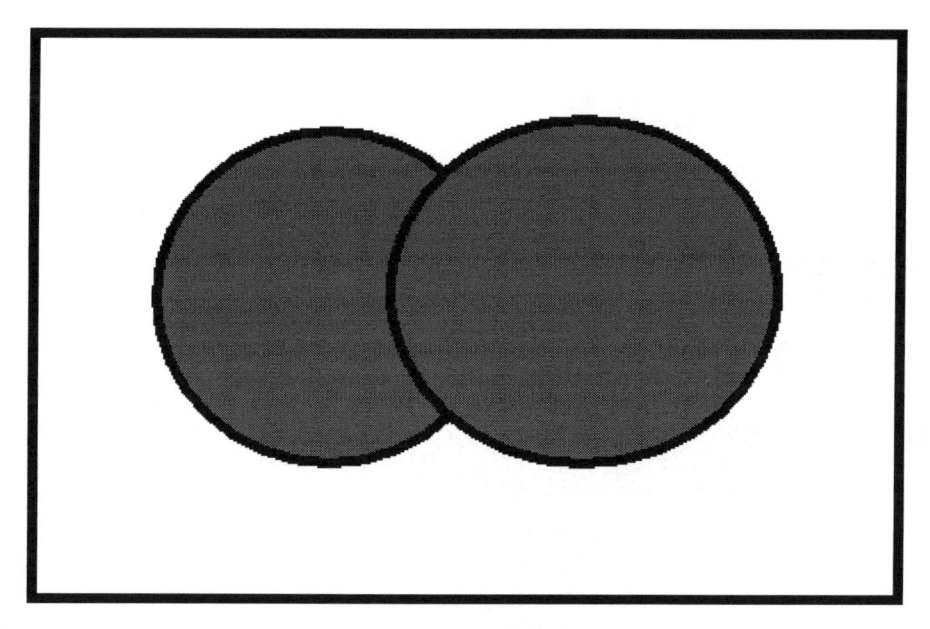

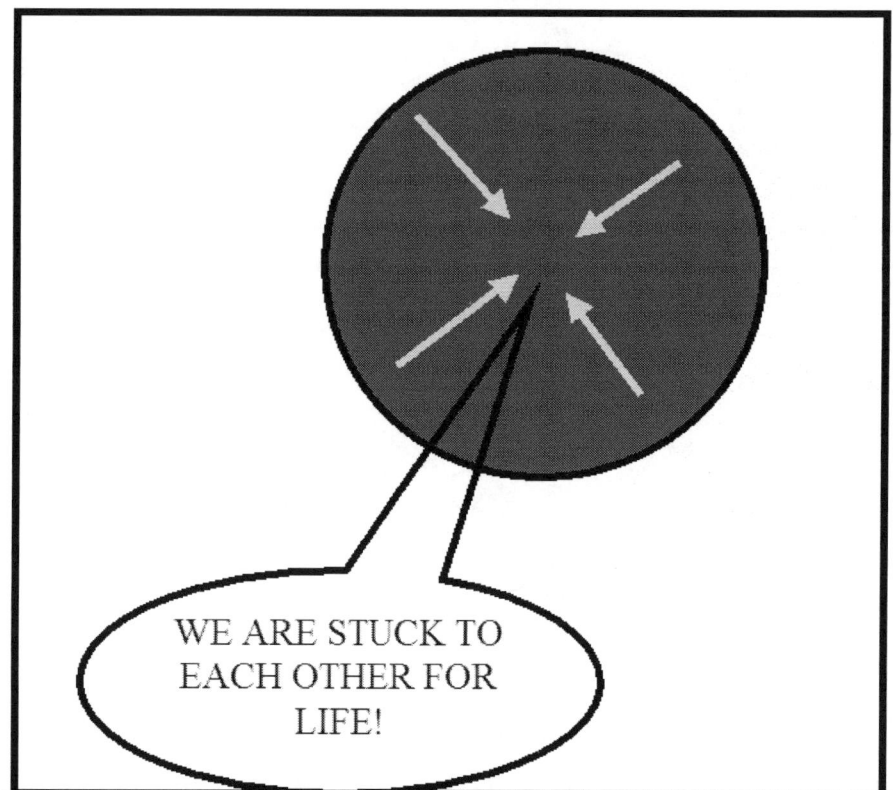

Someone, not, me playing with mercury:
https://www.youtube.com/watch?v=0j2X6HZrfdE
More mercury playing: https://www.youtube.com/watch?v=Bg8yfetAD8s
Still more: https://www.youtube.com/watch?v=pvdWOwG4aUY

If you want to try a cool experiment (without blowing anything up) try to see how many drops of water you can put on a clean penny. It is more than you think. The forces of surface tension, adhesion and cohesion help you.

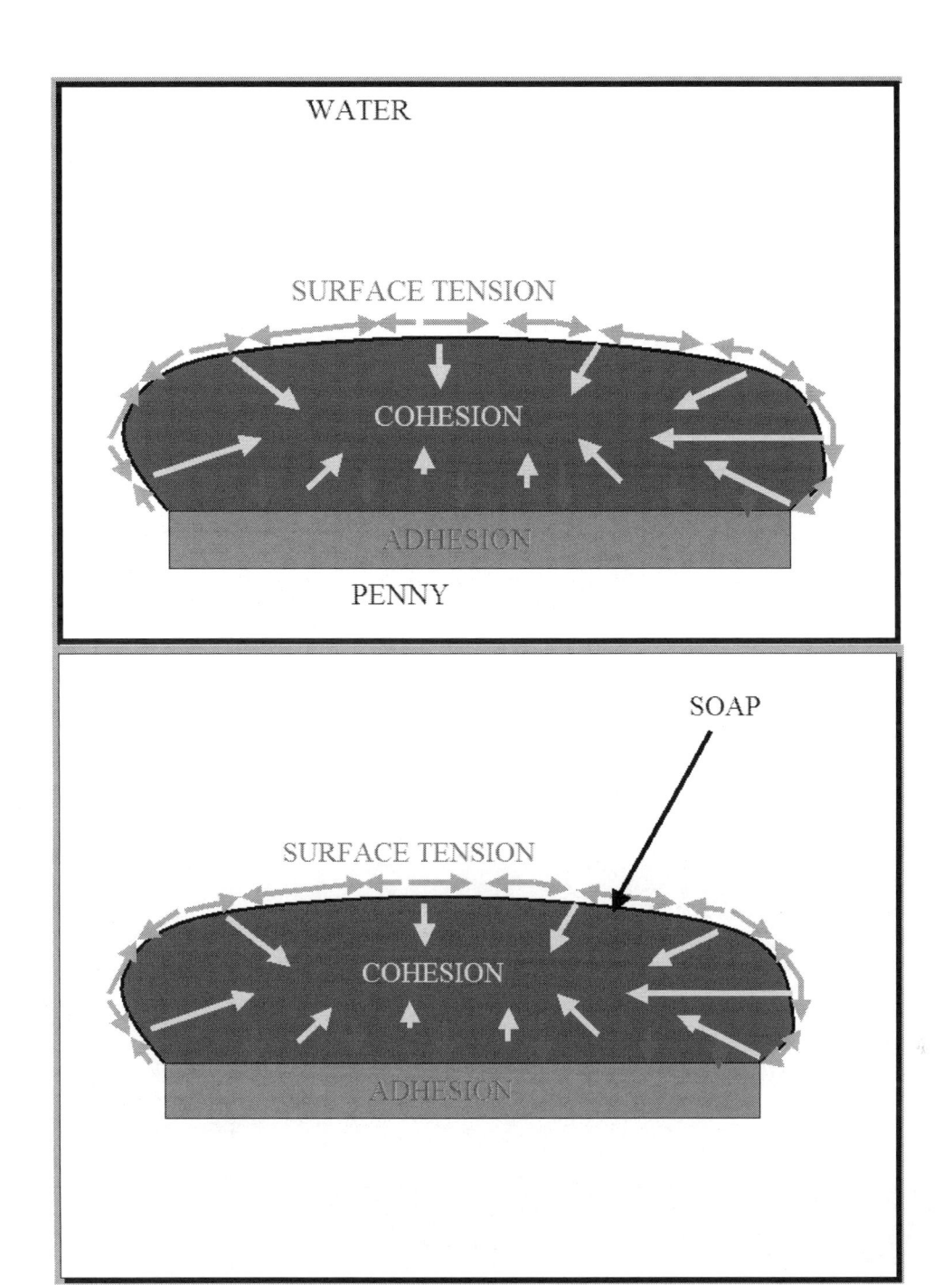

Add some soap.

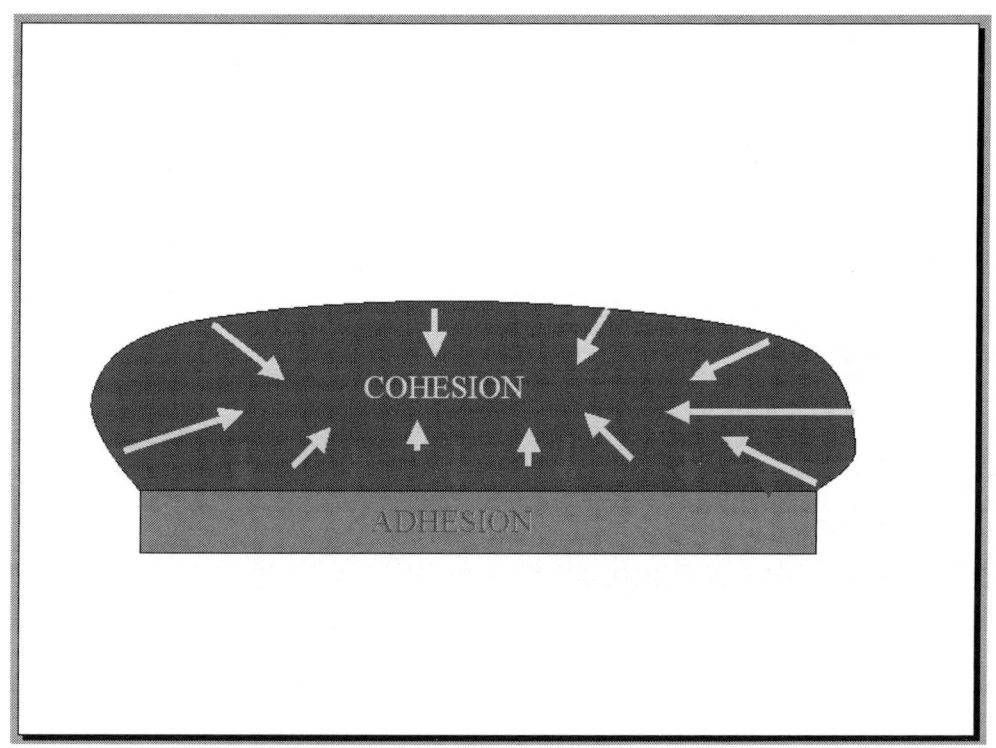

The surface tension is destroyed.

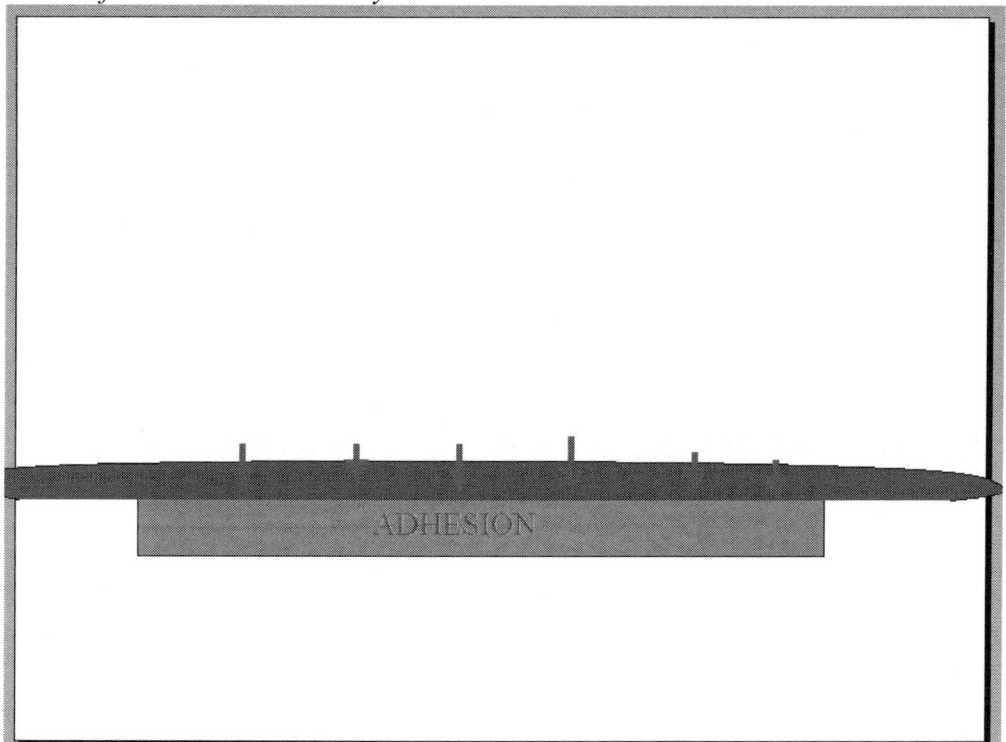

And it all spills.

Drops of water on a penny: https://www.youtube.com/watch?v=APllHgHmhRY
My favorite astronaut showing how water behaves in microgravity: this is beyond cool: https://www.youtube.com/watch?v=o8TssbmY-GM

More water in space: https://www.youtube.com/watch?v=s63JXdsL5LU

Review

The properties of liquids include surface tension, cohesion and adhesion. Surface tension is like a piece of elastic on the surface. Adhesion makes different things stick, and cohesion makes similar things stick.

Flashcard review from quizlet: https://quizlet.com/174919632/chapter-35-fluids-flash-cards/

A day in the life of Earwig Hickson III

The snow was gone and the Sasquatch hunters were gone. Otis and I had been good for a long time. We were ready to do a project. Earlier that day, in science class we did an experiment where we mixed cornstarch and water to make a non-Newtonian fluid. It was fun to play with and did all sorts of unique things. We were in the process of making a batch in my kitchen. We were going to put it on one of my Dad's big stereo speakers, to see if it would jump. My Mom caught us just in time, preventing a big mess and a potential grounding from my Dad. We were quickly chased outside. The other thing we learned in science class was how Quicksand formed. This idea had merit.

Quicksand is really just regular sand with an underground spring pushing water upward. The water lubricates the sand particles, and if you step on it, it acts like water and you sink. If you struggle as you try and get out, the sand gets pushed together and acts like a solid. You are stuck.

One of the neighborhood bullies who called himself "Big Criminal Scratch," but his real name was Oliver, was giving us a hard time about our lack of athleticism and our general nerdiness. We were kind of tired of it. He was a bit tired of me calling him Oliver, too. Our brains were our strongest muscle. His brain was his weakest, but he made up for it with his fists. As it turns out he did not pay attention in science class.

Otis and I had done an experiment. Since we could not do the cornstarch project in my kitchen, we decided to go big. First, we dug a big hole that was about 30 feet long and 30 feet wide. It was eight feet deep. At the bottom of the hole we put a fire hose (from a previous project) and attached it to a fire hydrant. We filled the hole with sand. We played with our quicksand experiment all the previous day. When the water was turned on, we could not walk on it, we just sank. When the water was off it was just like normal sand. We practiced getting out of the quicksand and getting stuck in the quicksand. We had fun, but now we had a problem. Little old Ollie was pestering us again.

He challenged us to fights, feats of strength, bravery, anything he could think of. He tried to embarrass us and insulted us. Then I had an idea. I challenged him to a volleyball game.

At first he laughed. Volleyball was a sissy game to him. I insinuated that he was afraid. He felt trapped. The Great Volleyball game was on. As we negotiated the rules, he was allowed to have his whole gang on his team and I had to play by myself. After all he claimed, I chose the game, he should choose the rules. I agreed.

The whole neighborhood was there, especially Alexia, who he was sweet on. As the crowd gathered around MY PERSONALLY made volleyball court, Otis snuck off to the fire hydrant. Ollie boy was exceptional annoying as he showed off for Alexia. Six players to one, did not matter to him, only winning seemed to be important. He made fun of me, Alexia smiled; he insulted me, Alexia smiled, he bragged before he won, a bad mistake.

When the game began, I got to serve first. I warmed up. Oliver and his five henchmen waited for my serve. I stretched, Ollie waited. I threw the ball up to start my serve. Otis hit the water valve. Oliver and all his gang instantly sank into the quicksand, up to their necks. They screamed like little girls. My serve landed right in the middle. Otis turned off the water. I had one point and I retrieved the ball. The gang was stuck, up to their necks in sand. The crowed cheered. I served again.

The final score was 21 – 0. Oliver "Big Criminal Scratch" was defeated in front of Alexia and 100 other kids. The spectators all "accidentally" stepped on Oliver and his gang's heads. Alexia stepped on his head too. I was a hero. Oliver never bothered me or anyone else again. Oliver had reached his peak; it was down hill from there. I did not feel bad at all.

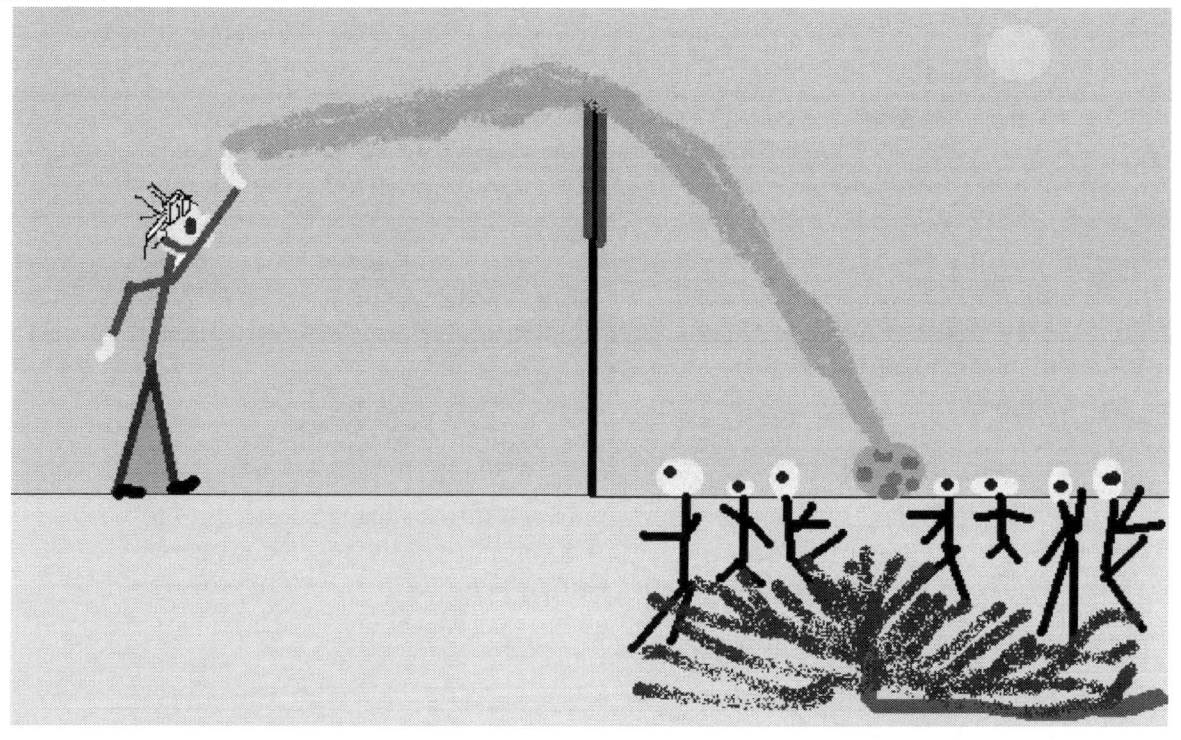

Fun things to Google
Surface tension experiment
Surface tension
Adhesion
Cohesion
Surface tension how to float metal
Drops on a penny
Quicksand
How to get out of quicksand
Non-Newtonian fluid

Silly putty
Slime
Gawk
Rain-x

Links
Try making some: http://www.pbs.org/parents/crafts-for-kids/homemade-silly-putty/

Five ways to make silly putty: http://www.wikihow.com/Make-Silly-Putty

How to make silly putty (video): https://www.youtube.com/watch?v=7KYv5s_yVm0

Three ways to make silly putty: http://www.instructables.com/id/Three-Ways-to-Make-DIY-Silly-Putty/

How to make Oobleck: http://www.wikihow.com/Make-Oobleck

Oobleck on a speaker video: https://www.youtube.com/watch?v=3zoTKXXNQIU

Oobleck: https://www.youtube.com/watch?v=FXkwv_zWL84

How to make slime: http://www.hometrainingtools.com/a/slime-recipes-project

More slime recipes: http://www.stevespanglerscience.com/lab/experiments/glue-borax-gak/

Bill Nye explains surface tension: https://www.youtube.com/watch?v=Hm52rkh68JA

Some cool and colorful surface tension experiments: https://www.youtube.com/watch?v=WsksFbFZeeU

Pepper and surface tension experiment: https://www.youtube.com/watch?v=vmilnEWwFPk

Water strider walking on water: https://www.youtube.com/watch?v=V-cXzZt2iVk

This has nothing to do with surface tension but is just too cool not to include. Lizards walking on water: https://www.youtube.com/watch?v=45yabrnryXk

A cool little experiment using adhesion to pour water down a string: https://www.youtube.com/watch?v=kUL_317rptE

Another version: https://www.youtube.com/watch?v=iJBHckmq_o0

Certain lizards can walk on walls and glass: https://www.youtube.com/watch?v=hoIk8G7ykDI

How do geckos defy gravity? By TED-Ed: https://www.youtube.com/watch?v=YeSuQm7KfaE

Someone, not, me playing with mercury: https://www.youtube.com/watch?v=0j2X6HZrfdE

More mercury playing: https://www.youtube.com/watch?v=Bg8yfetAD8s

Still more: https://www.youtube.com/watch?v=pvdWOwG4aUY

Drops of water on a penny: https://www.youtube.com/watch?v=APllHgHmhRY

My favorite astronaut showing how water behaves in microgravity: this is beyond cool: https://www.youtube.com/watch?v=o8TssbmY-GM

More water in space: https://www.youtube.com/watch?v=s63JXdsL5LU

Chapter 36
Pressure in fluids

Power point lesson for this chapter:
https://www.youtube.com/watch?v=CjoiapJDSZ0

The wonders of science

It was the first use of a submarine in war. The prefix "sub" means under and "marine" means sea, so "submarine" means under the sea. Under water is a great place to hide. The American Revolutionary war was in full swing and British war ships were everywhere. The *Turtle* was an egg shaped wooden vessel that used hand cranks to turn a screw for propulsion. The single person inside was supposed to maneuver across New York harbor and attach a bomb to a British ship, the HMS *Eagle*. The driver, Sergeant Ezra Lee, was not able to attach the bomb and was spotted and chased away by some British solders in a rowboat. The attack was so much of a failure that the British did not even know they had been attacked. Some historians do not think the event even happened and was a made up story to boost American moral.

The Turtle was not the first submarine; people had been experimenting with them since the late 1600's. The *Nautilus* built by American inventor Robert Fulton for the French navy is the first practical submarine built in 1797. In the end Napoleon was not interested.

In the Civil War, a submarine called the *H. L. Hunley*, became the first submarine to sink an enemy ship. Unfortunately it sunk itself too.

It was not until the development of Diesel Electric power that submarines became practical and it would be the German Navy in the Second World War, that would introduce submarines (U-boats) to the world. Germany used "wolfpacks" in the Battle of the Atlantic and nearly starved out the United Kingdom, which depended on ships from America for food and supplies.

Modern nuclear powered submarines can stay submerged for months, the limiting factor being the amount of food it can carry.

What makes submarines work is their ability to change their weight. They are made of three tubes, one for the people, and two for either water or air. When these ballast tanks are filled with air, the sub floats, when water, they sink. Using ballast tanks a submarine can control its buoyancy.

What you need to remember

Fluids include liquids and gases. Pressure is the amount of force per unit of area, such as pounds per square inch. Gravity causes all fluids to have weight (force). Density is how much mass per unit of volume, Density = mass/volume.

Pressure in fluids

Fluids can be heavy. They can squish you; in fact, a fluid, a very heavy fluid that you do not even notice, is squishing you right now. The fluid squishing you is the Earth's atmosphere and it weights about 11,700,000,000,000,000,000 pounds! Fortunately not all this weight is above you so you are safe. The only part of the atmosphere that is actually trying to crush you, is what is directly above your head. This is still a lot. You are at the bottom of a giant ocean of air 50 miles deep. Fifty miles of air is rather heavy and we call this *atmospheric pressure*. If you draw a 1-inch by 1-inch square on the top of someone's baldhead, that square has about 14.7 pounds of air pushing down on it. Imagine putting 15 pounds of weights on this square. That is what the atmosphere is doing. The atmosphere is not just pushing down though; it pushes everywhere, even up. You are surrounded by atmospheric pressure pushing on you from all directions, trying to make you smaller. It is a three-dimensional kind of thing. It works exactly the same way when under water.

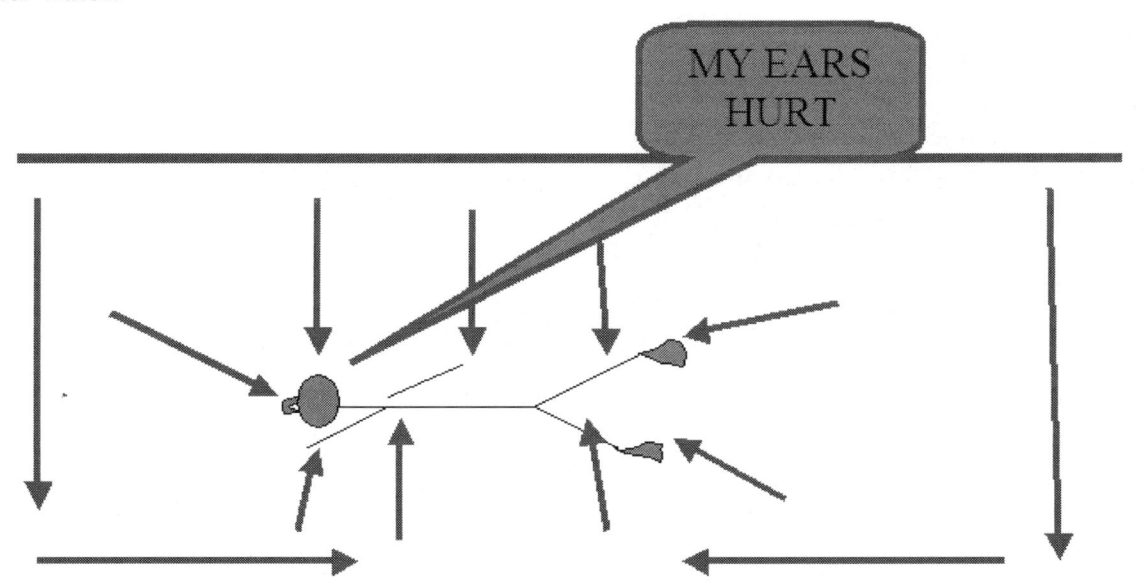

The atmospheric pressure at sea level is 14.7 pounds per square inch, but you are more than just one square inch. If I was to peal my skin off, lay it flat and measure how many square inches in area it is, I would get a large number. It turns out that my skin's surface area is 2.16 m^2, which converts to 3348 in^2. Each of these square inches has 14.7 pounds of air pushing on it so I multiply 3348 x 14.7 and get a grand total of 49215 pounds trying to crush me. That comes out to 24 tons. No wonder I am so tired all of the time. By the way, you have exactly the same pressure inside you pushing outward (trying to make you bigger) so you do not notice this giant weight.

A good video by TEDed explaining all this: http://ed.ted.com/lessons/how-heavy-is-air-dan-quinn

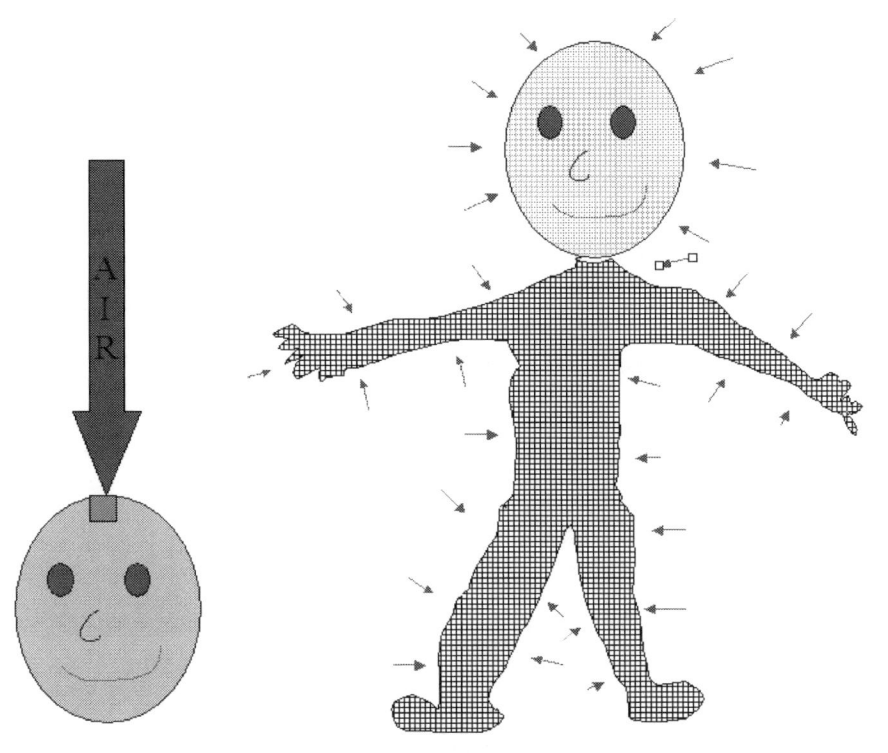

THINK WATER AND AIR

- **IN FLUIDS PRESSURE IS EXERTED EQUALLY IN ALL DIRECTIONS**
- **UP, DOWN, SIDEWAYS**

I think that was a cool story about the atmosphere but there is more to it than just that. Air isn't the only fluid that does this, all gases and all liquids are fluids, so all of them try and squish you when they are above you. Water is the easiest to understand, so from now on I will be discussing water, but everything I say about water is true for other liquids and all gases.

What causes pressure?
Pressure is caused by the Earth's gravity pulling fluids down. You may not notice air pressure but when you dive deep under water, you notice water pressure. It works exactly the same, as the atmosphere. The water above you is trying to squish you (from the top, bottom, and sides). The only difference is the water is denser, so there is more weight above you. Water is so much denser than air that you only need to dive 15 feet down to equal the pressure of 50 miles of air. So, pressure depends on two things, the depth of the fluid above, and how dense the fluid is.

223

PRESSURE DEPENDS ON

- DEPTH
- DEEPER = HIGHER PRESSURE
- DENSITY
- DENSER = HIGHER PRESSURE

Depth

The deeper under a fluid you go, the greater the pressure becomes. Every 32 feet of water increases the pressure by 15 lbs/in^2. This is why your ears may hurt when you dive as the pressure from the water tries to get into you, and your ears are the weak spot. Your ears can also hurt when you fly in an airplane or drive up a mountain for the same reason, usually your ears "pop" to relieve this pressure.

There are small submarines called submersibles or bathyscaphes that can go to the bottom of the ocean, seven miles deep, in a place called the Challenger Deep, in the Mariana Trench, where the water pressure is 15,750 lbs/in^2, these are built extremely strong. The passengers on these often put a Styrofoam cup on the outside as a souvenir, all the air gets squeezed out of the Styrofoam and the result is a very small cup.

A rather hilarious video of a poor critter diving under water: https://www.youtube.com/watch?v=Gbzvtxj-6hc

Life in the Mariana Trench: https://www.youtube.com/watch?v=6N4xmNGeCVU

National Geographic Documentary - Discovery Secret Mariana Trench – a rather long documentary but interesting: https://www.youtube.com/watch?v=cpxdT3KvlGY

Now let us imagine an inflatable rubber frog balloon. I am going to send it to the bottom of the ocean. At first it is a normal sized frog, but when I put it under a lot of water, it gets squished from all sides. The deeper it goes the smaller it gets.

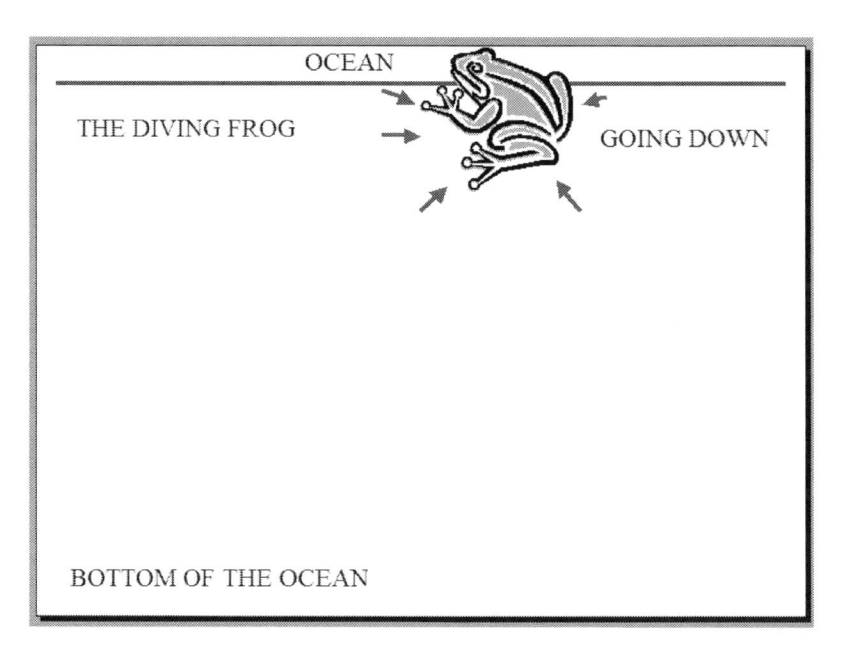

OCEAN

THE DIVING FROG   GOING DOWN

BOTTOM OF THE OCEAN

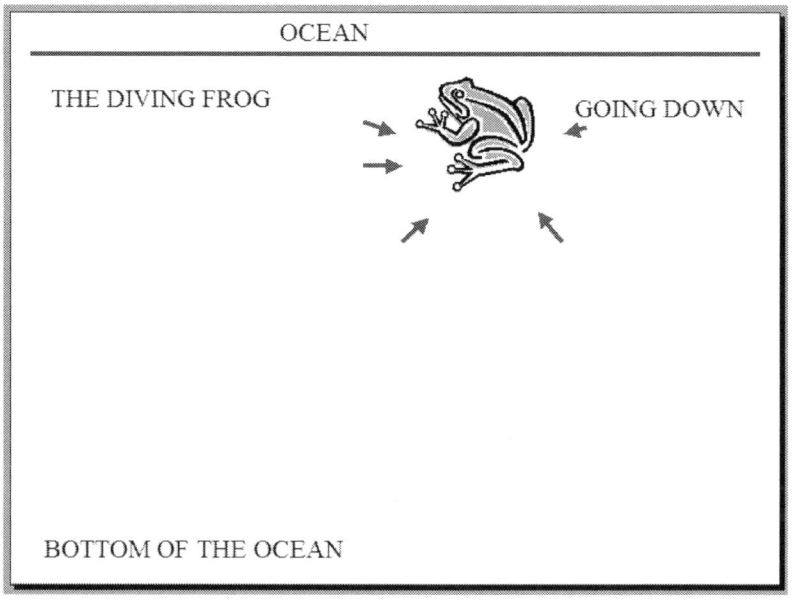

OCEAN

THE DIVING FROG   GOING DOWN

BOTTOM OF THE OCEAN

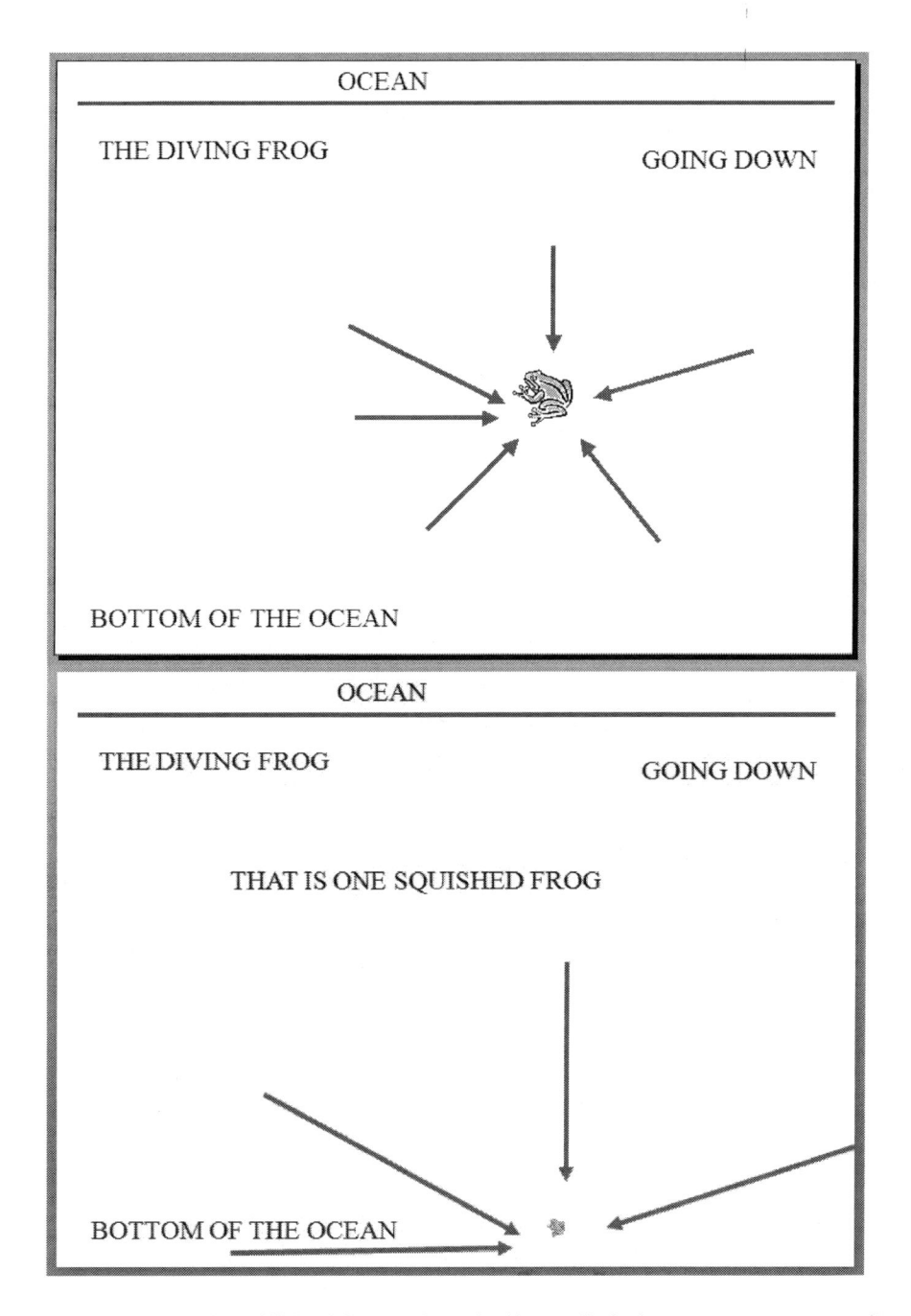

```
┌─────────────────────────────────────┐
│              OCEAN                  │
│  ─────────────────────────────────  │
│  THE DIVING FROG        GOING DOWN  │
│                                     │
│              ↓                      │
│         ↘    🐸   ↙                 │
│         →         ←                 │
│           ↗     ↖                   │
│                                     │
│  BOTTOM OF THE OCEAN                │
└─────────────────────────────────────┘

┌─────────────────────────────────────┐
│              OCEAN                  │
│  ─────────────────────────────────  │
│  THE DIVING FROG        GOING DOWN  │
│                                     │
│        THAT IS ONE SQUISHED FROG    │
│                                     │
│                  ↓                  │
│            ↘         ↙              │
│  BOTTOM OF THE OCEAN  →  •          │
└─────────────────────────────────────┘
```

Since the air is still inside my frog balloon (it is just very compact from the pressure). When I bring him back up, he gets big again as the air expands back to normal.

ON THE WAY BACK UP HE WOULD GET BIG AGAIN

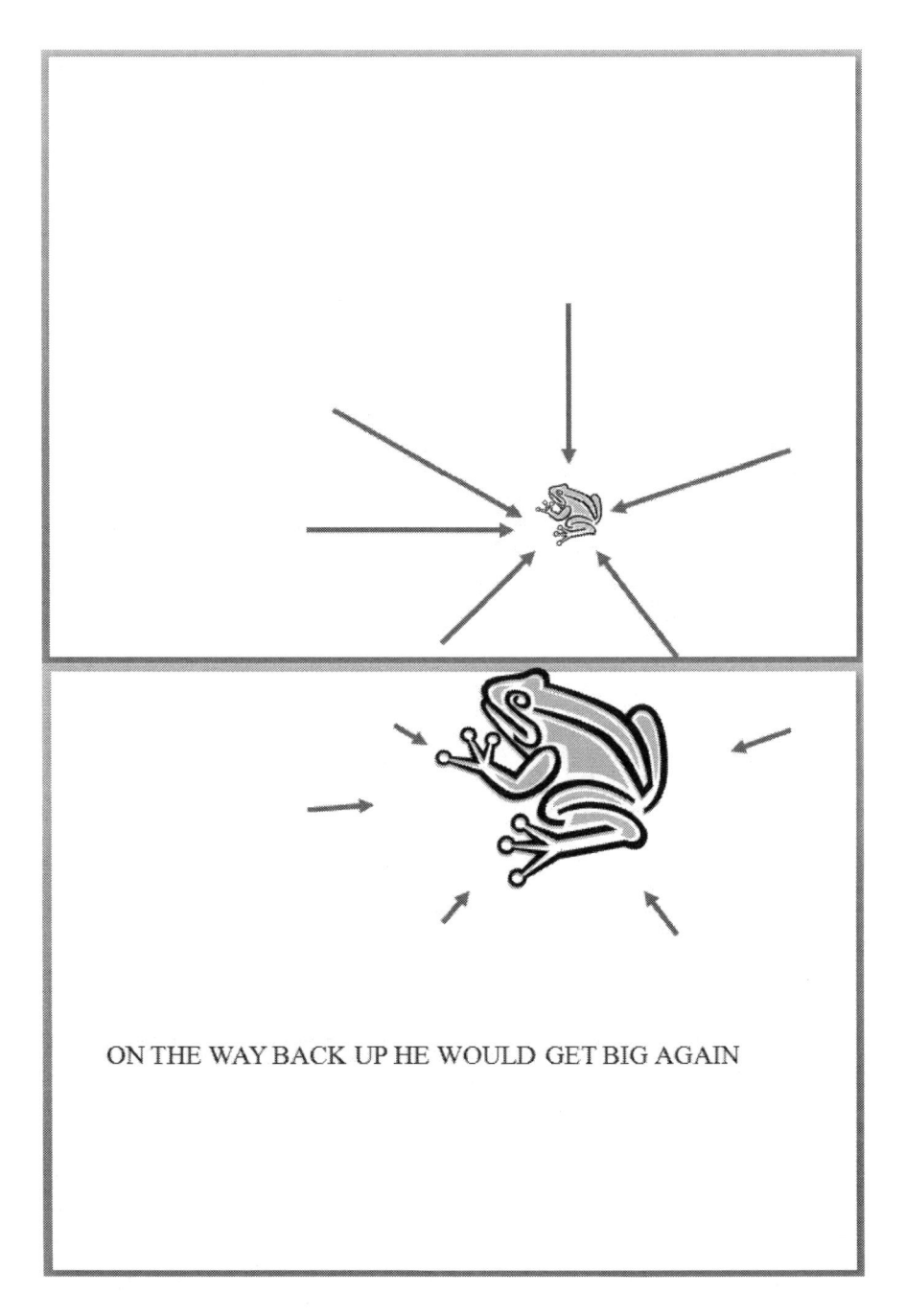

Density

We did density a long time ago in chapter 10. There you learned that density is a ratio between mass and volume, or how compact or fluffy something is, in other words, how many molecules are in a milliliter (or cubic centimeter). Obviously there are a lot more molecules crammed into a ml of water than a ml of air. The result is, water has a density of 1 g/ml, but air has a density of only 0.001225 g/ml, much less. As you read this you are under about 50 miles of 0.001225 g/ml density air, but what if you were under 1 g/ml water instead? The pressure would be a bit more than it is now, about 122,565 lbs/in^2; this is because water is roughly 800 times denser than air.

DENSITY

- SOME LIQUIDS ARE MORE DENSE THAN OTHERS
- WATER = 1 G/ML
- SALT WATER = 1.1 G/ML
- DENSE THINGS WEIGH MORE PER VOLUME
- DENSE LIQUIDS SINK IN LESS DENSE LIQUIDS

When you go swimming and dive 15 feet below the surface, the pressure is double what you are used to, at about 30 lbs/in^2, but what if you were swimming in a pool of mercury with a density of 13.5 g/ml? You would be very squished under 200 lbs/in^2. In fact, Mercury is so dense a human can float ON it, it would be like lying on your bed.

A cannon ball floating in mercury:
https://www.youtube.com/watch?v=Rm5D47nG9k4

Let's compare our inflatable frog in water to one in Mercury.

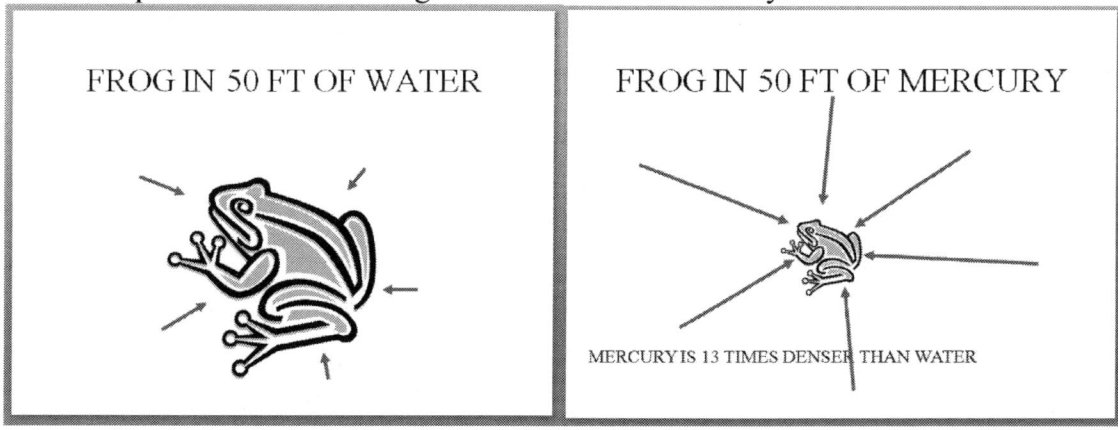

FROG IN 50 FT OF WATER

FROG IN 50 FT OF MERCURY

MERCURY IS 13 TIMES DENSER THAN WATER

A brief review of density
Density how much mass is in a certain volume.

228

DENSITY IS MASS PER VOLUME

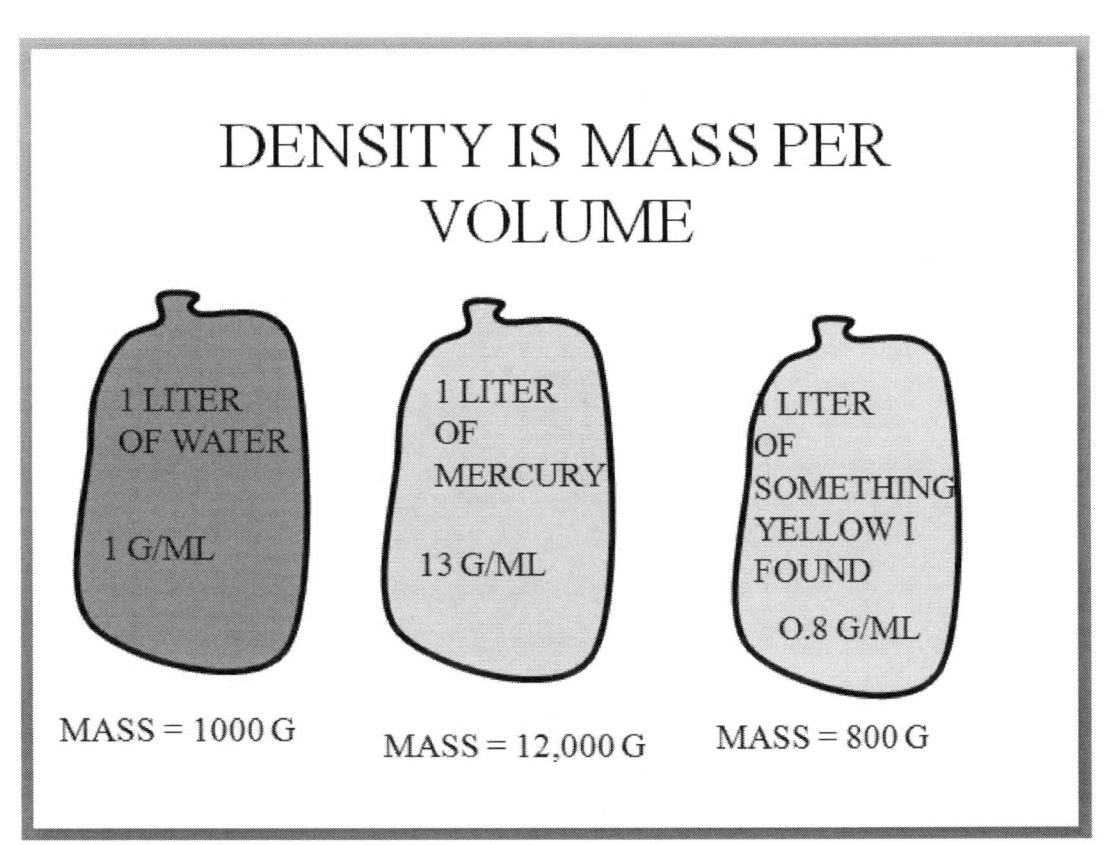

- 1 LITER OF WATER — 1 G/ML — MASS = 1000 G
- 1 LITER OF MERCURY — 13 G/ML — MASS = 12,000 G
- 1 LITER OF SOMETHING YELLOW I FOUND — 0.8 G/ML — MASS = 800 G

If I pour these liquids together, I get a density column.

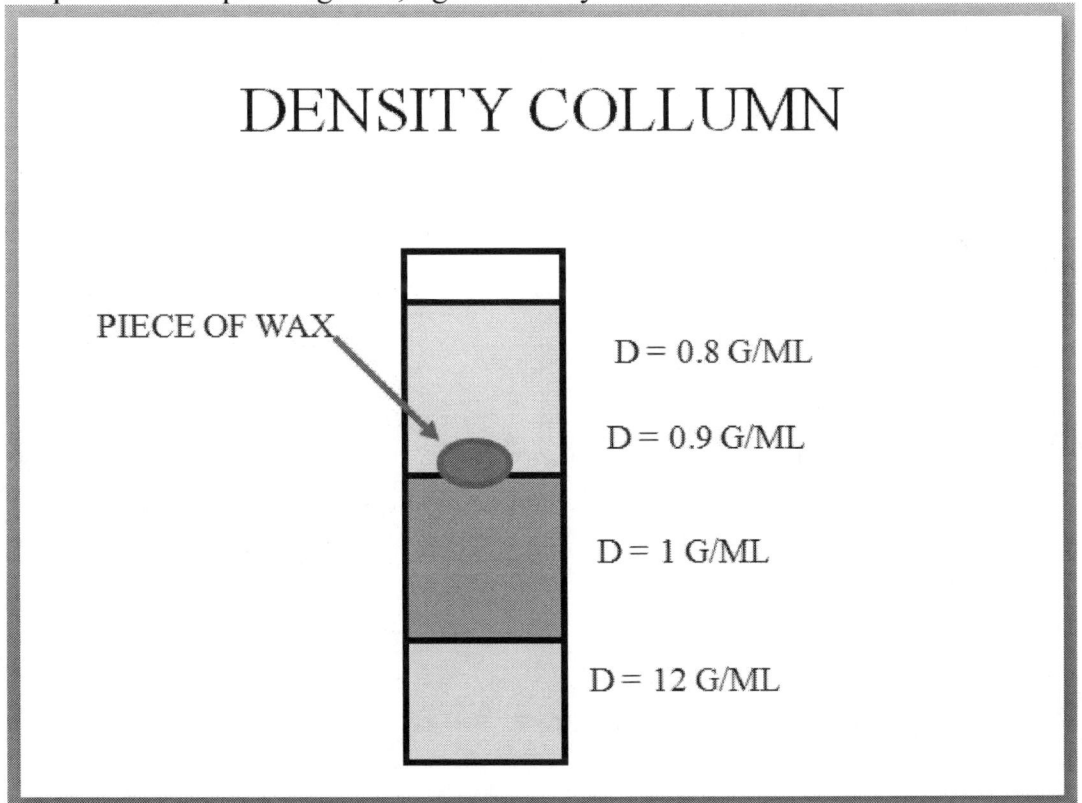

DENSITY COLLUMN

PIECE OF WAX

- D = 0.8 G/ML
- D = 0.9 G/ML
- D = 1 G/ML
- D = 12 G/ML

If I remember right, I told you a story about Arnold the duck and his trip to an oil spill in the Gulf of Mexico (back in Ch. 10).

REMEMBER ARNOLD THE DUCK?

OIL 0.7 G/ML

DUCK 0.8 G/ML

WATER 1 G/ML

It is interesting to note that the density of a solid floating in water tells you how much of the object is UNDER water. The Titanic was struck by an iceberg (with a density of 0.9 g/ml) floating in the ocean. This means 0.9 (90%) of the iceberg was under water with only 0.1 (10%) of it visible above the water. The results were not good.

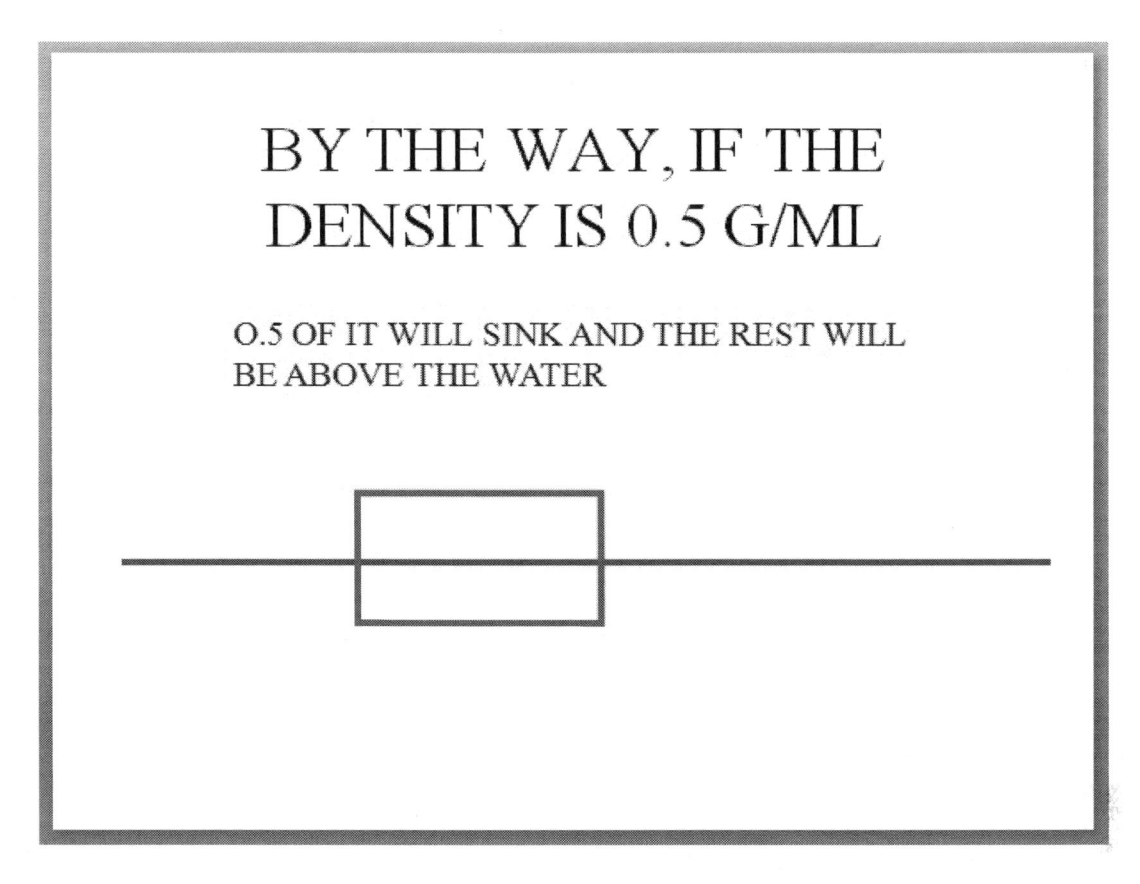

> # BY THE WAY, IF THE DENSITY IS 0.5 G/ML
>
> 0.5 OF IT WILL SINK AND THE REST WILL BE ABOVE THE WATER

Review
The pressure in fluids depends on how deep you are and the density of the fluid. As you go deeper the pressure increases. The denser the fluid is, the greater the pressure.

Flashcard review on Quizlet: https://quizlet.com/173972286/chapter-36-pressure-in-fluids-flash-cards/

A day in the life of Earwig Hickson III
Not all my ideas are good ones, and all of Otis' ideas are bad. This was one of Otis' famous ideas. One he actually deserved to be punished for. It all started with an airplane I wanted to build, not a paper or balsa wood one, but a real one. I had collected the materials for the wings but I needed a lightweight bike for the body. The store at the corner had the perfect bike for my needs, a racing bike with a lightweight Magnesium frame. It was ideal. The only problem was it cost money, none of which either of us had.

That is when Otis remembered something. Something I think he had been thinking of for quite a while. Something he knew was wrong, something he could blame on me. There was an old well in our town; it must have been hundreds of years old. No one knew who had originally dug it or how deep it was. If you dropped a rock into it, it fell for 4 whole seconds before a splash could be heard. I calculated how deep that was and came up with 78 meters or 256 feet. Not only that, Otis had taken in upon himself to drop a lead mass on a fishing line down to find out how deep the water was. He ran out of line. The well was a dangerous place. Falling into it was dangerous. Going into it on purpose was stupidity. This is what we decided to do. You see Otis has realized that for 200 years people had been making wishes over that well, and with every wish, they dropped a coin. MONEY was at the bottom of that well and it may have been a lot, and I needed that lightweight bike. I was to be dropped down into the wishing well.

We placed a new rope on the well crank; a harness was made to hold me. I had a bucket over my head to hold in air once I reached the water. Otis lowered me to the riches. Suddenly I was falling! Otis was not slowing me down! I splashed down into the cold dark water, my bucket trapping the air I needed to breathe. I sank very fast and hit the bottom.

It turned out that the local police deputy was curious as to why a small boy named Otis was standing beside the town wishing well. Otis said he did not know. The deputy sent him home.

I on the other had had made a discovery. As I sank the air bubble in my bucket started getting smaller. The pressure of the deep water was squeezing the air in my bucket and making it smaller. The water was rising in my bucket. Finally I hit the bottom and there was money! All kinds of money, modern money, old money, I could see gold and silver, I was going to be rich!

Apparently police deputies are a curious lot. This one wanted to know why a new rope was tied to the windlass of the wishing well, and he began to turn the crank.

I reached down to scoop up my riches when suddenly I was pulled from above. My hands barely grazed the piles of money when I was hoisted upward.

The deputy was surprised when he pulled up an angry young boy with a bucket over his head out of the well. He took me home.

The city decided the well was too dangerous, and it was filled with stones and covered with cement. It was all but forgotten. Forgotten by almost everyone, but not by me. My fortune waits.

Fun things to Google
Oil spill
The Titanic
Icebergs
USS turtle
Submarines
Submarine surfaces at the North Pole

Links

A good video by TEDed explaining air pressure: http://ed.ted.com/lessons/how-heavy-is-air-dan-quinn

A rather hilarious video of a poor critter diving under water: https://www.youtube.com/watch?v=Gbzvtxj-6hc
Life in the Mariana Trench: https://www.youtube.com/watch?v=6N4xmNGeCVU

National Geographic Documentary - Discovery Secret Mariana Trench – a rather long documentary but interesting: https://www.youtube.com/watch?v=cpxdT3KvlGY

A cannon ball floating in mercury: https://www.youtube.com/watch?v=Rm5D47nG9k4

Lesson 14 - Pascal's Principle - The Properties of Liquids - Demonstrations in Physics: https://www.youtube.com/watch?v=8ma4kW3xVT0&t=68s

Chapter 37
Pressure in gases

Video of power point lesson for this chapter:
https://www.youtube.com/watch?v=Qce4FBD9b_g

The wonders of science
Hot air balloon have been around for quite a long time. In China small Kongming lanterns were used for military signaling. In 1783 the first tethered manned balloon flight was done by Jean-François Pilâtre de Rozier. The first free flight was done by Rozier and Marquis François d'Arlandes in France the same year. As with many new technologies, balloons were first used for military purposes. Before the invention of the airplane, the best way to map or see a battlefield was with a hot air balloon. They were first used in 1794 but it was not until the American civil War and The Union Army Balloon Corps and its leader, Thaddeus S. C. Lowe, that they were an actual part of the Army. In this war reconnaissance over the battlefield was very important. In fact the Civil War saw the first use of an aircraft carrier, The *USS George Washington Parke Custis,* not for airplanes but for balloons. In WWII the Japanese launched Hydrogen balloons (fire balloons) over the Pacific Ocean and actually bombed the west coast of the United States.

In modern times these balloons are used as weather balloons. This is because balloons can go very high, 60,000 to 120,000 ft (11 to 23 mi), higher than an airplane. These balloons go to a place called near space, high above where airplanes fly but below where satellites can orbit.

Balloons are used by modern adventurers such as Richard Branson and Bertrand Piccard who made attempts to circle the Earth and cross the Atlantic and Pacific Oceans. In 2012 Felix Baumgartner jumped from a balloon at an altitude of 24 miles. He even broke the sound barrier as he fell.

What you need to remember
Gases easily expand and compress, they do not have a definite shape or volume.

Pressure in gases
You may remember from chapter 9 that gases are free to expand or compress, where liquids and solid were not. Gases, when they get hot (more energy) can expand, since there are not really any chemical bonds holding them. Hot air expands, cool air condenses. So, gases can change volume with temperature, but they can change with pressure also. If a bunch of gas changes temperature or pressure, its volume is going to change too, that is the way they are.

Boyle's law
Boyle's Law is about the relationship between pressure and volume. Pressure squeezes a gas into a smaller volume, and when the pressure is lowered the gas increases in volume.

When you fill a bike or car tire with air, you are using energy (a pump) to force a room-full of air into a smaller volume. The air in this smaller volume tire wants to get out so it pushes very hard creating a high pressure inside the tire. Boyle's law says that high-pressure gas (the air pump) makes the volume of a gas smaller (in the tire). The high-pressure air in the tire wants to get out and if the tire gets a hole in it, it will fill the room. Boyle's law also says that lower pressure (hole in the tire) causes the volume of the gas to increase (as the air escapes).

GASES CAN EXPAND OR COMPRESS

- BOYLES LAW
  - PRESSURE SQUEEZES GASES TO A SMALLER VOLUME
  - COMPRESSED GAS EXERTS A LOT OF PRESSURE ALSO
  - $P^\uparrow = V^\downarrow$
  - $P^\downarrow = V^\uparrow$

This is why containers of high-pressure gas can be dangerous, that gas wants to get back out.

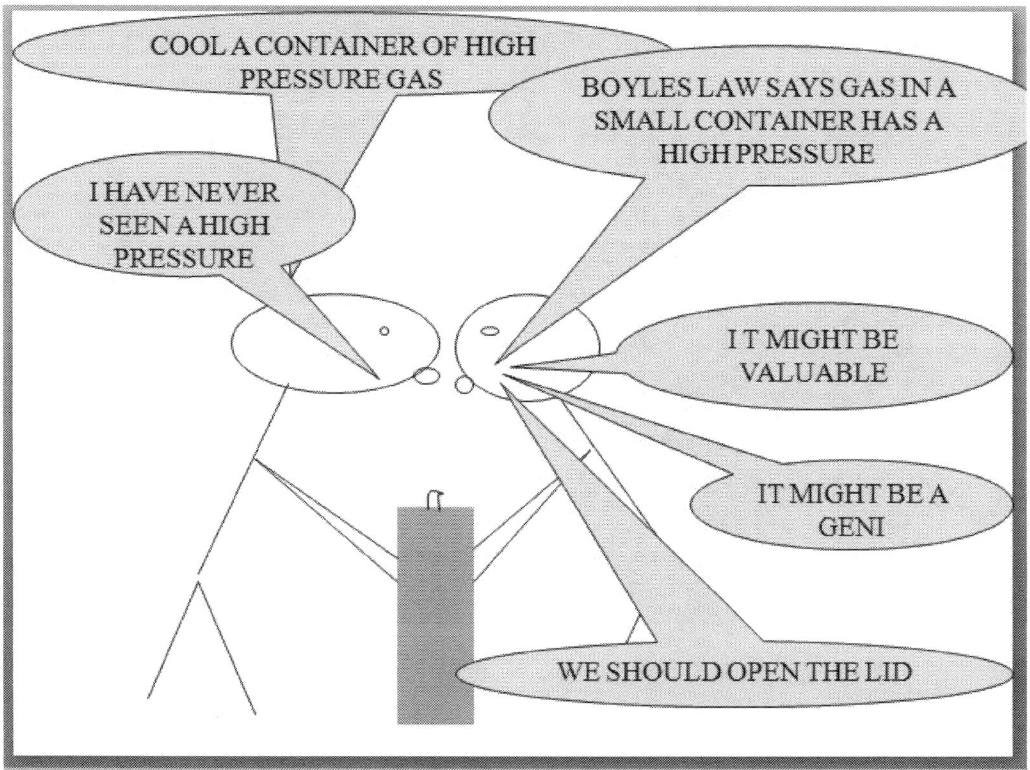
COOL A CONTAINER OF HIGH PRESSURE GAS
BOYLES LAW SAYS GAS IN A SMALL CONTAINER HAS A HIGH PRESSURE
I HAVE NEVER SEEN A HIGH PRESSURE
I T MIGHT BE VALUABLE
IT MIGHT BE A GENI
WE SHOULD OPEN THE LID

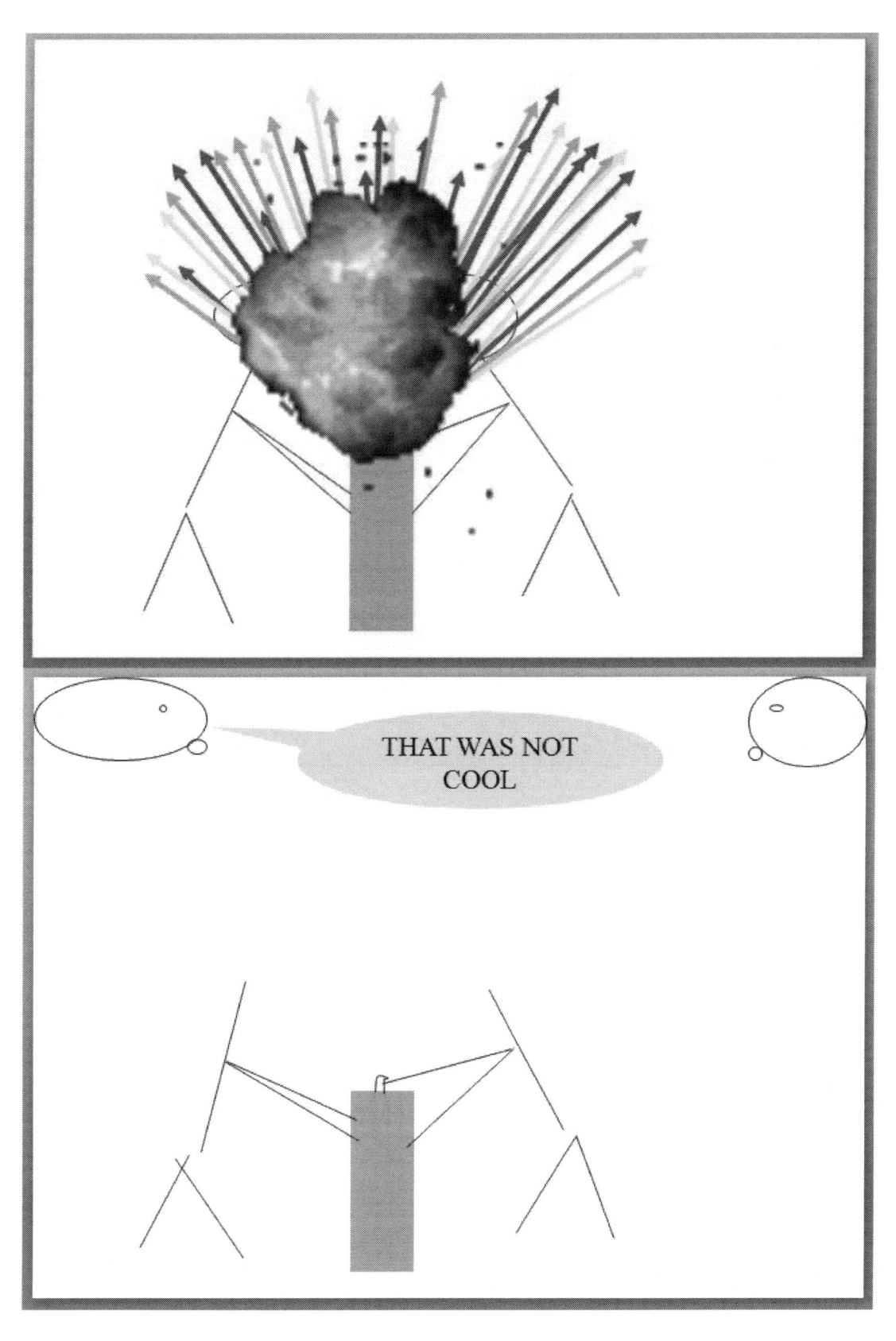

Charles's Law

Charles's Law is about the relationship of temperature and pressure. As a gas gets hotter, the molecules go faster and the gas tries to expand, but if it cannot, the pressure increases instead. This is what happens when you heat up a closed container. When a container of compressed gas (like a spray paint can) gets cold, the pressure becomes lower and the paint can will not work properly.

Did you ever notice that when you pump up a bike or car tire it feels warm? The higher pressure results in a higher temperature. Did you notice that when you let the air out of a tire it feels cold? The lower pressure air results in a lower temperature. This is Charles's Law too.

Another time Charles's law is noticed is when someone does something stupid, like throwing a spray can of compressed gas (spray paint can) into a fire. The molecules in the can go faster, resulting in a higher pressure, until POOF, no eyebrows.

A paint can in a fire: https://www.youtube.com/watch?v=D61wFfJMwYQ

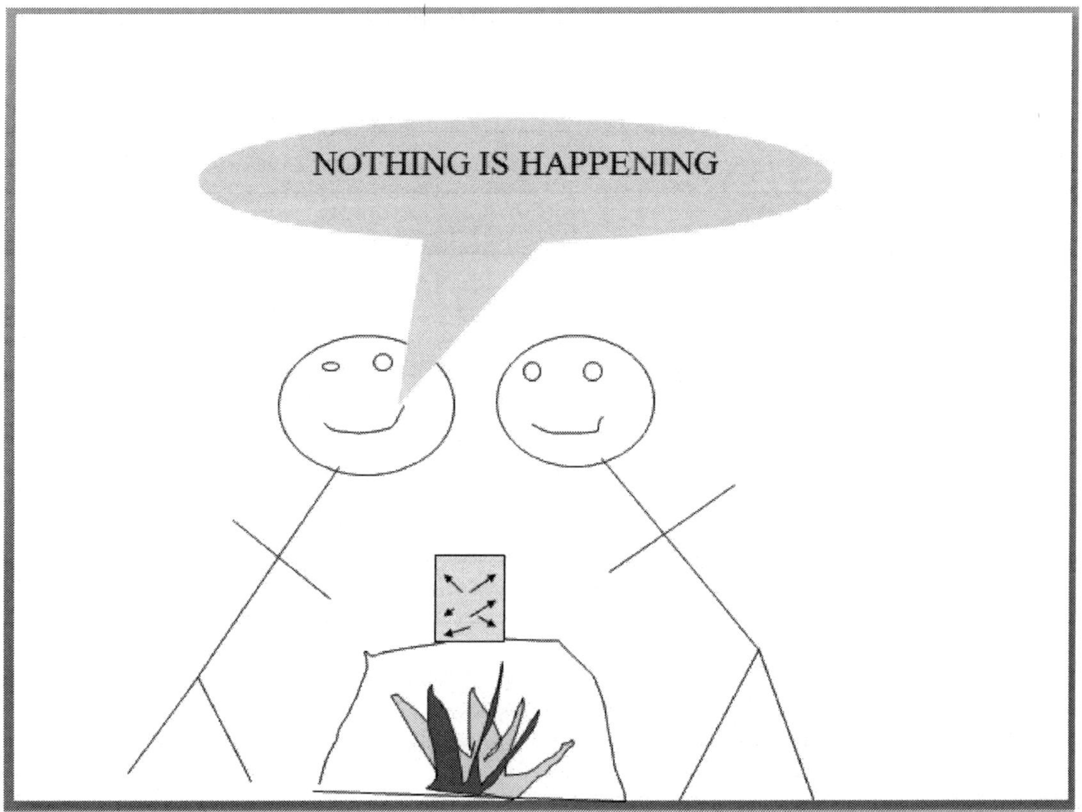
NOTHING IS HAPPENING

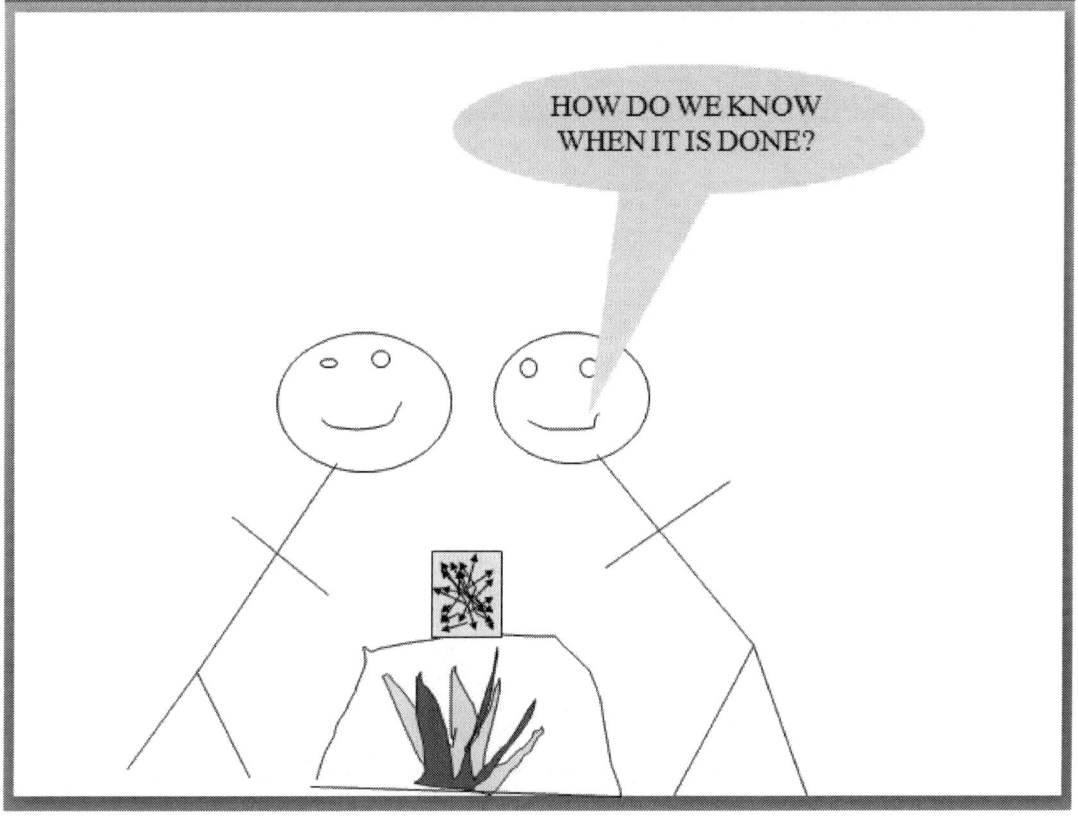
HOW DO WE KNOW
WHEN IT IS DONE?

OH GOOD, IT IS DONE

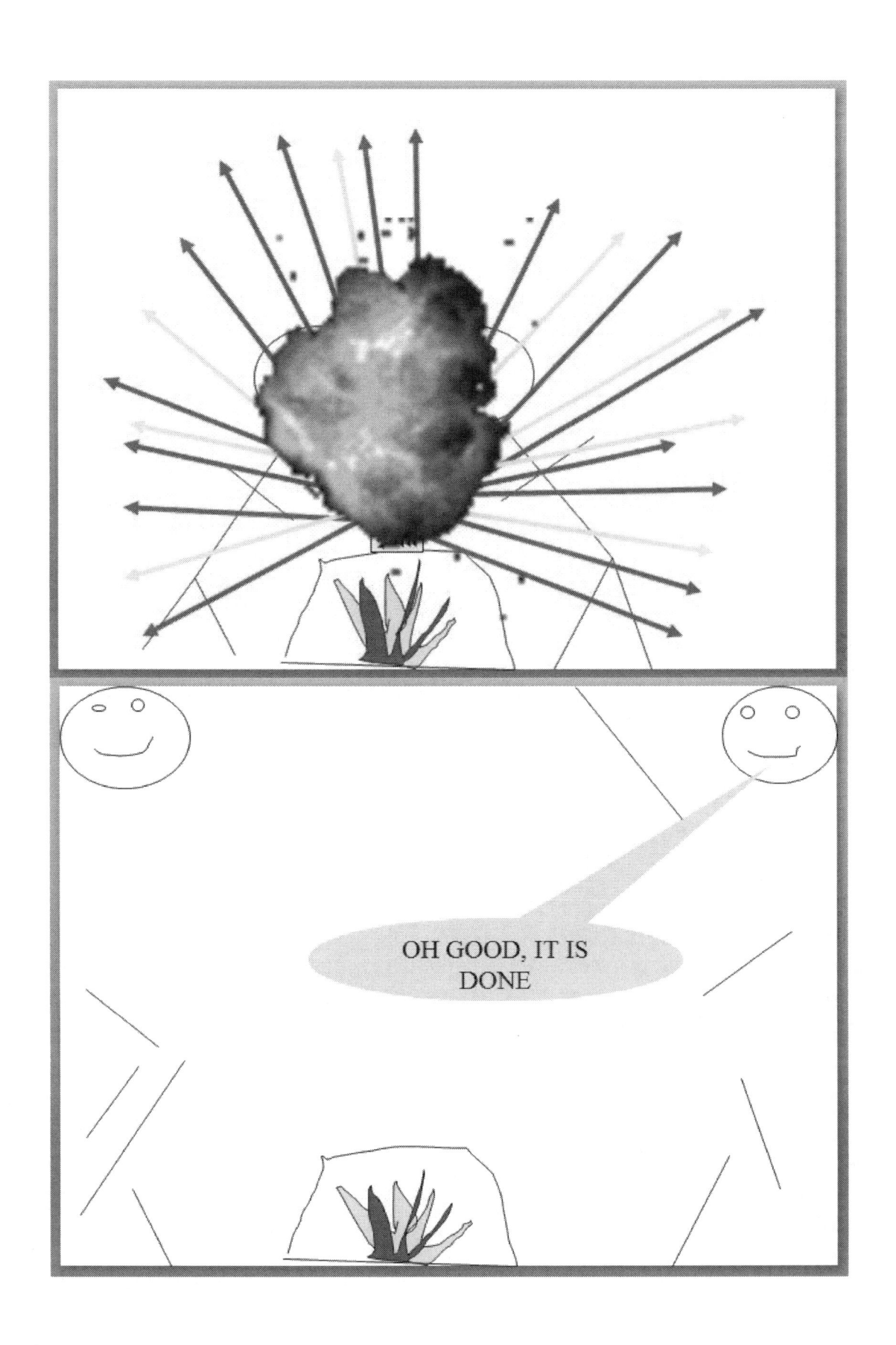

Atmospheric pressure

> # ATMOSPHERIC PRESSURE
>
> - THE WEIGHT OF THE EARTH'S ATMOSPHERE
> - THE MOLECULES MOVE AND HIT THINGS
> - AIR PRESSURE AT SEA LEVEL IS ABOUT 15 LBS/IN^2

Earlier I talked about atmospheric pressure, now is time to look at it more closely. You know that at the bottom of the Earth's ocean of air you are being squished, but luckily you have the same pressure pushing outward and it cancels everything out, but what happens if you were to go up in altitude towards outer space? The higher you go, the less air is above you, the lower the pressure. The pressure inside you would push out and you would expand. One day my school decided to release hundreds of helium balloons for a publicity stunt. Where did those balloons go? They went up. As they went up, they expanded (or got bigger) because the atmospheric pressure was lower. Eventually they got so big that they popped and rained back down all over the Earth. Not nice for the environment. Try not to let helium balloons go and don't let anyone tell you latex balloons are safe for wildlife, they are not. Sure they decompose eventually after MONTHS. I wonder if we killed any Sea Turtles that day.

Let us see what an inflatable Gumby man will do as he goes up into the atmosphere.

GUMBY MAN IS BEING SQUISHED BY 15 LBS/IN^2

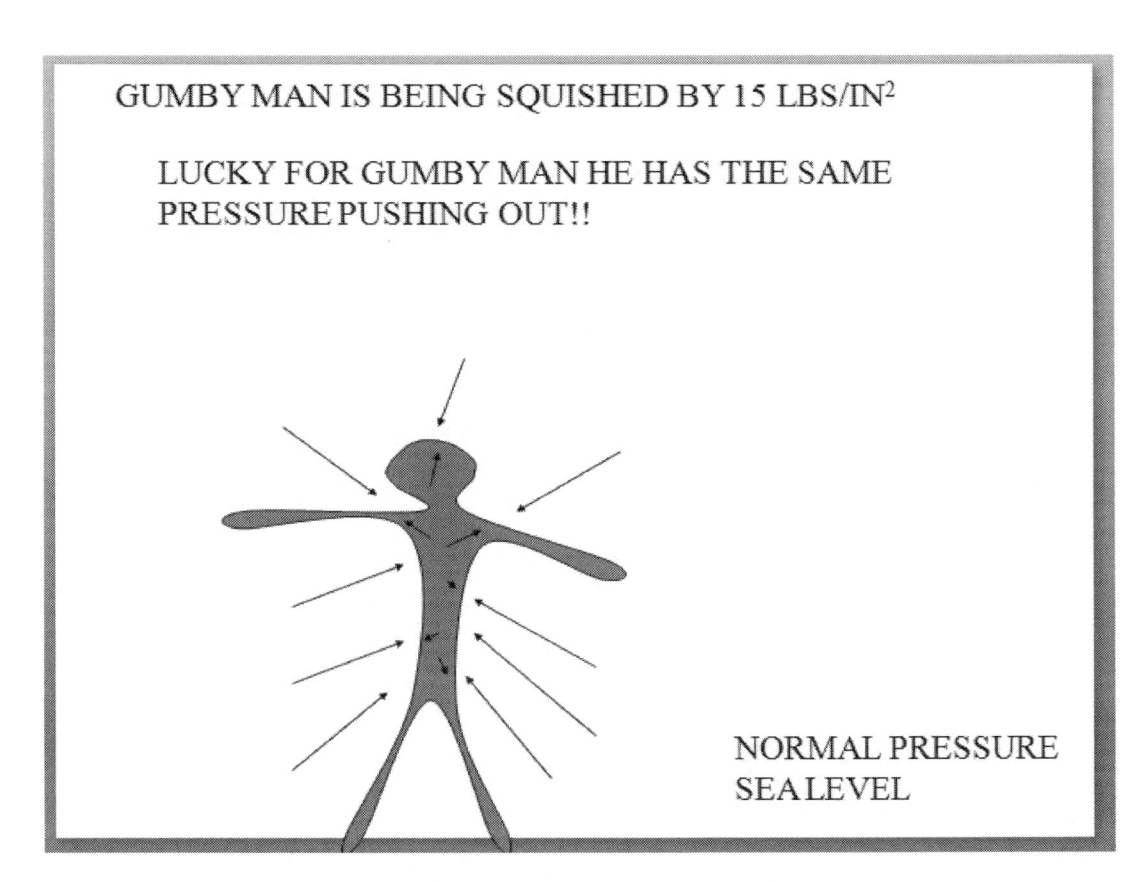

LUCKY FOR GUMBY MAN HE HAS THE SAME PRESSURE PUSHING OUT!!

NORMAL PRESSURE
SEA LEVEL

GUMBY MAN IS GOING UP IN ALTITUDE!!!!!!

BUT THERE IS STILL PRESSURE INSIDE HIM TRYING TO GET OUT

AS GUMBY MAN GOES HIGHER

THE INSIDE PRESSURE PUSHES OUT

HE GETS BIGGER

GOOD-BYE GUMBY MAN

BOOM

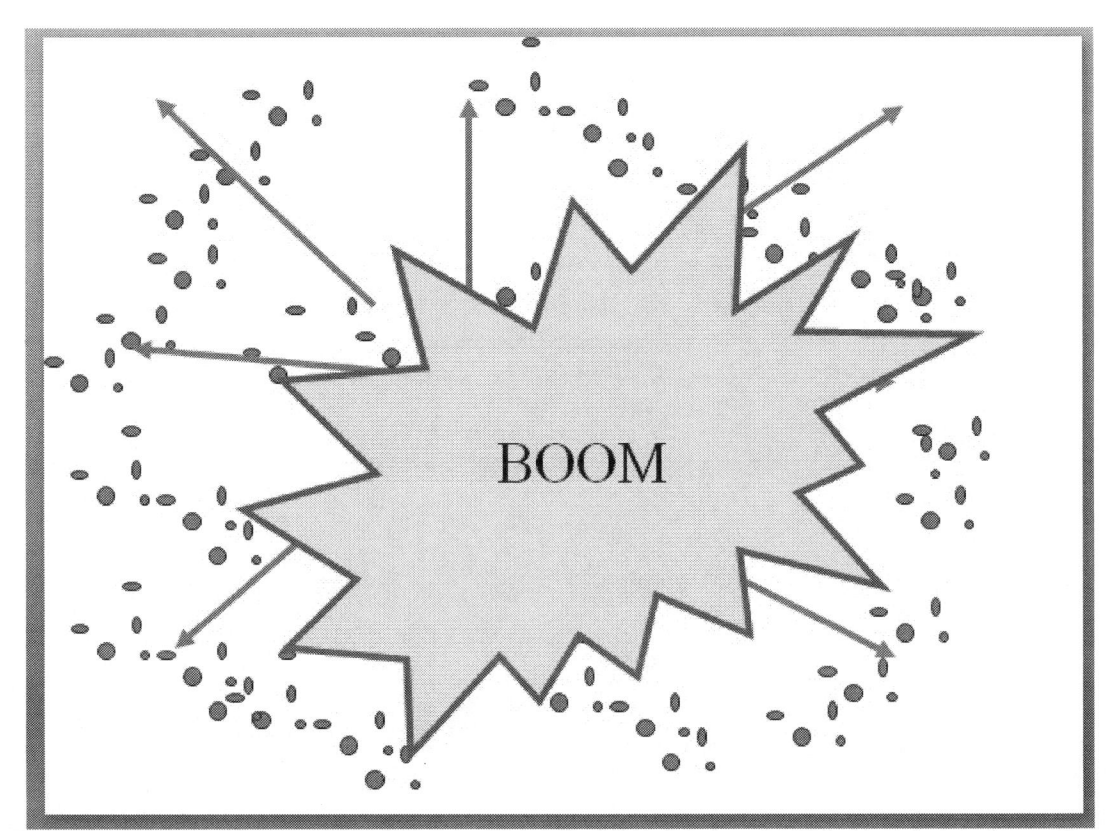

As you go higher in altitude the air becomes less dense, the pressure becomes lower, until there is no air or air pressure left. This is outer space. There is no air in space. It is a vacuum.

Remember when I said the Earth's atmosphere is like a giant ocean of air? Imagine a bunch of aliens coming to this ocean of air and going fishing. If they catch someone what would happen when they bring them up?

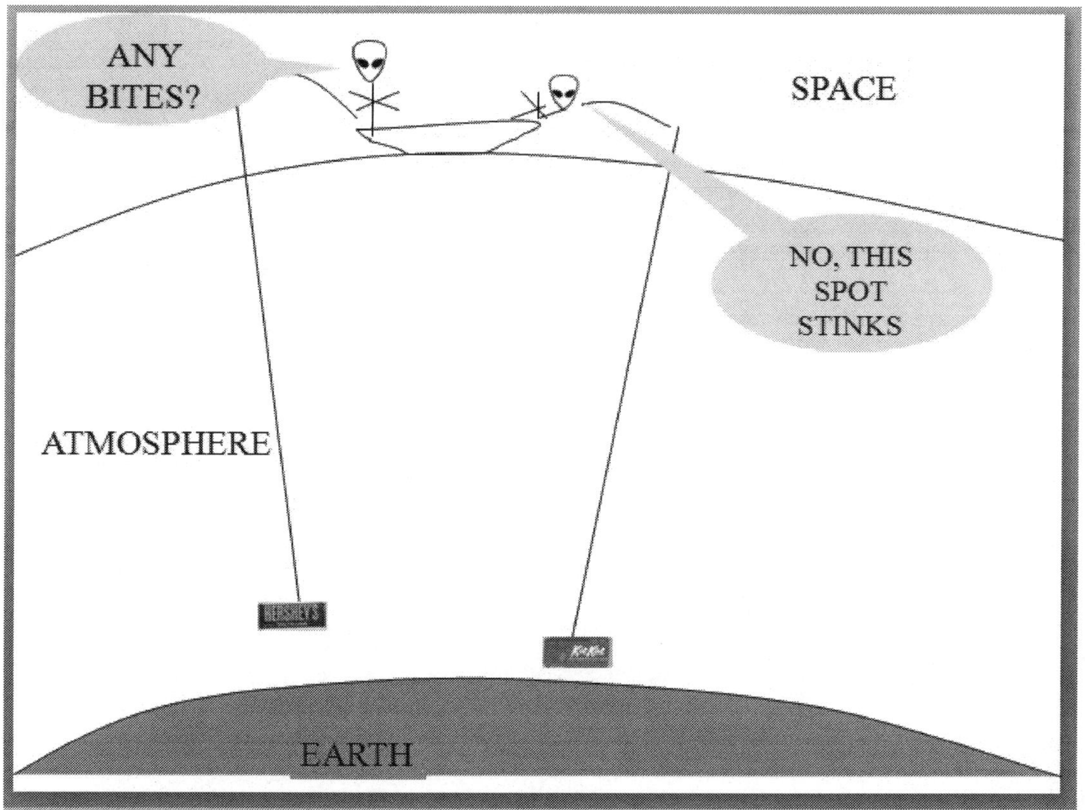
ANY BITES?
SPACE
NO, THIS SPOT STINKS
ATMOSPHERE
EARTH

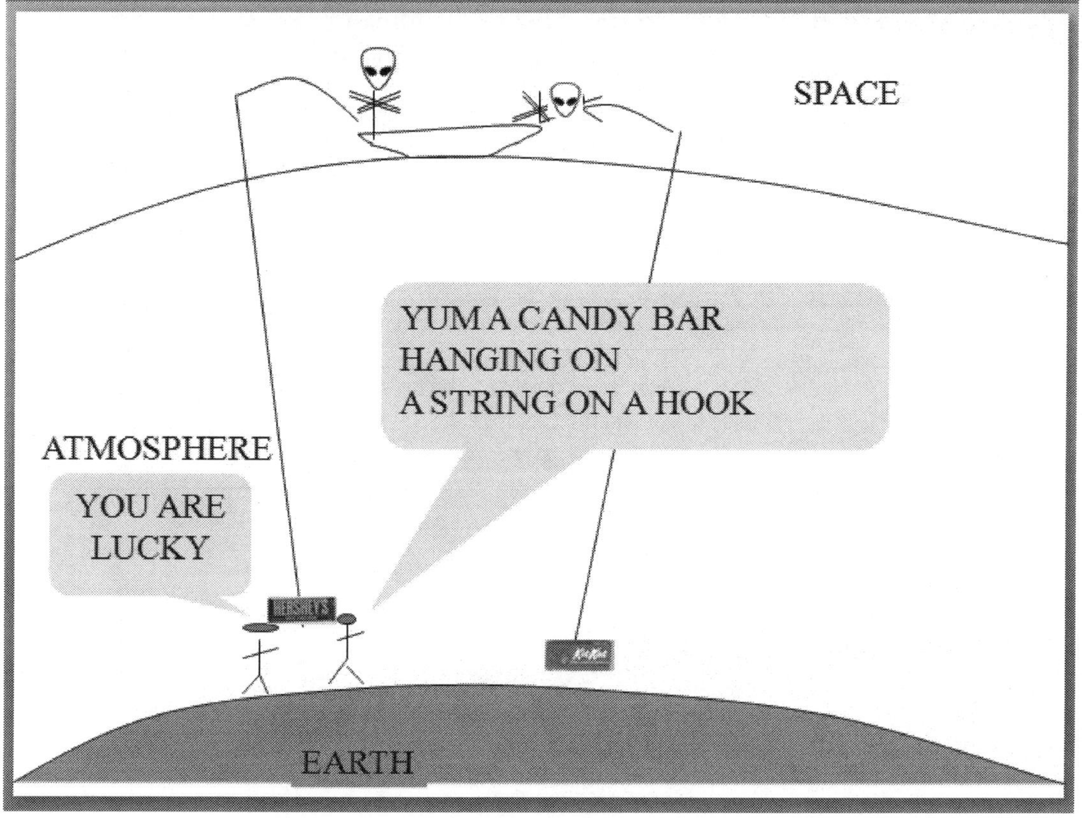
SPACE
YUM A CANDY BAR HANGING ON A STRING ON A HOOK
ATMOSPHERE
YOU ARE LUCKY
EARTH

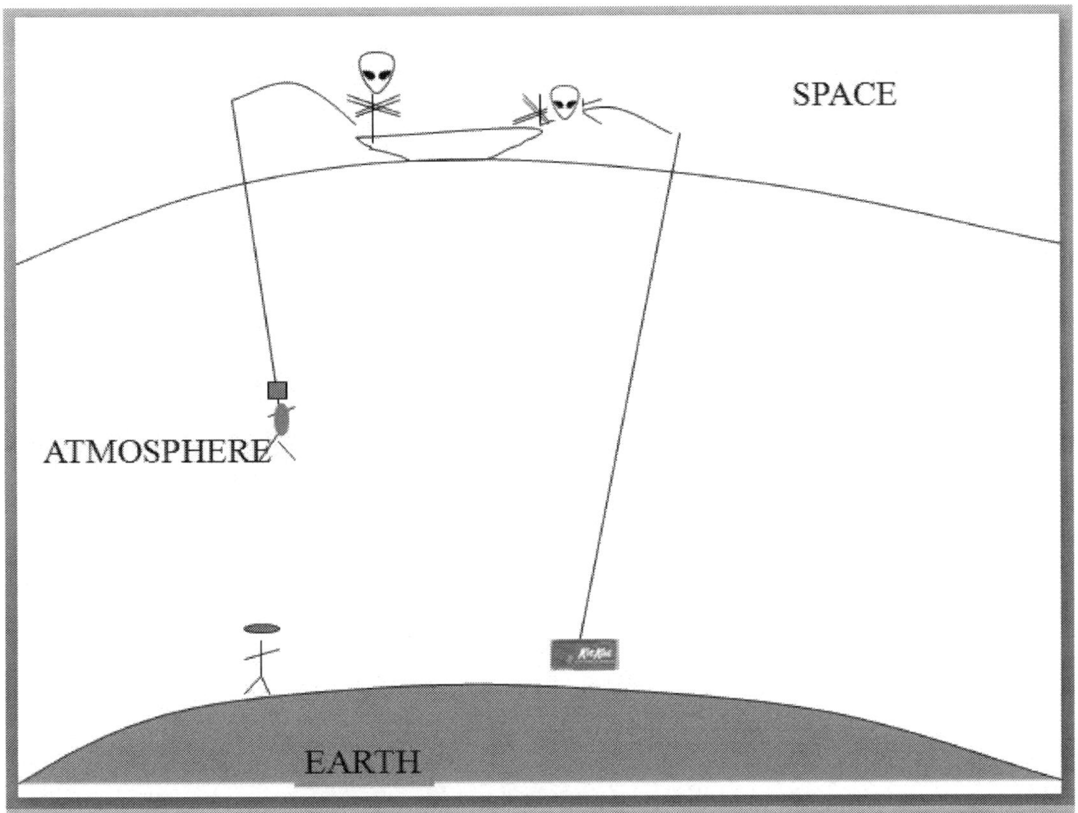
SPACE
ATMOSPHERE
EARTH

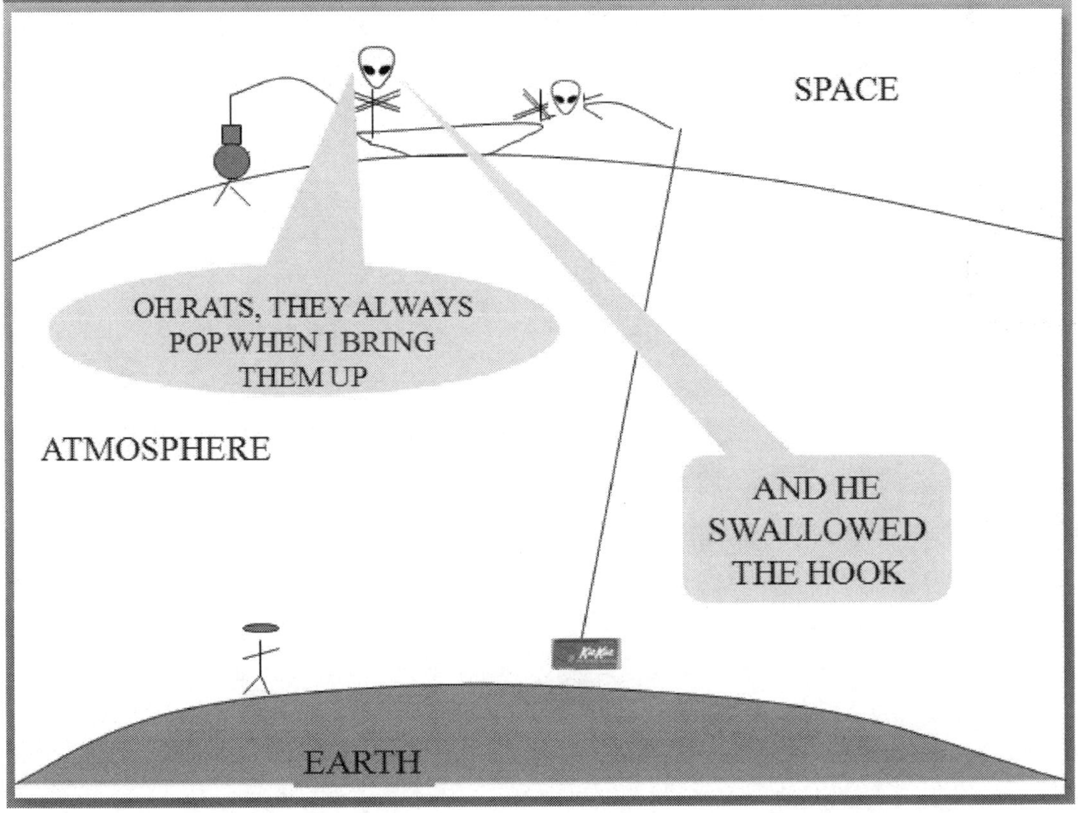
SPACE
OH RATS, THEY ALWAYS POP WHEN I BRING THEM UP
AND HE SWALLOWED THE HOOK
ATMOSPHERE
EARTH

Vacuum

As the aliens bring their catch to the surface of the atmosphere, it reaches a point where there is no air left. A hard vacuum is a space with nothing in it. Outer space is a total vacuum

Check out these videos of objects in a Ultra-high vacuum chamber, which would imitate what they would do in outer space

Peeps in a vacuum chamber: https://www.youtube.com/watch?v=fxLY1SGXV_E
Marshmallows in vacuum chamber: https://www.youtube.com/watch?v=bWd31AefKns
Popping Corn In a Vacuum: https://www.youtube.com/watch?v=aiMseDBTp7s
Egg in a Vacuum Chamber: https://www.youtube.com/watch?v=odiCKmM0WWE
What Happens When You Put Shaving Cream Balloons In A Huge Vacuum Chamber? : https://www.youtube.com/watch?v=Cs7tdHT9NmU
Boiling Water Until It Freezes: https://www.youtube.com/watch?v=y4BGV7-1lhs
Boiling Water at Room Temperature: DIY Vacuum Chamber: https://www.youtube.com/watch?v=jn1X_I8-9h8
Freeze Water by Boiling: https://www.youtube.com/watch?v=8oCjj8iDB9I
5 Awesome Vacuum Chamber Experiments with The King of Random: https://www.youtube.com/watch?v=SwqneULbJcs
Magdeburg Sphere: https://www.youtube.com/watch?v=OnDej9tfdZg
Magdeburg Hemispheres- amazing alternate method- no pump // Homemade Science with Bruce Yeany: https://www.youtube.com/watch?v=Q66DxZB6plE

..

The weight of the atmosphere

Atmospheric pressure is a lot more powerful than you realize. I once decided to shoot a ping pong ball faster than sound. I used only atmospheric pressure (no high pressure)

and reached 1100 miles/hour. Some day I want to build one that reaches 1600 miles/hour (mach 2). I have the design, I just have to build it.
(https://www.youtube.com/watch?v=zIwz6XcndSk)
Atmospheric pressure can be used to crush a can too.
https://www.youtube.com/watch?v=I4Znh5RLSyM
Here is a 55 gal drum being crushed. https://www.youtube.com/watch?v=c5_ho2sc0fc

Review

Gases can expand or compress. High pressure causes the volume of a gas to get smaller. As the temperature of a gas goes up, so does the pressure in a closed container. The weight of the Earth's atmosphere is atmospheric pressure. As you go up in altitude, the pressure becomes less.

Flashcard review on Quizlet: https://quizlet.com/173977383/chapter-37-pressure-in-gases-flash-cards/?new
https://quizlet.com/101743553/gas-laws-review-flash-cards/

A day in the life of Earwig Hickson III

I was not the first to think of it. In fact I did not think of it at all. I saw it on the news. Some guy decided to tie a bunch of helium weather balloons to a lawn chair and lift off into the sky. His name was Larry Walters and he invented cluster ballooning, and he was my hero. Otis thought it was a stupid idea but agreed to watch from the ground, as a good friend should. I started with my comfy camping chair and tied a few ropes to it. I tied the ropes to stakes I had hammered deep into the ground. The ropes were about 20 feet long and I figured a 20-foot flight was high enough. I wore a seat belt and put I put a heavy barbell on my lap to hold me down until I was ready to fly. Otis was in charge of filling the hundreds of party balloons with helium and tying them to my chair. It took a lot of balloons. Otis was busy at work while I took a short nap, actually a long nap. I had a dream about monsters chasing me and when I woke up I tried to run away but instead knocked the barbell off my lap. Just before I shot up into the air, I saw hundreds of balloons above me and a big Otis head looking at me with a look of fear on his face. Otis ran away, I flew up until the ropes holding me to Earth became tight. Then the stakes pulled out. Up I went, my safety ropes dangling below me. It was not fun at first as I saw my little town getting smaller and smaller. Soon I was enjoying myself. I was flying! The gentle breeze was carrying me over farms and towns. The view was wonderful!

I saw birds flying by me with looks of confusion on their little faces. I saw a plane flying below me. Below me! That was not good. I was too high. It was getting hard to breathe. The air seemed thin, I gasped for breath. The balloons above and holding me up seemed to be getting bigger. I started to worry.

As I gained altitude the balloons really started to expand. I knew the air pressure around them was getting weaker. I had to do something. I did not think to bring anything along to pop the balloons; I did not think to bring anything at all.

I searched my pockets and found the only thing my father trusted me with, a pair of fingernail clippers. I started cutting the many strings with them. My assent slowed, then stopped, then I started going down. I thought it best to stop cutting the strings.

My decent took quite a while. Darkness came. Then the sun came up. My balloons began to get small again. The Earth was getting closer. I landed in a cornfield three states away from where I began. The farmer let me use his phone but kept a sharp eye on me. He let me stay in the barn until my father showed up to take me home, where I got to play with some of his cows. I was having a grand old time.

I did not have a grand old time when I got home. Added to the list of things I could not own were chairs, fingernail clippers and balloons. I had to sit in the corner again but I had to sit on the floor since I could not be trusted with chairs.

Fun things to Google
Hot air balloons
Hot air balloon records
Atmospheric pressure
Boyle's law
Charles's law
High altitude
Weather balloons
Helium balloons and sea life
Cluster ballooning
Larry Walters

Links
A mach 1 ping pong ball using a vacuum cannon:
https://www.youtube.com/watch?v=zIwz6XcndSk

A small vacuum cannon shot: https://www.youtube.com/watch?v=pvMWJB9q3fc

Making a cloud by reducing air pressure: https://www.youtube.com/watch?v=OCmhvC27bto

Crushing a soda can with air pressure: https://www.youtube.com/watch?v=I4Znh5RLSyM

A pressure demonstration using water and air: https://www.youtube.com/watch?v=Tn89IBl_LOw

The old atmospheric pressure and egg trick (OK a water balloon): https://www.youtube.com/watch?v=6JBIvlFvw-4

Flaming floating bubbles: https://www.youtube.com/watch?v=zN3GbcGZQGU

Crush a 55 gallon drum with air pressure: https://www.youtube.com/watch?v=c5_ho2sc0fc

Mythbusters implode tanker train car AWESOME! https://www.youtube.com/watch?v=T9bpUfWy8Wg

Bill Nye The Science Guy on The Atmosphere: https://www.youtube.com/watch?v=gGNxYtT_36I

Bill Nye - Atmospheric Pressure: https://www.youtube.com/watch?v=QeAp3CuGjk8

Atmospheric Pressure: https://www.youtube.com/watch?v=xJHJsA7bYGc

Lesson 10 - Atmospheric Pressure - Properties of Gases - Demonstrations in Physics: https://www.youtube.com/watch?v=P3qcAZrNC18

Peeps in a vacuum chamber: https://www.youtube.com/watch?v=fxLY1SGXV_E

Marshmallows in vacuum chamber: https://www.youtube.com/watch?v=bWd31AefKns

Popping Corn In a Vacuum: https://www.youtube.com/watch?v=aiMseDBTp7s

Egg in a Vacuum Chamber: https://www.youtube.com/watch?v=odiCKmM0WWE

What Happens When You Put Shaving Cream Balloons In A Huge Vacuum Chamber? : https://www.youtube.com/watch?v=Cs7tdHT9NmU

Boiling Water Until It Freezes: https://www.youtube.com/watch?v=y4BGV7-1lhs

Boiling Water at Room Temperature: DIY Vacuum Chamber: https://www.youtube.com/watch?v=jn1X_I8-9h8

Freeze Water by Boiling: https://www.youtube.com/watch?v=8oCjj8iDB9I

5 Awesome Vacuum Chamber Experiments with The King of Random: https://www.youtube.com/watch?v=SwqneULbJcs

Magdeburg Sphere: https://www.youtube.com/watch?v=OnDej9tfdZg

Magdeburg Hemispheres- amazing alternate method- no pump // Homemade Science with Bruce Yeany: https://www.youtube.com/watch?v=Q66DxZB6plE

Chapter 38
Buoyancy and Archimedes principle

Video of the power point lesson:
https://www.youtube.com/watch?v=MGRIj_Rf5Qs

The wonders of science
It is not often that a single person can save millions of human lives. He is said to have saved more lives than the work of any other human, and he only had one good idea. It was a great idea though. Back in his day, the 1700s many people died of a disease called smallpox. Smallpox was a nasty disease that covered a person with maculopapular rash or many large liquid filled blisters, 30–35 percent of those infected died (mostly children).

His name was Edward Jenner and he was a doctor. His one good idea was to actually inject healthy people with a disease called cowpox, a disease much like smallpox but not as serious. He had noticed something. People who worked on farms often caught cowpox (from cows) but never caught smallpox (the bad one). He reasoned that the cowpox disease gave these people immunity to smallpox. Others had noticed this before Jenner and even gave cowpox to people to protect them from smallpox (Benjamin Jesty), but he took it one step farther.

Jenner not only believed that giving cowpox to people would protect them from getting smallpox, he proved it. He scraped the fluid from people sick with cowpox and injected it into an eight-year-old boy named James Phipps. A few weeks later he injected the fatal smallpox into the boy to see if he lived. He did. Jenner then repeated his experiment and wrote a paper about why it worked. He had developed the first vaccine, the smallpox vaccine. His ideas about vaccines resulted in others developing vaccinations against other fatal diseases such as measles and Polio. The result is million of people are alive today who would have died from one of these diseases (maybe you are one).

Vaccines are the most effective means to fight and eradicate serious infectious diseases. They have been a great success. Recently there has been an anti-vaccine movement in the United States led by people with no medical training or knowledge. They have managed to convince some people not to vaccinate their children. Since the germs are still in the environment, some of these children are getting these awful diseases. Do not fall for the anti-vaccine movement.

What you need to remember
Less dense materials float in more dense fluids. Helium and hot air balloons float in more dense air. Less dense oil floats on more dense water.
Net force is the sum of all opposing forces on an object.
Weight is a force of gravity.

Buoyancy
Did you ever find yourself swimming in a pool with a younger brother, sister, or cousin, (or maybe you were the younger brother, sister, or cousin), and found it necessary to give them a wedgie? You may have noticed that you could lift them up out of the water

much higher than if they were not submerged in water. This is because the water helped you lift them up. It applied an upward force on them called buoyancy, or buoyant force. Buoyancy is the force that makes boats and other things float. Buoyancy pushes thing up against gravity. Buoyancy makes submerged objects weigh less, even if they sink.

BUOYANT FORCE

- THE UPWARD FORCE THAT A <u>FLUID</u> EXERTS ON A SUBMERGED OBJECT
- GAS OR LIQUID
- BUOYANCY MAKES <u>SUBMERGED</u> OBJECTS WEIGH <u>LESS</u>
- <u>FLOATING</u> OBJECTS WEIGH <u>NOTHING</u>

The <u>buoyant force</u> is a force; it pushes against anything submerged in a fluid. If it is strong enough it makes an object float. A 200-pound person in a pool is being pulled by gravity with a force of 200 pounds. Gravity wishes to drown us. Fortunately the water pushes the person up with a force too. If the buoyant force is 200 pounds up, the person floats.

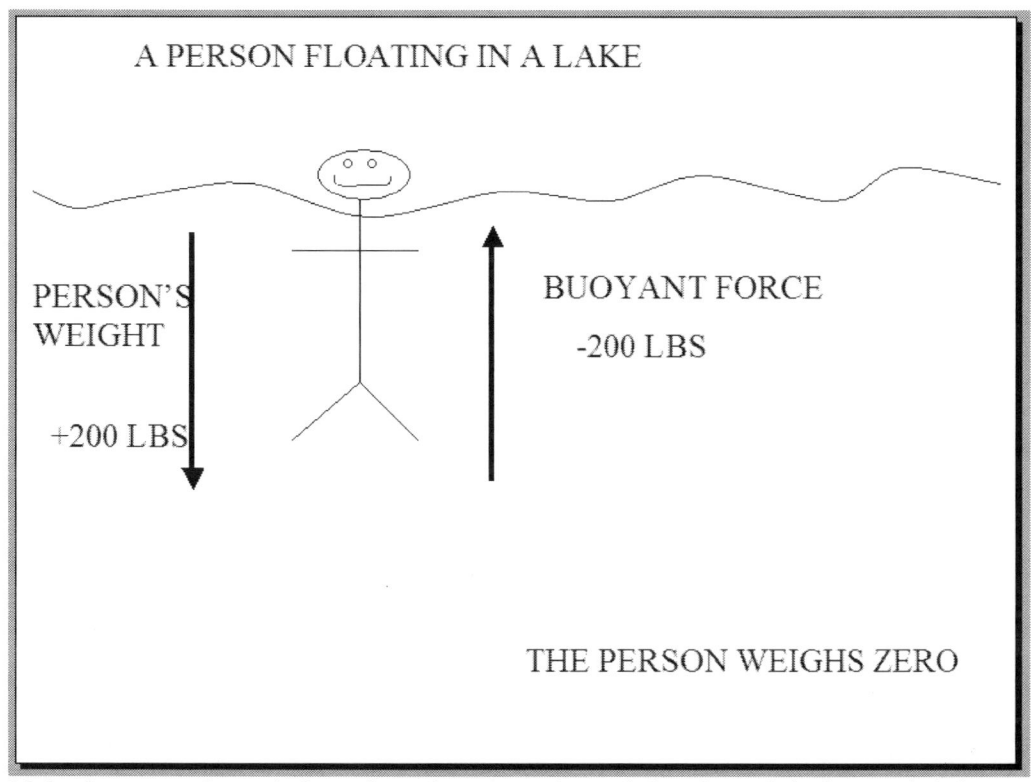

```
┌─────────────────────────────────────────────────┐
│           A PERSON FLOATING IN A LAKE           │
│                                                 │
│                       ☺                         │
│    ↓              ╱│╲           ↑               │
│  PERSON'S          │         BUOYANT FORCE      │
│  WEIGHT            │         -200 LBS           │
│                   ╱ ╲                           │
│  +200 LBS        ╱   ╲                          │
│    ↓                                            │
│                                                 │
│                        THE PERSON WEIGHS ZERO   │
└─────────────────────────────────────────────────┘
```

If the buoyant force is less than the person's weight, they sink, but weigh less.

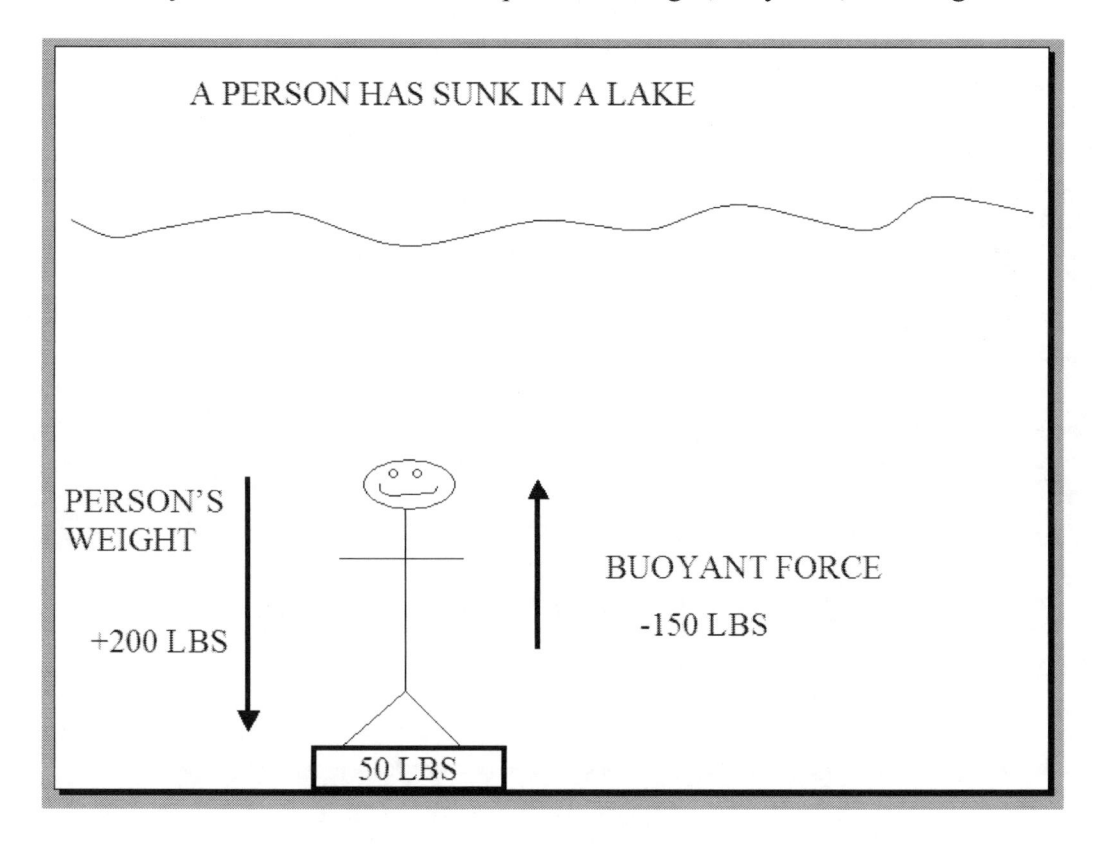

```
┌─────────────────────────────────────────────────┐
│           A PERSON HAS SUNK IN A LAKE           │
│                                                 │
│  ～～～～～～～～～～～～～～～～～～～～～         │
│                                                 │
│                                                 │
│                       ☺                         │
│   ↓                  ╱│╲           ↑            │
│  PERSON'S             │         BUOYANT FORCE   │
│  WEIGHT               │         -150 LBS        │
│                      ╱ ╲                        │
│  +200 LBS           ╱   ╲                       │
│   ↓              ┌─────────┐                    │
│                  │ 50 LBS  │                    │
│                  └─────────┘                    │
└─────────────────────────────────────────────────┘
```

Anything submerged in any fluid always weighs less that it did before.

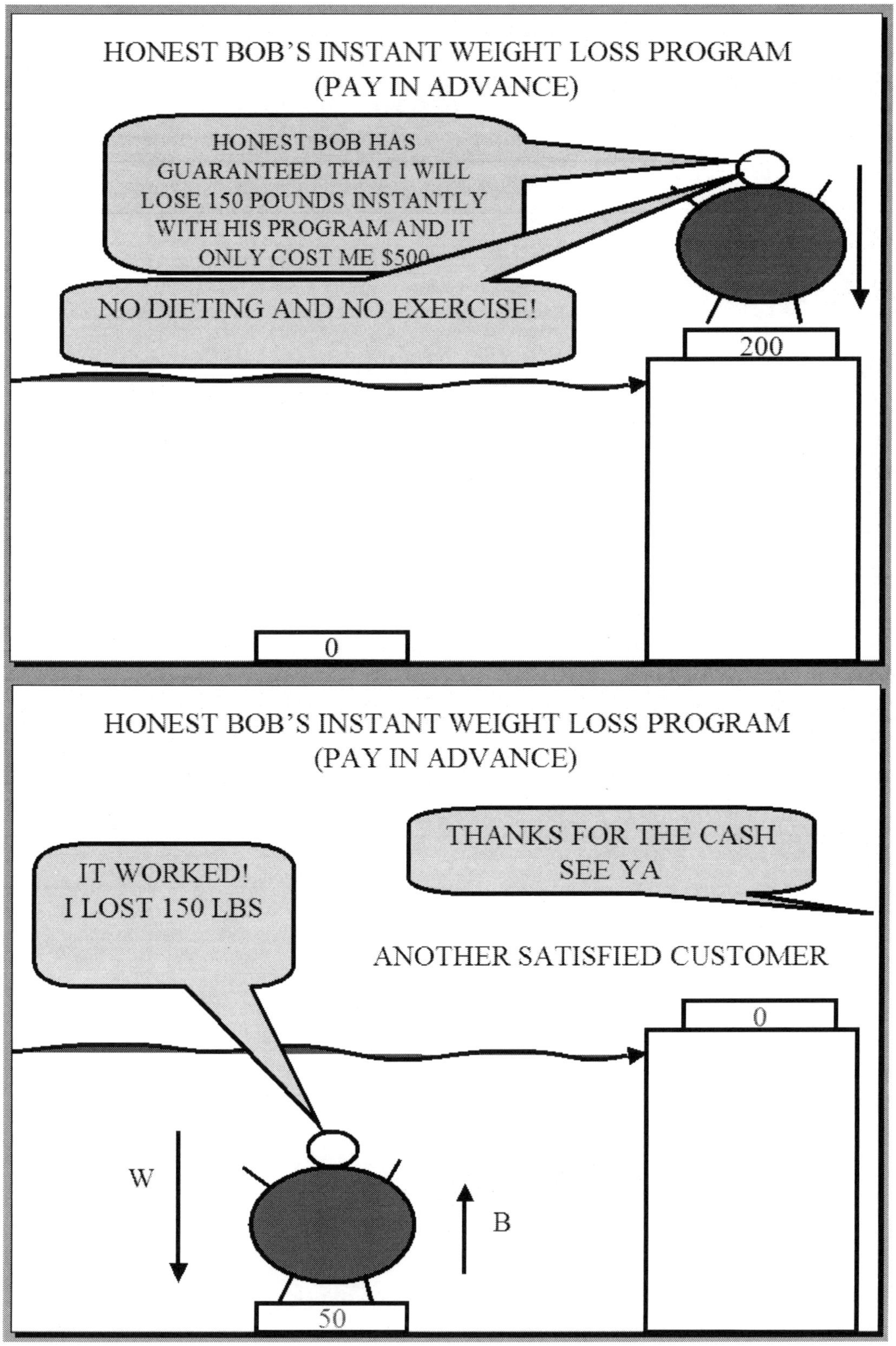
HONEST BOB'S INSTANT WEIGHT LOSS PROGRAM
(PAY IN ADVANCE)
HONEST BOB HAS GUARANTEED THAT I WILL LOSE 150 POUNDS INSTANTLY WITH HIS PROGRAM AND IT ONLY COST ME $500
NO DIETING AND NO EXERCISE!
200
0
HONEST BOB'S INSTANT WEIGHT LOSS PROGRAM
(PAY IN ADVANCE)
IT WORKED! I LOST 150 LBS
THANKS FOR THE CASH SEE YA
ANOTHER SATISFIED CUSTOMER
0
W
B
50

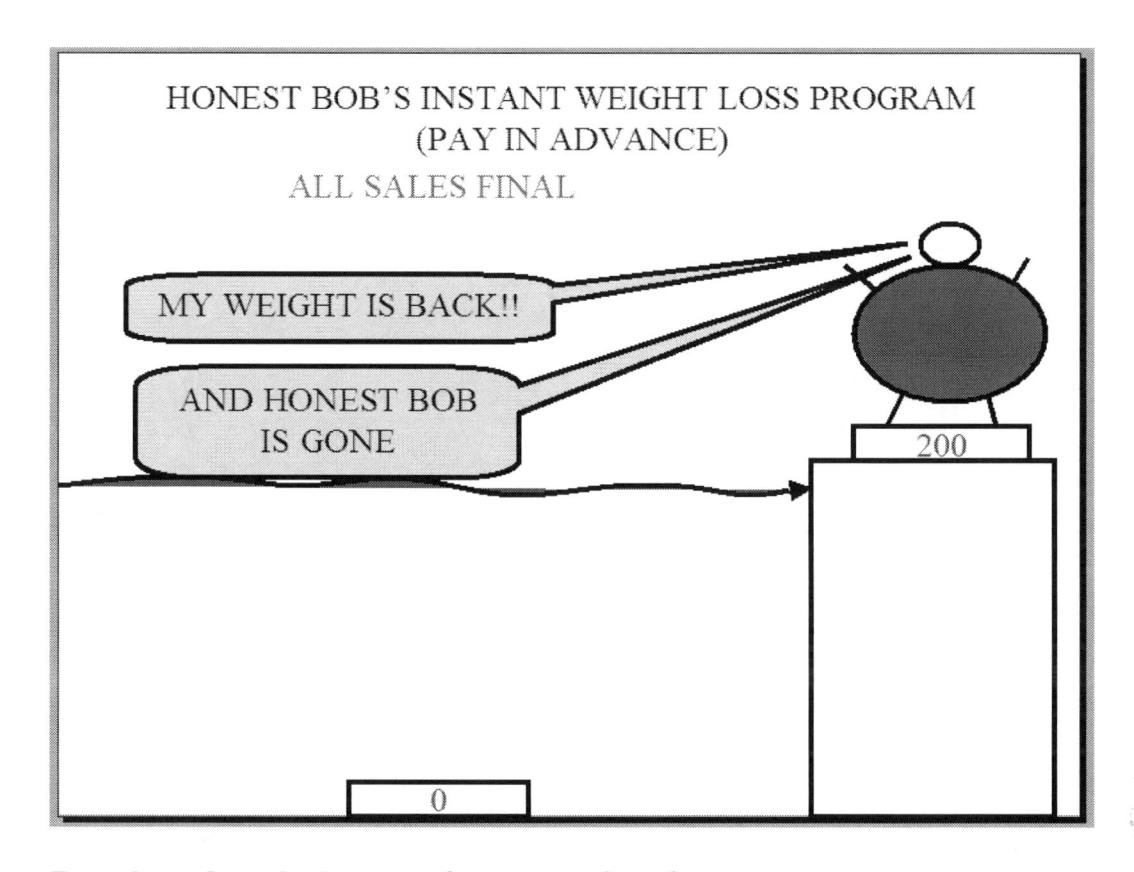

```
HONEST BOB'S INSTANT WEIGHT LOSS PROGRAM
          (PAY IN ADVANCE)
         ALL SALES FINAL

   MY WEIGHT IS BACK!!

   AND HONEST BOB
       IS GONE                    200


                 0
```

But where does the buoyant force come from?

It actually is a side effect of gravity pulling a fluid down. That is right, a side effect of gravity that causes something to be pushed up. Buoyancy can only exist because of gravity. When you jump in a pool, your body pushes some water out of the way to make room for you. This is called displacement, so you displace some water. This water tries to push back against you to get where it was. It turns out that the force the water pushes back on you is exactly the same as the weight of the water you pushed out of the way. So, as you sink, more, and more water is displaced until its weight is the same as you. Then you stop sinking since buoyancy cancels out your weight. You actually sink until you float.

Let's imagine a 200-ton boat being lowered into the ocean.

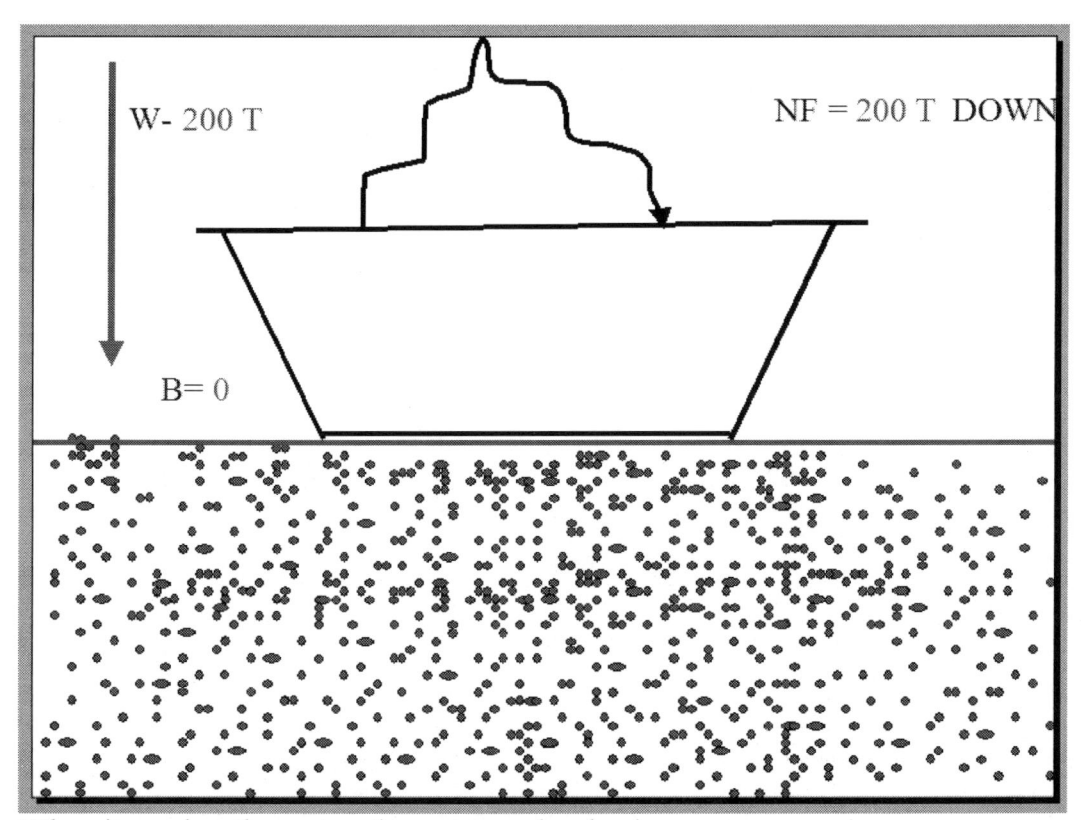

When it reaches the water, the water pushes back.

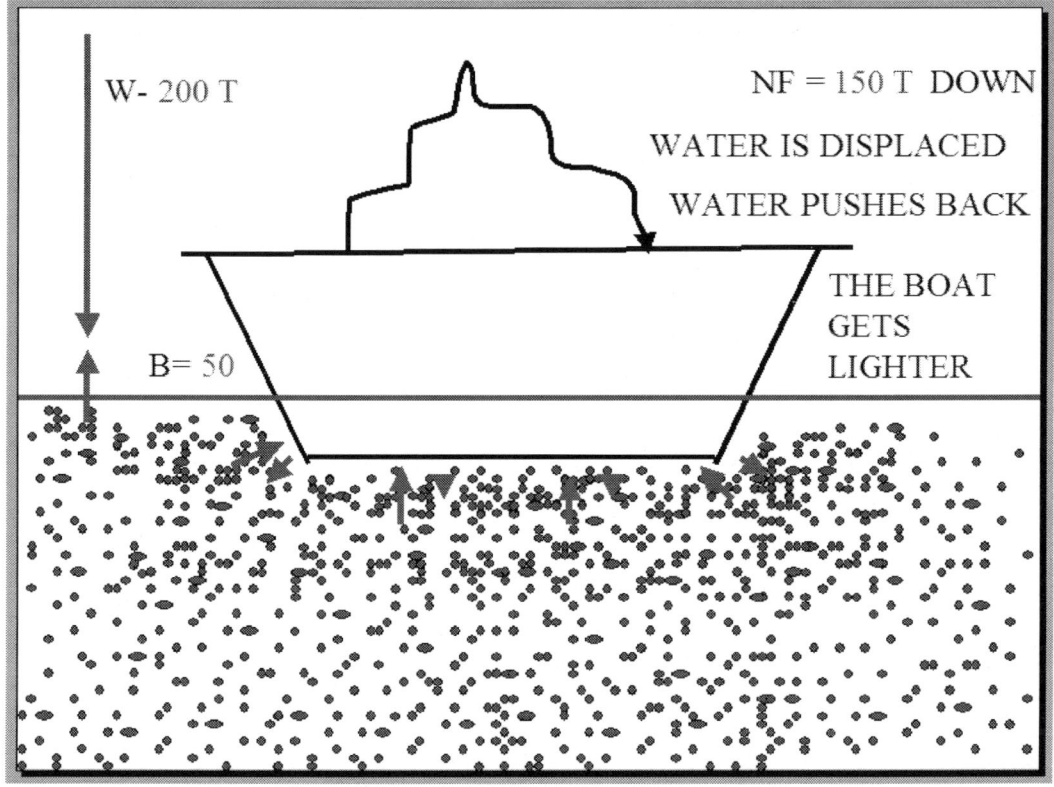

The boat gets lighter as the buoyant force pushes back, but something else is happening too. The ocean is getting deeper! Not much only a billionth of an inch, but deeper nonetheless, this is from the displaced water. This means the pressure under the boat is a bit more (remember depth causes pressure) gravity pulls the water molecules a bit harder too.

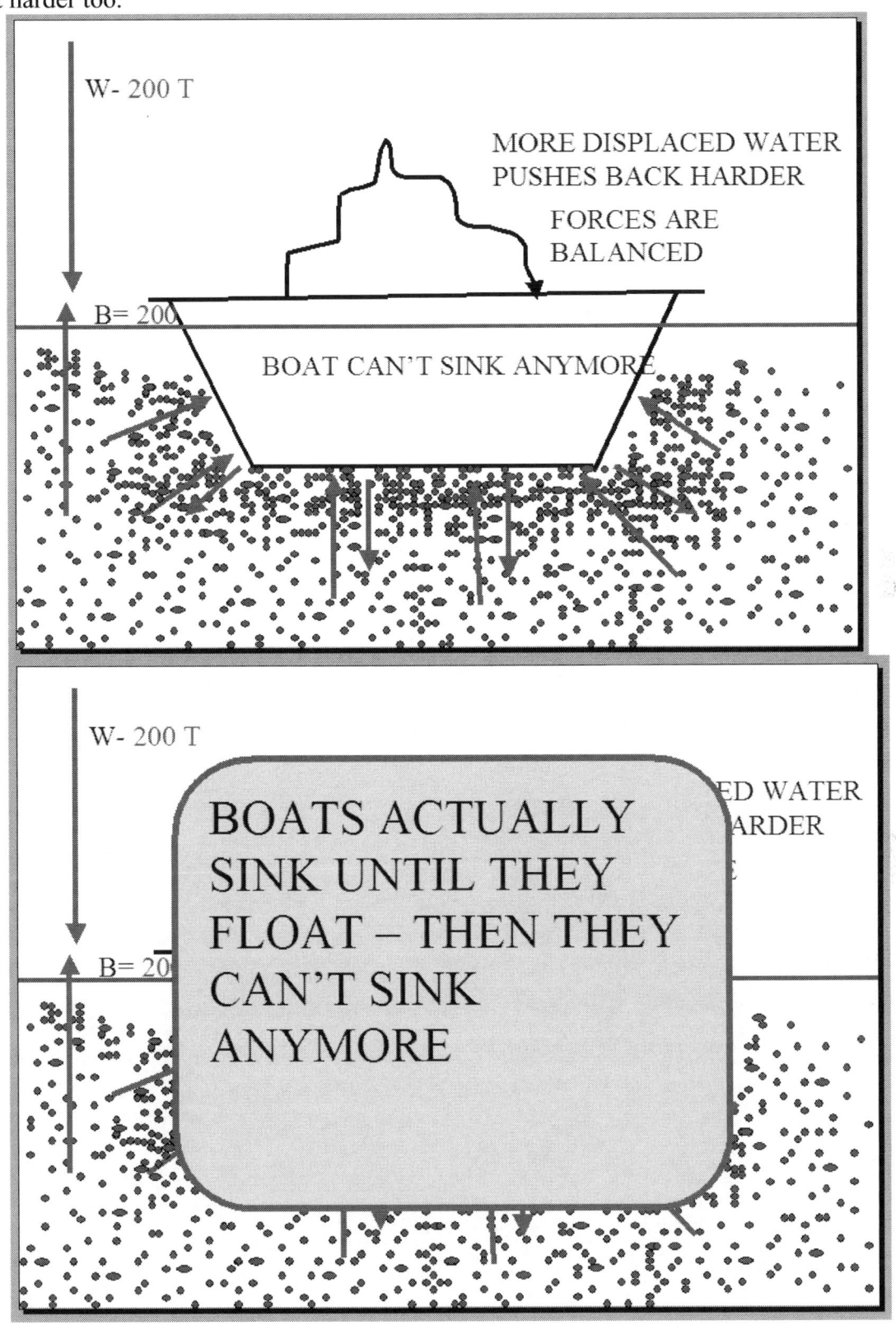

W- 200 T

MORE DISPLACED WATER
PUSHES BACK HARDER

FORCES ARE
BALANCED

B= 200

BOAT CAN'T SINK ANYMORE

W- 200 T

BOATS ACTUALLY SINK UNTIL THEY FLOAT – THEN THEY CAN'T SINK ANYMORE

The density of the object makes a difference too. A Helium filled balloon is less dense than the surrounding air, and floats, but it is the buoyant force from the displaced air that pushes it up. If the balloon pushes a bigger weight of air out of the way than the balloon (and helium) weighs, the net force is up.

THE AIR PUSHES
BACK (BUOYANCY)

LESS DENSE He

DISPLACES THE
HEAVY AIR

AND IT PUSHES
THE BALLOON UP

JUST AS IF IT WAS UNDER WATER AND WAS LET GO

Archimedes principle

This brings us to the scientific principle that causes buoyancy. It was discovered by our old friend Archimedes, remember, the guy who figured out how to tell if the kings crown was made of gold or not?

Archimedes' principle states that the buoyant force is equal to the weight of the fluid displaced.

ARCHIMEDES' PRINCIPLE

- THE BUOYANT FORCE IS EQUAL TO THE WEIGHT OF THE FLUID DISPLACED
- IF 5 LBS OF WATER ARE DISPLACED IT PUSHES BACK WITH EXACTLY 5 LBS OF FORCE

FLOATING OBJECTS

- DISPLACE AN AMOUNT OF WATER EQUAL TO THEIR WEIGHT
- WEIGH NOTHING

500 LB BOAT

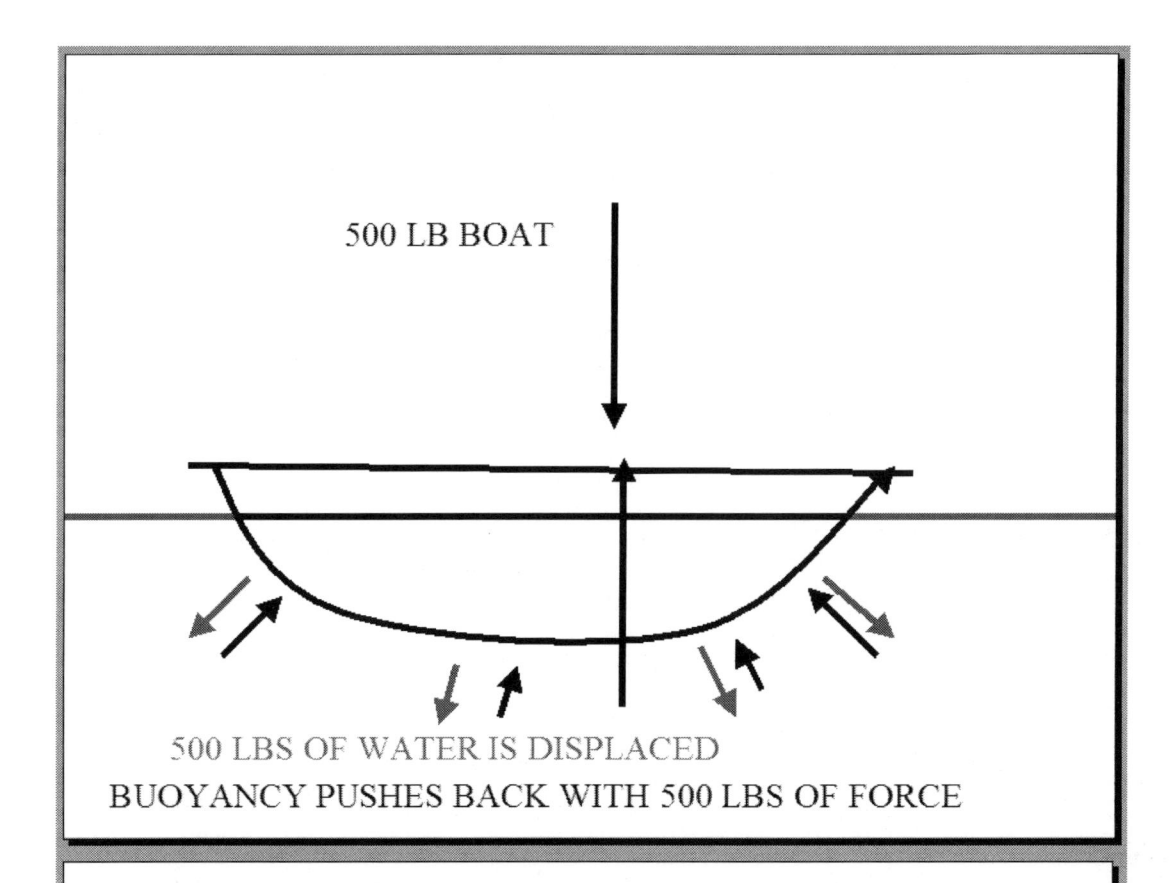

500 LBS OF WATER IS DISPLACED
BUOYANCY PUSHES BACK WITH 500 LBS OF FORCE

SUBMERGED OBJECTS

- LOSE WEIGHT EQUAL TO THE WEIGHT OF THE FLUID DISPLACED
- THINGS WEIGH LESS UNDER WATER

THE BOAT WEIGHS 100 LBS

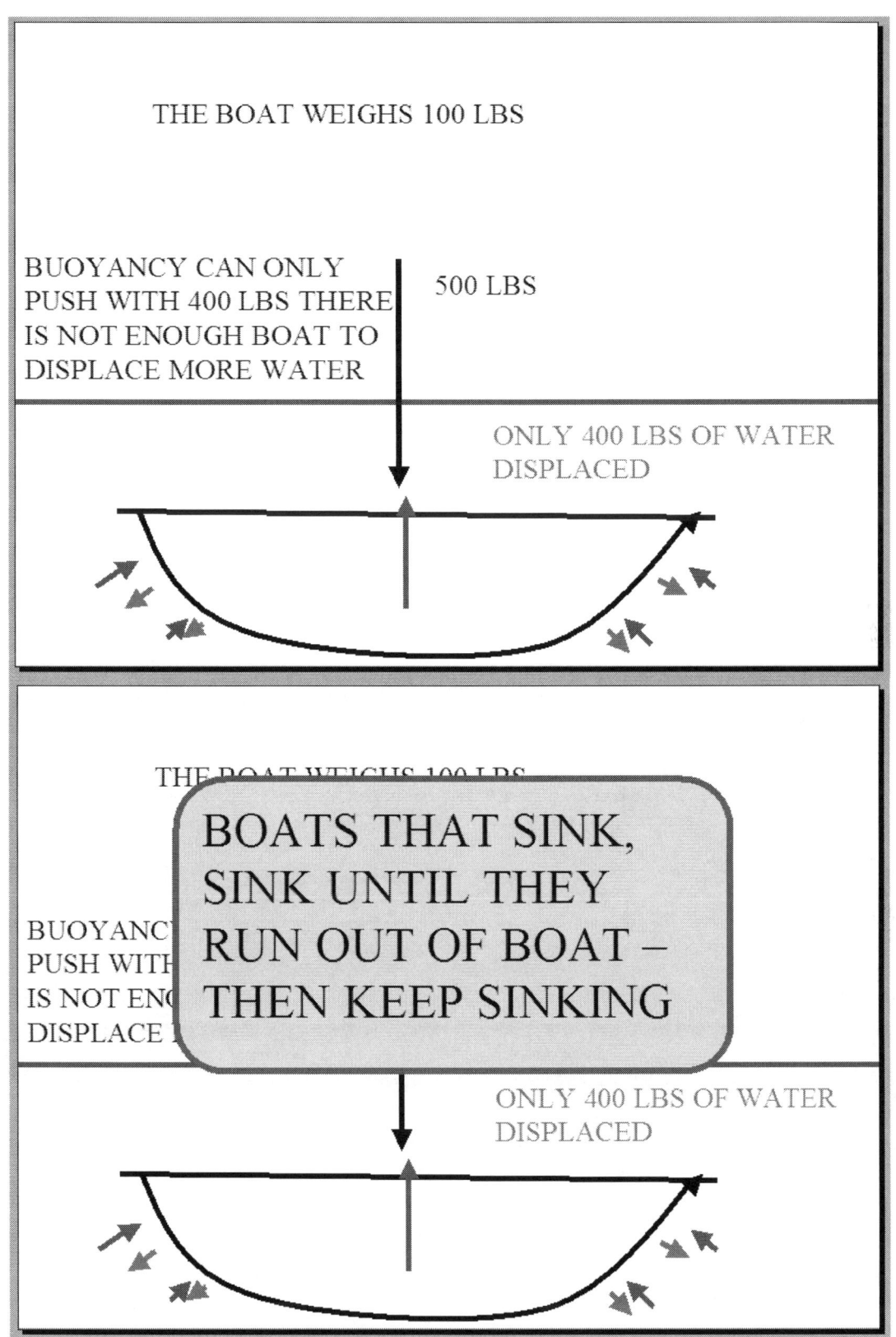

BUOYANCY CAN ONLY PUSH WITH 400 LBS THERE IS NOT ENOUGH BOAT TO DISPLACE MORE WATER

500 LBS

ONLY 400 LBS OF WATER DISPLACED

BOATS THAT SINK, SINK UNTIL THEY RUN OUT OF BOAT – THEN KEEP SINKING

Just for the fun of it. You realize you are at the bottom of an ocean of air, and you displaced some of that air, and that air is pushing back on you with the force of buoyancy. This means that you actually weigh more than you thought you did. I displace

24 gallons (3.3 cubic feet) of air. Each cubic foot of air weighs 0.0807 pounds for a grand total of 0.26631 pounds of air (and buoyancy). So to find my true weight I should add 0.26631 pounds to it. The next time your Mom complains about her weight, mention that is could be worse if air was not making her lighter, OK maybe you shouldn't.

Bill Nye the Science Guy S01E05 Buoyancy:
https://www.youtube.com/watch?v=v3Kc0ahGGMU

Submarines

Submarines are special; they can float or sink in water at will. They do this by changing their weight. They cannot change their buoyancy because they cannot change their volume. They change their weight by adding water to their ballast tanks, or forcing the heavy water out with compressed air.

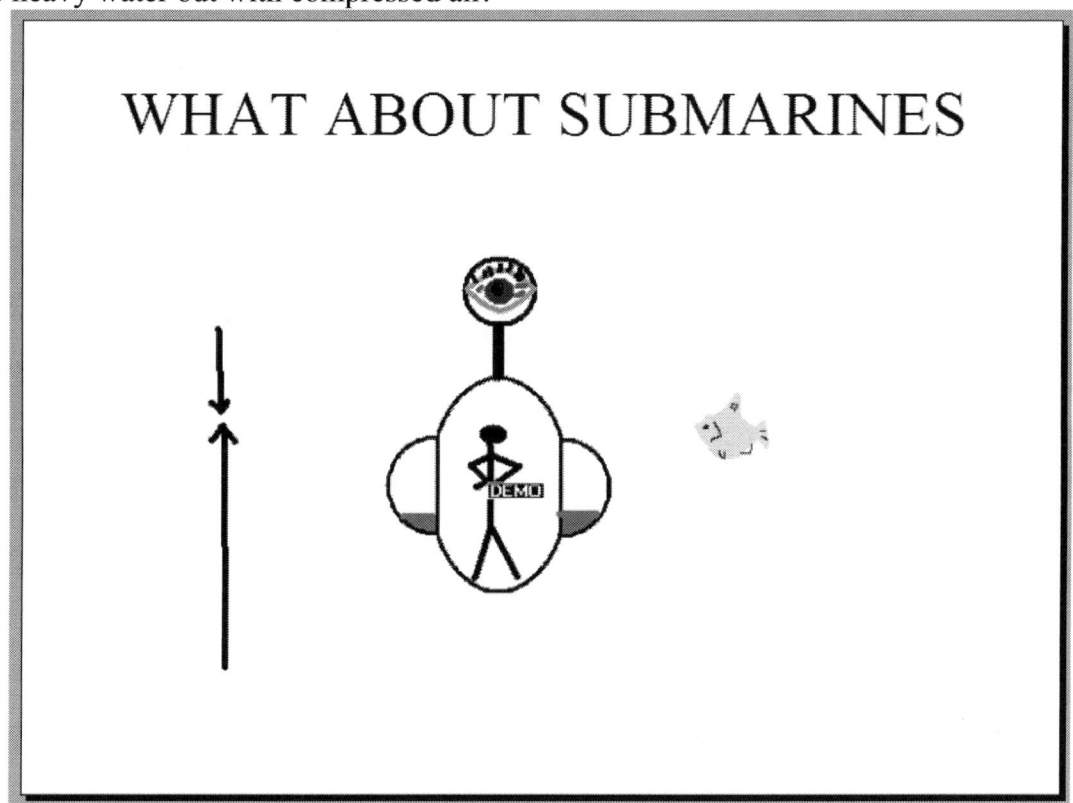

WHAT ABOUT SUBMARINES

WHAT ABOUT SUBMARINES

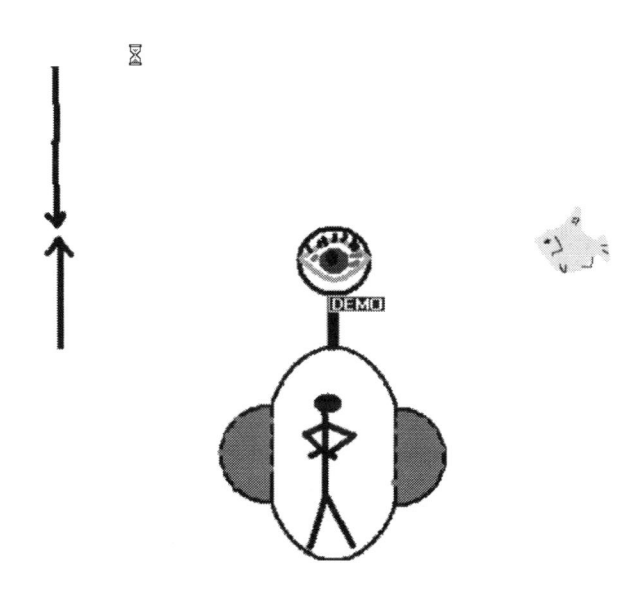

Review

The buoyant force pushes objects up against gravity. The buoyant force is caused by displaced fluid. Archimedes' principle states that the buoyant force is equal to the weight of the fluid displaced.

How submarines work: https://www.youtube.com/watch?v=yb3e4IegeJ0

Flashcard review on quizlet: https://quizlet.com/175187788/chapter-38-archimedes-principle-flash-cards/?new

A day in the life of Earwig Hickson III

I wanted a boat of my very own. I asked my dad for one and he immediately put it on the list of things I could not be trusted with, along with chairs. Immediately I went to work. I was going to build a pirate ship. I had heard that long ago people had built small boats with hollow reeds of grass tied together. I did not have the thick hollow plants but I did have pool noodles and I had Otis who was a Boy Scout and could tie knots. We tied a bunch of noodles together and made a small sail. Off to the local lake we went. It was great. The ship had a lot of buoyancy, the wind was fair, and our boat scooted over the lake like a feather. It was great fun for quite awhile.

The controversy of what went wrong is debated to this day. Was it true that Otis did not pay attention during knot tying lessons or was my kite string too weak to hold a hundred pool noodles together? Either way our boat started to peel. Noodles were popping off as I steered toward shore. Otis had panicked and jumped overboard.

Apparently he did pay attention during swimming lessons. Soon the lake was covered with colorful pool noodles with me in the middle. It was a great shipwreck.

Later, there was a big pool party at the lake but I could not go. Pool noodles and lakes were added to the *Not For Earwig List*.

Fun things to Google
Density diver experiment
Buoyancy diver experiment
Submarine
Why do boats float?
Bill Nye buoyancy
Submarines
How vaccines save lives

Links
Calculate your volume: http://www.aqua-calc.com/calculate/weight-to-volume

Lesson 13 - Archimedes' Principle - Demonstrations in Physics:
https://www.youtube.com/watch?v=uIbX4TSguTI

Conceptual Physics: Demo of Archimedes' principle:
https://www.youtube.com/watch?v=g6aErhwFXsg

Archimedes Principle - Class 9 Tutorial:
https://www.youtube.com/watch?v=2RefIvqaYg8

Why Don't Big Ships Sink? https://www.youtube.com/watch?v=Yvvyj41njBs

Science - Archimedes' Principle: https://www.youtube.com/watch?v=OrpeXFpHLmw

Chapter 39
Bernoulli's principle

Video of this chapters power point lesson:
https://www.youtube.com/watch?v=PgcXfon0Kng

The wonders of science
It was way back in 1799 when George Cayley began experimenting with human flight. He is known as the father of aviation but did not build a powered airplane. He could not. The technology he needed had not been invented yet. He did build a glider capable of carrying a passenger but that was all he could do. He was an expert in aeronautics before there was aeronautics.

It was the Wright brothers, Wilber and Orville who would follow his ideas 100 years later (December 17, 1903) and build the first airplane that could take off from the ground, gain altitude and land again. Up until this point pilots could fly from the top of a hill and land at a lower place but no one could actually gain altitude. The Wrights achieved the first sustained and controlled heavier-than-air powered flight. But that is not all; they invented one thing that no one before had considered. What do you do once you are flying? It was one thing to get into the air but they actually figured out a way to control a plane once it was air born. They used the idea of wing warping. Wing warping was a way to bend the wings and control a plane, this is called three-axis control. They were able to control the pitch (up and down), roll (turning) and yaw (side to side) of a plane. They could correct the plane in flight. This in fact is the only patent the Wright brothers received from the U.S. Patent office. They did not actually invent flight but had learned to control it.

What made the Wright brothers successful was their experience in their shop with printing presses, bicycles, motors, and other machinery. They are famous for being bicycle repairmen but that was because bicycles were the most common thing people needed fixed back then (there was a bicycle craze back then). Another thing they had in their favor was the invention of the gasoline engine. They built their own or one of their employees, Charlie Taylor did. It was light in weight and powerful. Before that the only engines available were steam engines, which were much too heavy.

The brothers were serious about building an airplane and researched the problem extensively. They built their own wind tunnel to test different wing shapes. They experimented with different shaped propellers. They practiced in gliders so they could learn the problem of flight control.

The next step was to find a place to fly. The engine for their first airplane was good but not great. They needed a good head wind. They searched the Weather Bureau data for the windiest places in the USA. They decided on Kill Devil Hills near Kitty Hawk, North Carolina.

In 1903, after many tests and failures they finally flew in their 1903 Flyer (or Wright Flyer). Wilbur got to go first after winning a coin toss. They took turns and on the last flight of the day covered 800 feet in 59 seconds. Since then many people have claimed to have flown before the Wright brothers but the fact is they were the first to take off from level ground, in a heavier-than-air aircraft, and fly in a controlled and sustained flight. Quite an achievement.

What you need to remember
Pressure is a force (P=F/A) and pushes on things.
Fluids are liquids and gases, and can flow.
Fluids move from high pressure towards low pressure.
Wind blows from high pressure to low pressure.

Bernoulli's principle
　　Sometimes common sense is not enough, which is how Bernoulli's principle works that is why it is so cool. It acts the opposite of what you might think. You might think that when air blows really fast it would push things around like a large force, but it doesn't. You might think that a tornado's high winds knock things down, but they do not, at least not exactly. You may think that airplanes fly because of …. Wait, how do airplanes fly? The wings do not flap, the plane does not always point upwards, and some planes do not even have wings.
　　The hints are around you if you look closely. Did you ever notice that when you are driving in a car and a big truck passes you close in the opposite direction very fast you do not get pushed away from the truck you get pulled into it. In the same car with the windows open, the trash in the back of your car will blow out the window into the fast wind. If you have a shower curtain at home and you turn the shower on, where does the shower curtain go? Away from the speeding water or towards it? You might expect it is blown away but it is actually pulled into it. Money can be made with the understanding of Bernoulli's Principle because knowledge can be converted into money. Let me explain.
　　Let's say you have a ping-pong ball in a funnel and you blow air up from the bottom of the funnel. Where will the ping-pong ball go?

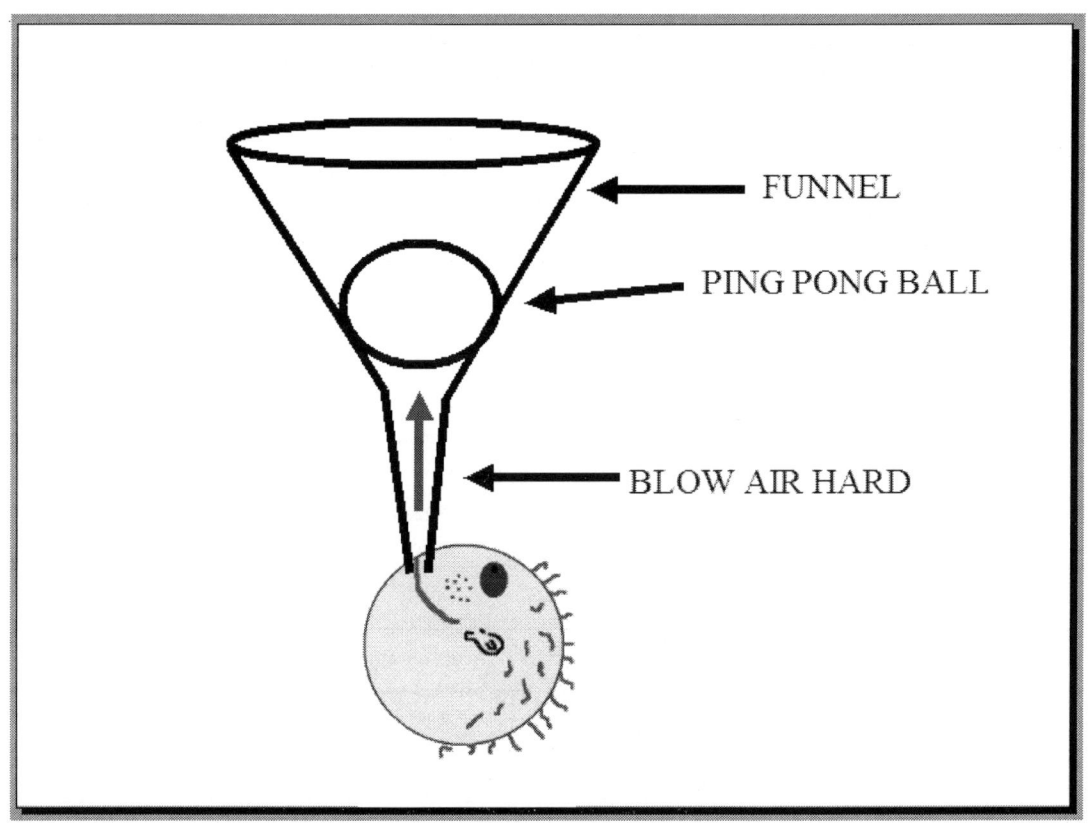

What did you guess? That the ball would fly upward? Nice try but it does not. In fact the harder you blow, the harder the ball is pushed <u>DOWN</u>. Go ahead and try it, then make a bet with the boss at home for double or nothing on your chores.

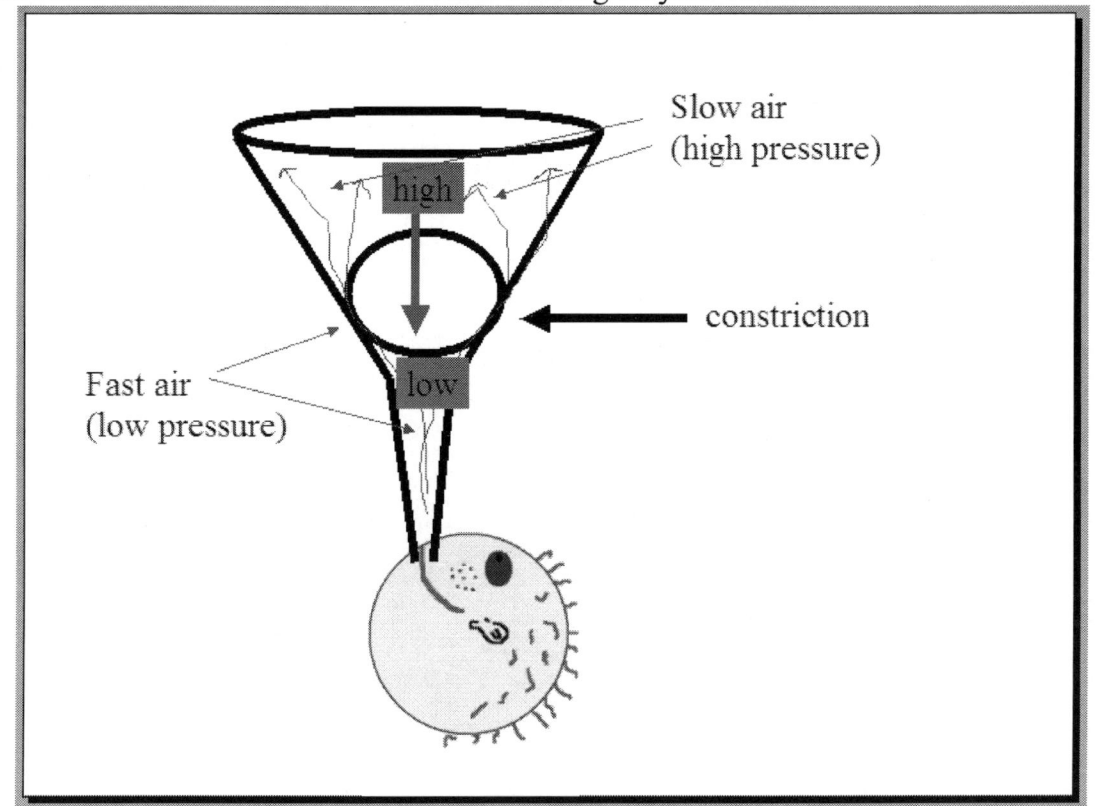

272

What is really wild though is that if the funnel was not there, you could levitate the ball on a stream of air from a hair dryer, and could even tilt it sideways a bit. Go ahead and try it.

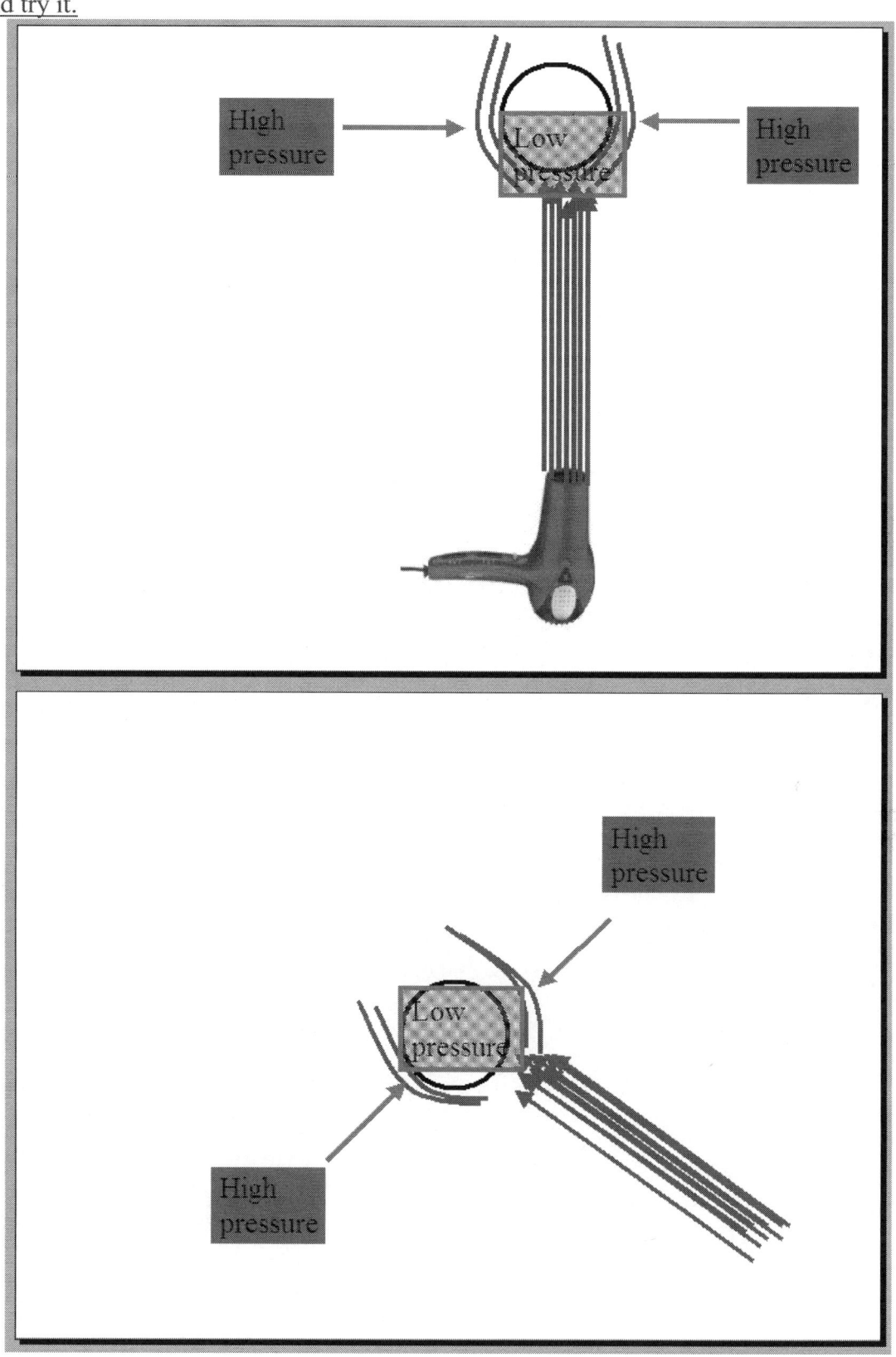

Bernoulli discovered that the faster a fluid (air or liquid) moves the lower the pressure it produces. Slow moving fluid causes high pressure. This means that the harder you blow on the funnel, the lower the pressure. The air above the ball is much slower so it produces high pressure. Wind always goes from high pressure to low pressure, so there is a wind pushing the ball down. This wind is called lift.

So, Bernoulli's principle says the pressure in a moving stream of fluid is less than the pressure in the surrounding fluid. In simpler terms, fast air makes low pressure and slow air makes high pressure.

BERNOULLI'S PRINCIPLE

- THE PRESSURE IN A <u>MOVING</u> STREAM OF FLUID IS <u>LESS</u> THAN THE PRESSURE IN THE SURROUNDING FLUID
- <u>FAST</u> FLUID = <u>LOW</u> PRESSURE
- <u>SLOW</u> FLUID = <u>HIGH</u> PRESSURE
- WIND (FORCE) GOES FROM HIGH PRESSURE TO LOW PRESSURE

Now imagine a roll of toilet paper free to spin, and you blow fast air over the top, what does the paper do? If you put the air compressor over the top, the paper is pushed in and it shoots out, a fun thing to do. https://www.youtube.com/watch?v=RdQeyy9skDA

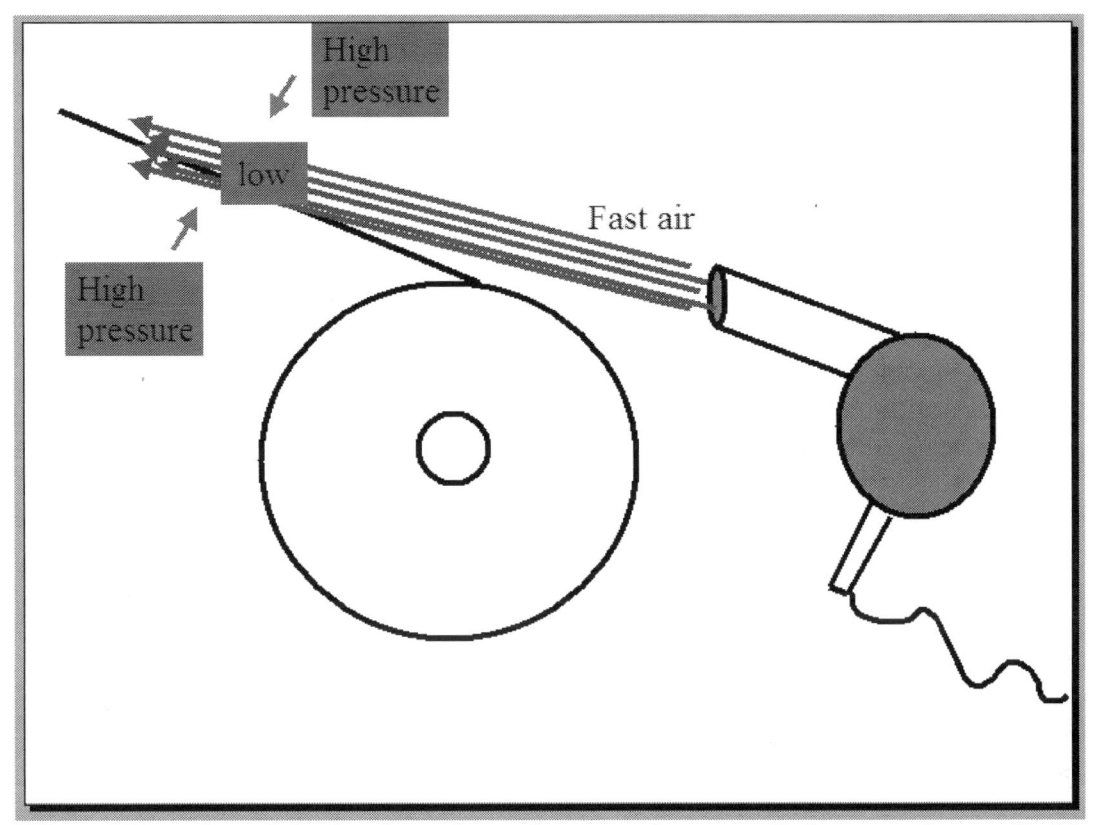

Where I live a lot of the local creeks and streams have small dams on them called low head dams. Water goes over the dam and forms a small waterfall. The fishing is always good below these dams. If the water flow is slow there is generally not a problem, but in the spring when the water flow is a little more than normal bad things can happen. People do not respect the power of Bernoulli's principle and do not realize that when they bring a boat too close to the waterfall (and it can be a couple of yards) the high pressure behind them pushes them into the waterfall and sinks the boat. This is not the worst part though. Behind the waterfall the water is going super fast and causes a very low-pressure area, the people get trapped and drown. Respect Bernoulli's principle, and listen to science teachers. Ping pong ball and faucet demo

While I am on the subject of keeping you alive, there is another way Bernoulli's principle can "get" you. Obviously the swift current in the middle of a river is hazardous, but what about the slower moving water near the bank. Bernoulli would warn you to be careful. The slow moving water has high pressure pushing to the fast flowing water in the middle. If you wade out past a certain point, the force will push you into the deeper water. Sometimes small children are not strong enough to fight this force (if the water is deep enough) and get pushed into the center and swept away. Very sad.

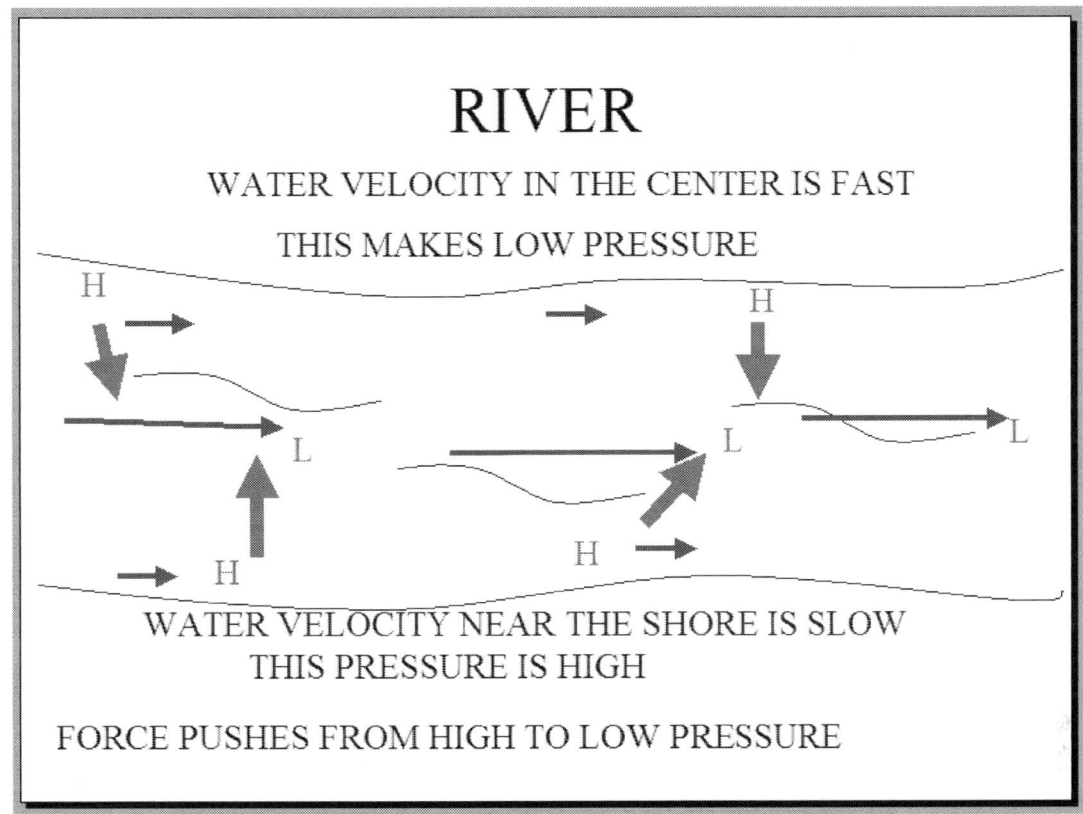

```
┌─────────────────────────────────────────────────┐
│                    RIVER                        │
│        WATER VELOCITY IN THE CENTER IS FAST     │
│             THIS MAKES LOW PRESSURE             │
│   H                           →        H        │
│   ↓→                                   ↓        │
│  →         L         →        L        →  L     │
│            ↑                                    │
│      H           H →                            │
│   →                                             │
│        WATER VELOCITY NEAR THE SHORE IS SLOW    │
│                THIS PRESSURE IS HIGH            │
│                                                 │
│        FORCE PUSHES FROM HIGH TO LOW PRESSURE   │
└─────────────────────────────────────────────────┘
```

    Now let's talk about a windy day. Did you ever notice that when a storm is coming and the wind picks up and your window is open, that air sometimes goes out of your house to the outside? Your curtains may be pushed out of the window or an inward swinging door will slam shut. If the wind is blowing at a certain angle to your house this happens. The slow-high pressure air inside your house goes out into the faster low-pressure air.

    On an even windier day, a tornado, the air is so fast, and the pressure so low, that the roof can be pushed off a house from the inside and land on the neighbor's yard.

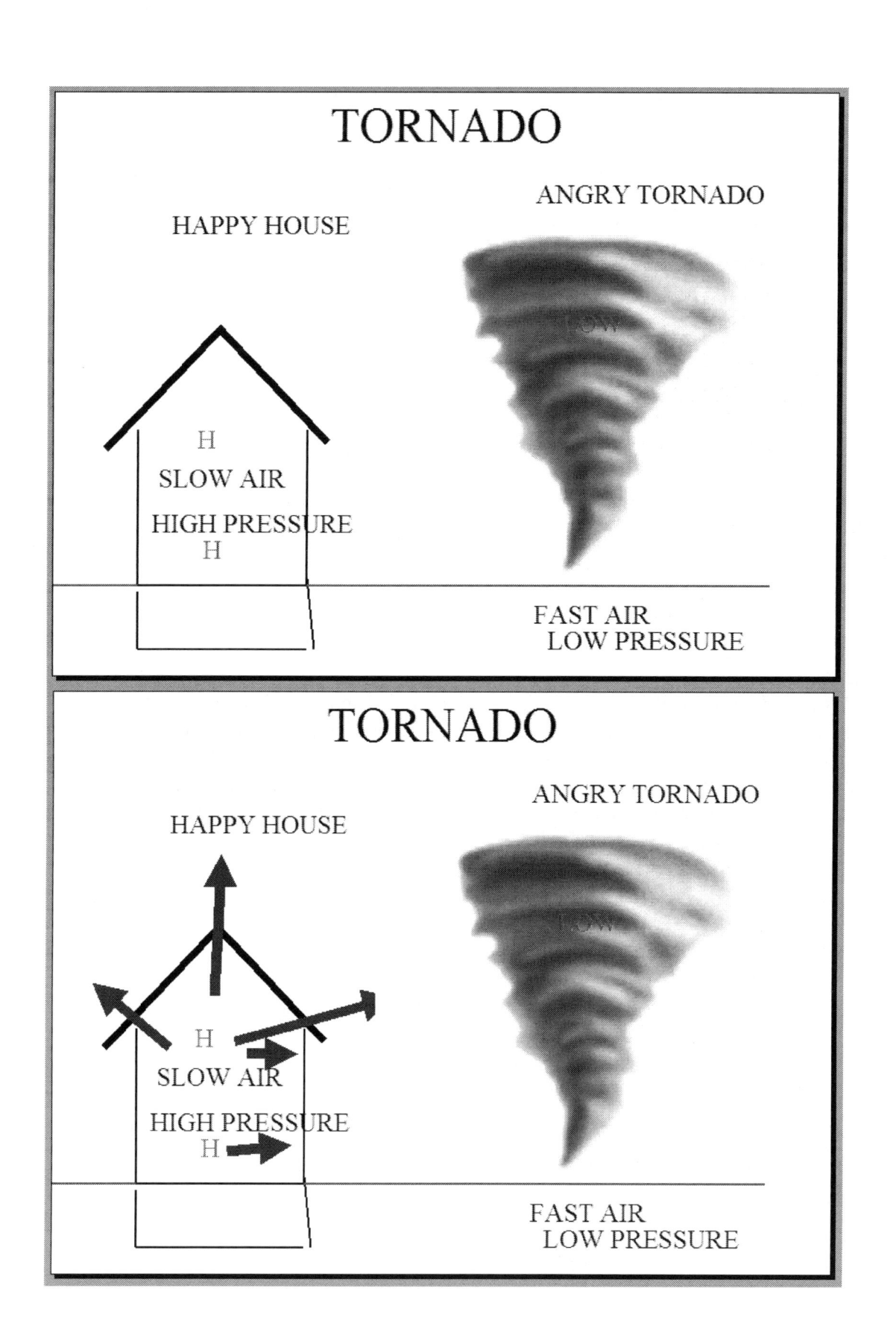

TORNADO

HAPPY HOUSE

ANGRY TORNADO

H
SLOW AIR
HIGH PRESSURE
H

FAST AIR
LOW PRESSURE

TORNADO

HAPPY HOUSE

ANGRY TORNADO

H
SLOW AIR
HIGH PRESSURE
H

FAST AIR
LOW PRESSURE

TORNADO

ANGRY TORNADO

HAPPY HOUSE

H →
SLOW AIR
HIGH PRESSURE
H →

FAST AIR
LOW PRESSURE

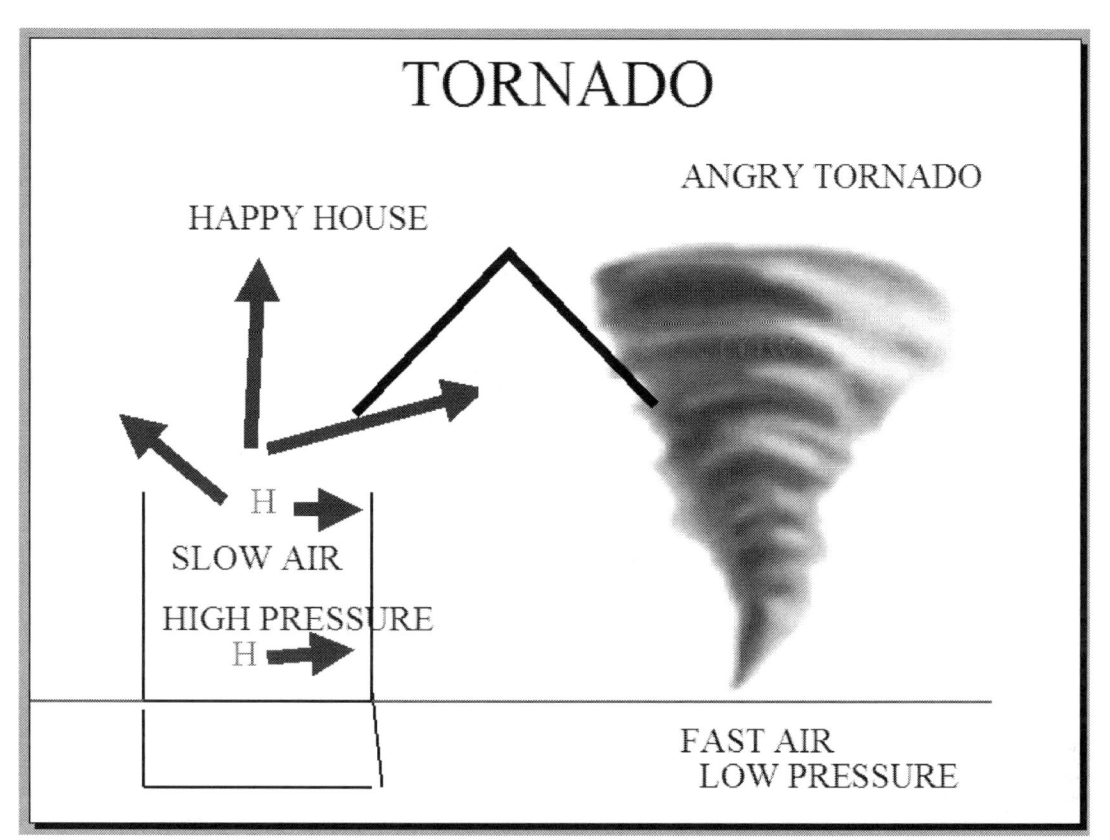

Where I live we do not have many tornados and the ones we do have are rather small, other parts of the country have giant tornados. The safest place to hide in your house is as low as possible (in the basement) on the tornado side of the house. If you do not have a basement the next best place is a small closet or bathroom with no windows (the smaller the better). If you are really afraid, hide in the bathtub with a bed mattress covering it, bathtubs are very strong.

TORNADO

OR A SMALL CLOSET OR BATHROOM WITH NO WINDOWS

SAFEST PLACE TO BE

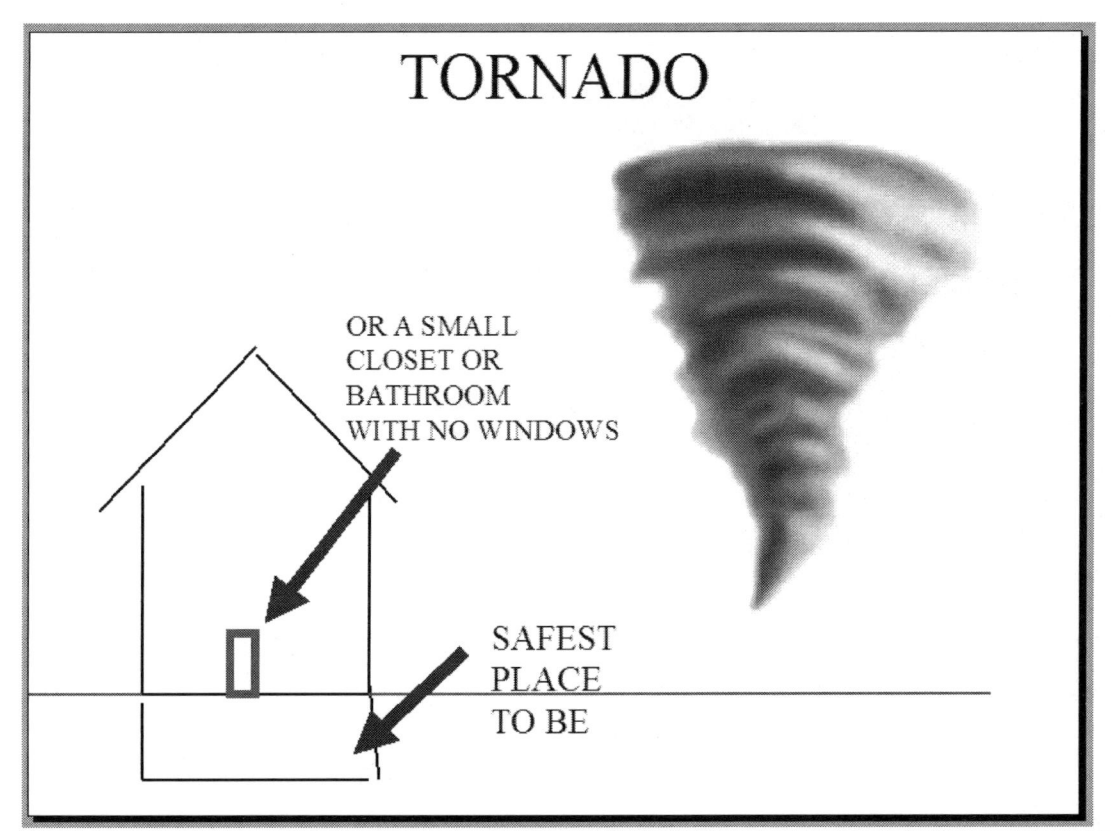

Even though we do not have many tornados where I live, my school has tornado drills just in case. Everyone goes into the hall and faces the wall. I am on the top floor and our hallway has skylights, a terrible spot. Right beside my room is a very small room with no windows. As a scientist I know the safest place is in that small room, but I am not allowed to send my students there and can get into trouble if I do. Maybe I will send them there when the glass breaks and the roof flies off.

Have you ever been attacked by a shower curtain? You are taking a shower, you have soap in your eyes and suddenly you feel someone tickling your leg. Do not worry. It is only Bernoulli, unless you do not have a shower curtain and have a door instead, then be worried.

SHOWER
CURTAIN
WALL
IT IS OK, NONE OF MR. RITTS' PICTURES HAVE CLOTHES
DON'T LOOK
SHOWER
CURTAIN
WALL
I HAVE SOAP IN MY EYE
SLOW AIR
HIGH PRESSURE
FAST WATER
LOW PRESSURE

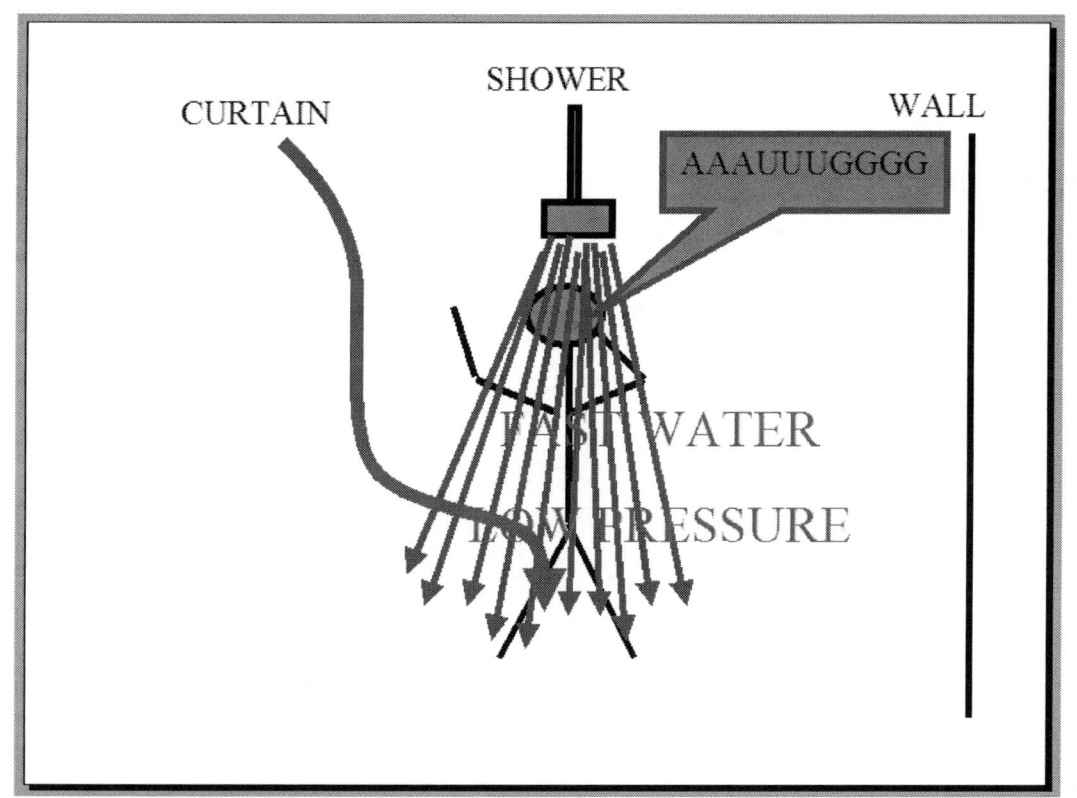

[Figure: Diagram showing a shower with curtain being pulled in. Labels: CURTAIN, SHOWER, WALL, AAAUUUGGGG, FAST WATER, LOW PRESSURE]

Long ago I told you about convection and how hot air rises and cold air sinks. If you have a fire place the hot air above the fire pushes the smoke out your chimney, but that is not when a chimney works the best. It works the best on a windy day. In fact the draft of a fire can be so strong that the fire gets bigger and the inside of a dirty chimney can catch on fire. This is why people clean the soot (and creosote) from inside chimneys. There was once at a hunting camp, I knew of, where this happened on a cold windy day. The chimney (a metal pipe) from the woodstove caught fire. No more hunting camp.

    A venturi is a T-shaped tube. It is used to make paint go up a tube in an airbrush, or perfume in an old fashion perfume bottle. It turns the liquid into a mist. I use a venturi to crush a metal can or a plastic bottle. I hook it up to my water faucet and take the air out from the inside.

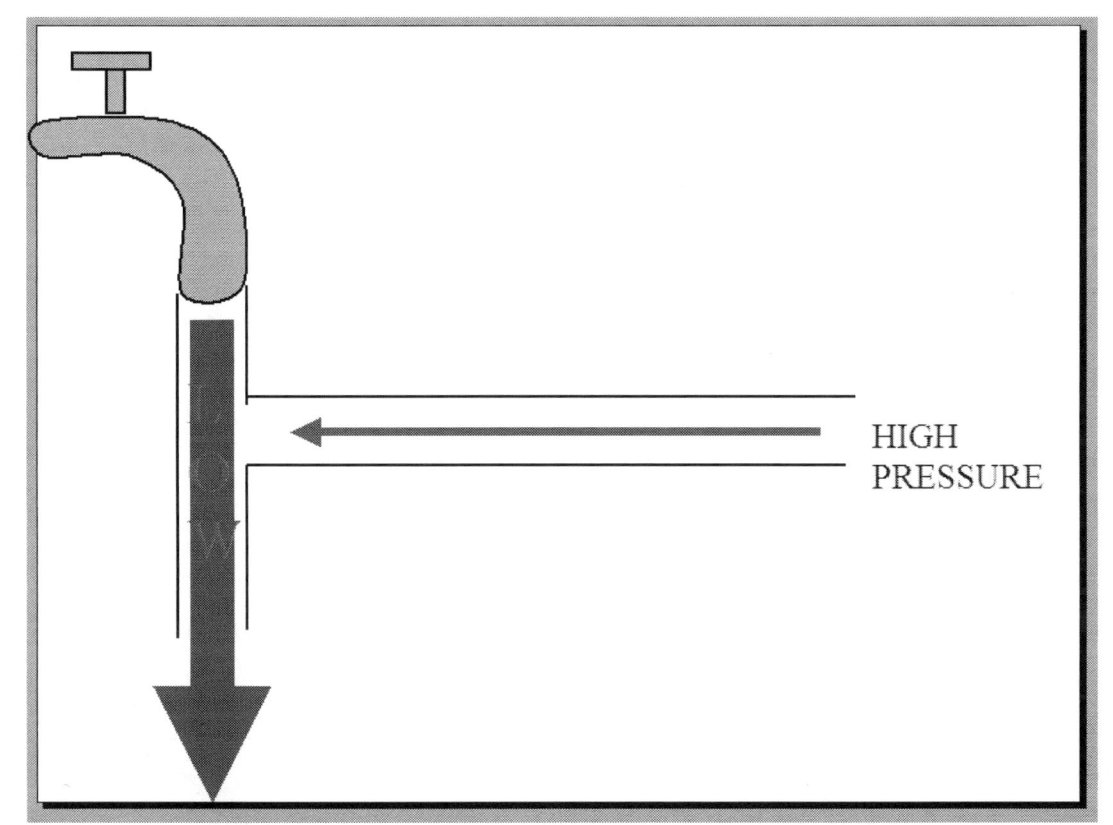

An excellent video by Julius Sumner Miller showing some cool demonstrations of Bernoulli's principle: https://www.youtube.com/watch?v=7XHohWDIUB0
https://www.youtube.com/watch?v=wwuffpiYxQU
https://www.youtube.com/watch?v=KCcZyW-6-5o

Flight

This brings us to how airplanes actually fly. There are a few reasons having to do with angle of attack and some other factors, but it is mostly Bernoulli's principle. To make this simple I am going to start with wind. It is important to remember that wind always blows from high pressure to low pressure. You may notice this on a weather map on T.V.

WEATHER

LOW PRESSURE AREA

★ HARRISBURG

HIGH PRESSURE AREA

WHAT DIRECTION DOES THE WIND BLOW?

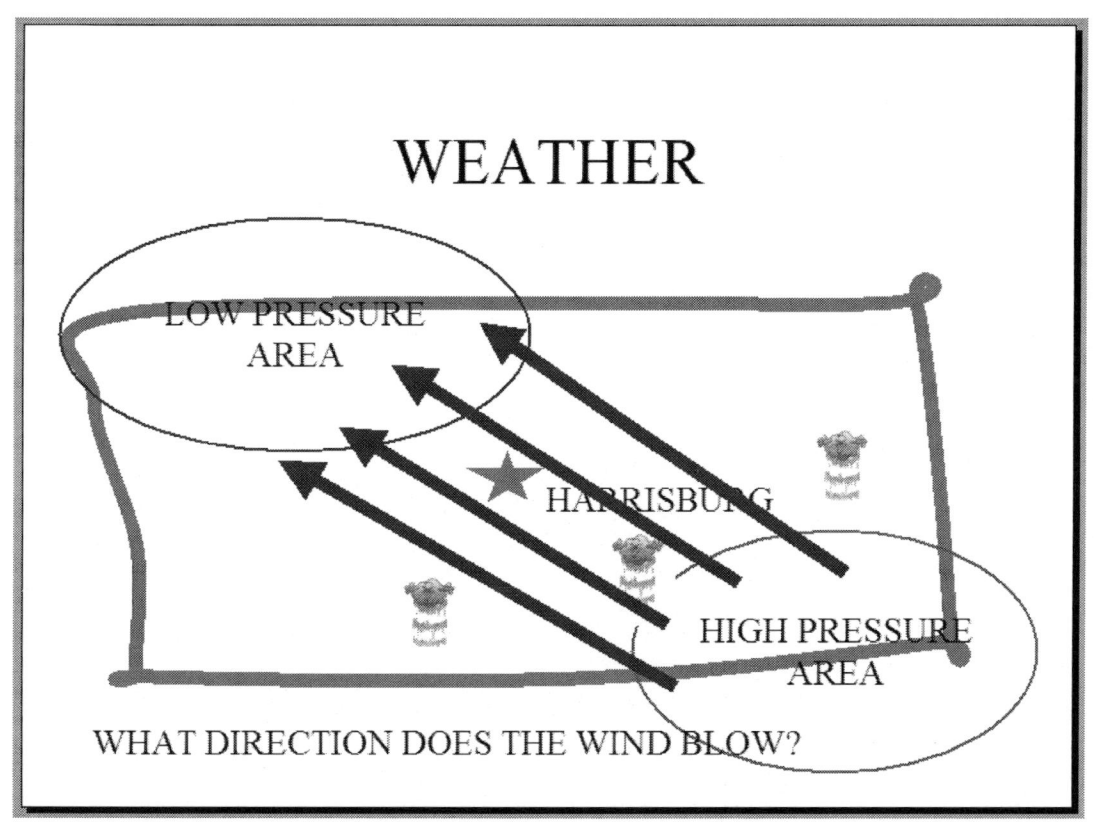

In flight, the trick is to make the wind blow UP, to push the plane wing up. The force of wind pushing up on an airplane wing is called <u>lift</u> (as in it *lifts* the plane up).

LIFT

- THE FORCE MOVING FROM AREAS OF HIGH PRESSURE TO AREAS OF LOW PRESSURE
- PUSHES A WING UP

In 1903 the Wright Brothers figured out how to do this with a wind tunnel they built. They realized that to make the wind push up, they had to make the pressure above the wing lower than the pressure below. To make the pressure on top lower, they had to make the air flow faster over the wing than the bottom. They shaped the wing to split the air. That is why an airplane wing is called an airfoil (air cutter). Of course birds figured this out millions of years earlier but get no credit.

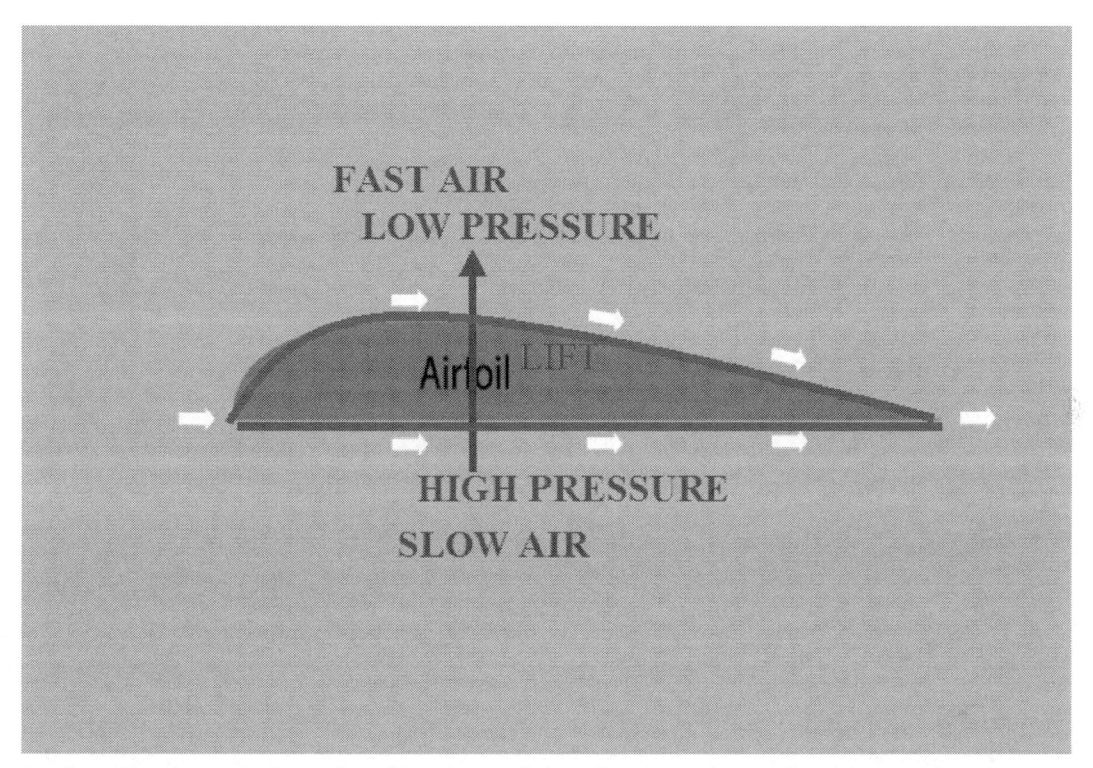

As the wing moves into the air, some of the air goes above the wing and some goes below. The air going over the top has farther to go and has to speed up compared to the air underneath. The pressure difference creates lift.

FLIGHT
- AIRPLANE AND BIRD WINGS ARE SHAPED SO AIR FLOWS FASTER
- OVER THE TOP OF THE WING
- THAN THE BOTTOM
- THIS IS CALLED AN AIR FOIL

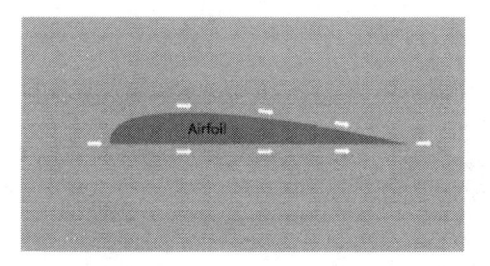

Since the first airplane, people have built hydrofoils, lifting bodies, flying cars, flying wings, and even wing suits. But not feathers glued to their arms, that does not work.

Review:
Fast moving fluid causes low pressure.
Slow moving air causes high pressure.
Lift and wind push from high to low pressure.
This is called Bernoulli's principle.

Bernoulli's Principle states that fast fluids cause lower pressure. An airfoil causes the air pressure below a wing to be higher than the pressure above a wing. Fast fluid causes low pressure. Slow fluid causes high pressure; Lift is a wind that pushes from high to low pressure.

Flashcard review on Quizlet: https://quizlet.com/175194163/chapter-39-bernoullis-principle-flash-cards/?new

A day in the life of Earwig Hickson III
Otis and I were at a going-out-of-business sale for a party supply store. We had birthday money and were hoping to find something interesting. We searched through the piles of different shaped balloons (none of which I was allowed to buy since it was on the

not for Earwig list, same with the Helium, but I searched anyway. There were decorations, games, small toys, costumes, carnival games and other things that did not interest me at all. Otis had found a clown nose he liked and a rubber frog that squirted water. I kept looking. In the back I found a box. I did not even know they made such a thing. I could not believe someone had not bought it earlier. It was amazing. On the box it said "Giant Human Hamster Ball". Apparently it was a giant inflatable ball of clear plastic that a person could get into and run around. It would go on water, on land, on snow, anywhere. I quickly showed Otis and he was excited. He imagined rolling around in the hamster ball in the back yard. Poor Otis, he had no imagination.

In science class my teacher showed us a cool trick where he floated a beach ball on the blower end of a vacuum cleaner. The ball floated in the blowing air like magic. He said it was an experiment that showed how Bernoulli's principle worked. He said if you had a strong enough blower you could float any round object, and I had a strong enough blower.

Once we were back home I dug out the old hurricane wind generator I had build to clean the garage. Otis was messing with the hamster ball. I set the hurricane wind generator in the back yard so it pointed up. Otis had gotten himself in the hamster ball and was zipping it closed. I plugged in the power to my wind machine. Otis slowly rolled into the yard. The zipper on the ball had gotten stuck and he could not get out. Sadly I realized what that meant. Otis would get to go first.

I rolled him onto the launch platform. Otis still thought he was going to roll around the yard. I hit the power switch.

It worked wonderfully. Up went Otis in the clear plastic ball. He was suspended in the air column just like my teacher said he would be. After his initial shock he actually was having fun. Then he began to spin. Faster and faster he spun, the ball and Otis. At first he tried to run just like a real hamster, but then the speed of the spinning ball got too fast. Down he went but he kept going only now he was doing summersaults. Faster the ball spun. Soon he looked like an Olympic gymnast doing cartwheels, handsprings and triple sow cows, like skaters do. Soon he was not only at the bottom of the ball but stuck to the side so he went around with it. Still the ball spun faster.

Apparently Otis had spaghetti, beef stew and corn, a lot of corn for lunch. I know this because Otis got sick and threw it all up all over the inside of my science experiment. It turns out Otis can eat a lot of food in one sitting. The liquid sloshed, Otis splashed, and the ball kept spinning faster. I tried to push him out of the air stream with a long stick but the low pressure kept pulling him back. Accidentally, I poked a hole in the big ball and out shot all the liquid, drenching me with Otis lunch. As the ball deflated, Otis descended back to Earth.

It was a long time but eventually Otis forgave me and became my friend again. My father never found out about this experiment and quite often played with our Human Hamster ball himself. He thought it was cool, but he complained that it smelled funny.

Fun things to Google
The Wright Brothers
The Wright Flyer
Lifting bodies
How to survive a tornado
TP cannon
Tornado
How airplanes fly
Lift
Low head dam

Links
An excellent video by Julius Sumner Miller showing some cool demonstrations of Bernoulli's principle: https://www.youtube.com/watch?v=7XHohWDIUB0

Steve Spangler TP experiment: https://www.youtube.com/watch?v=gtzVIXO7zh8

DT3 - Lifting Body Aircraft - https://www.youtube.com/watch?v=3JEU0QtEw-w

NASA Dryden -The Lifting Bodies - https://www.youtube.com/watch?v=PdVlMwlz6hc

VSKYLABS | Northrop M2-F2 Lifting Body Vehicle | X-Plane 10
https://www.youtube.com/watch?v=mxyxUED9BLE

Wright brothers 1908 France - https://www.youtube.com/watch?v=cKDepNYlbRA

THE WRIGHT BROTHERS' FLYING MACHINE - NOVA - Discovery History Science (full documentary) - https://www.youtube.com/watch?v=7XDE9WAH-RI

Airplane History (Full Video) - https://www.youtube.com/watch?v=NySGMQeBgBU

Dallas Tornado Video Shows Massive Twisters in Texas, Tractor Trailers Thrown Around Like Toys - https://www.youtube.com/watch?v=WABqwKjQM_c

How to Survive a Tornado - https://www.youtube.com/watch?v=yB1P7FAOU54

Emergency Preparedness: Tornados - https://www.youtube.com/watch?v=h3z50ZX_RMM

Bernoulli's Principle Demonstration: Bernoulli's Ball - https://www.youtube.com/watch?v=sIrJOrTAJjg

Bernoulli's principle - physics experiment - https://www.youtube.com/watch?v=P-xNXrELCmU

Bernoulli's Principle - https://www.youtube.com/watch?v=WDGNcmEOjs4

Bernoulli's Principle: Ping-pong Ball and Funnel - https://www.youtube.com/watch?v=1TQL1ju3RoQ

Ping pong ball and Funnel - https://www.youtube.com/watch?v=nsnMt8erxH8

STEMbite: Shower Curtain Mystery - https://www.youtube.com/watch?v=zeOXzXi1wlY

Dangerous Lowhead Dams are cause of Deaths to Canoe, Swimmers, tubing and Kayak - https://www.youtube.com/watch?v=lLZo0m1RDUI

Low Head Dams - A Dangerous Current (Presentation) - https://www.youtube.com/watch?v=XsYgODmmiAM

Tire stuck in a low head dam - https://www.youtube.com/watch?v=z_19V2tJ904

Raft stuck in dam - https://www.youtube.com/watch?v=5Ldn23yu_38

Over The Edge, dangers of low head dams - https://www.youtube.com/watch?v=JF7vB5aHnBc

The Venturi effect - https://www.youtube.com/watch?v=Na9ORhYjvJU

Make A Simple & Powerful Pump - The Venturi Pump -

https://www.youtube.com/watch?v=7can2eKgXS0

Bernoulli's Principle Demonstration: Bernoulli's Ball: https://www.youtube.com/watch?v=sIrJOrTAJjg

Bernoulli's Principle - physics experiment with ping pong ball: https://www.youtube.com/watch?v=epesI-fWvrY

Julius Sumner Miller - Physics - Bernoulli pt. 1: https://www.youtube.com/watch?v=KCcZyW-6-5o

Julius Sumner Miller - Physics - Bernoulli pt. 2: https://www.youtube.com/watch?v=wwuffpiYxQU

Quadrofoil - Amazing Electric Hydrofoil of the Future: https://www.youtube.com/watch?v=ooAAnZIgj8o

Chapter 40
Pascal's principle and Hydraulics

Power point video of this lesson: https://www.youtube.com/watch?v=hScTbyb3uRg

The wonders of science
One of the things that make echinoderms special is that they have radial symmetry. Echinoderms include the sea stars, sea urchins, sand dollars, and sea cucumbers. What is really special about them is not their symmetry but their water vascular system or hydraulic system. They have no muscles like us, but a hydraulic system, like a machine. Muscles use oxygen and sugar for energy and get tired. Hydraulics never get tired. Starfish eat bivalves like clams and mussels. Mussels are called mussels because they are very strong when it comes to holding their shells shut, clams are just as strong. Not much is strong enough to open them, but starfish can. Starfish have little tube feet all over their arms and on the end of each is a small suction cup. The tube feet are part of a series of connected tubes where water pressure can be reduced. When a starfish wants to eat a clam, it simply sticks its suction cup feet to the shells, sets this hydraulic system to pull the shells apart, then waits until the muscles holding the shells closed gets tired and opens a tiny bit. The Starfish then sticks its stomach into the clam and digests it. Its stomach is actually inside out when it eats the clam. The hydraulic system of starfish is amazing.

What you need to remember
Pressure = Force/Area or   P = F/A
Liquid take the shape of the container they are in.

Pascal's Principle
Blaise Pascal was a genius, so were most of his relatives. He was studying fluid behavior and came up with a very important observation. This observation eventually became Pascal's law (or Pascal's Principle) and it is very important in modern society. It is because of Pascal's law that a bulldozer, a backhoe, or a frontend loader can lift or move tremendous weight with very little force. It is why lowrider cars can be made to jump at the driver's command. It is why you can lift a car to change a flat tire by just pushing a lever a few times. Pascal's law led to hydraulics.

Pascal discovered a very simple relationship. In a closed container, the pressure is always the same. If one part of this closed container has higher pressure, the pressure in the whole container increases. This does not sound like much, but the implications are phenomenal. If the pressure in one end of a closed container increases, the pressure in the entire container increases! What makes this useful is that it does not matter the *shape* of the container. Do you understand this? The container can be any shape! This is important, a container could be thin or thick or a tube or hose. I know it does not sound important yet but the implications are actually rather extraordinary.

What Pascal said was that if *the pressure in a closed container changes, the pressure changes throughout the container*. The implications are amazing if you know where to look. Well, some people did know where to look.

PASCAL'S PRINCIPLE

- IF THE PRESSURE OF A FLUID IN A CLOSED CONTAINER CHANGES
- THE PRESSURE CHANGES THROUGHOUT THE CONTAINER

- IF ONE END OF A CONTAINER IS COMPRESSED THE OTHER END EXPANDS
- THE PRESSURE INSIDE IS THE SAME EVERYWHERE

Imagine a garden hose hooked to a faucet. When you turn on the water it creates high pressure at that end of the hose. The pressure at the other end of the hose becomes high also and water is forced out.

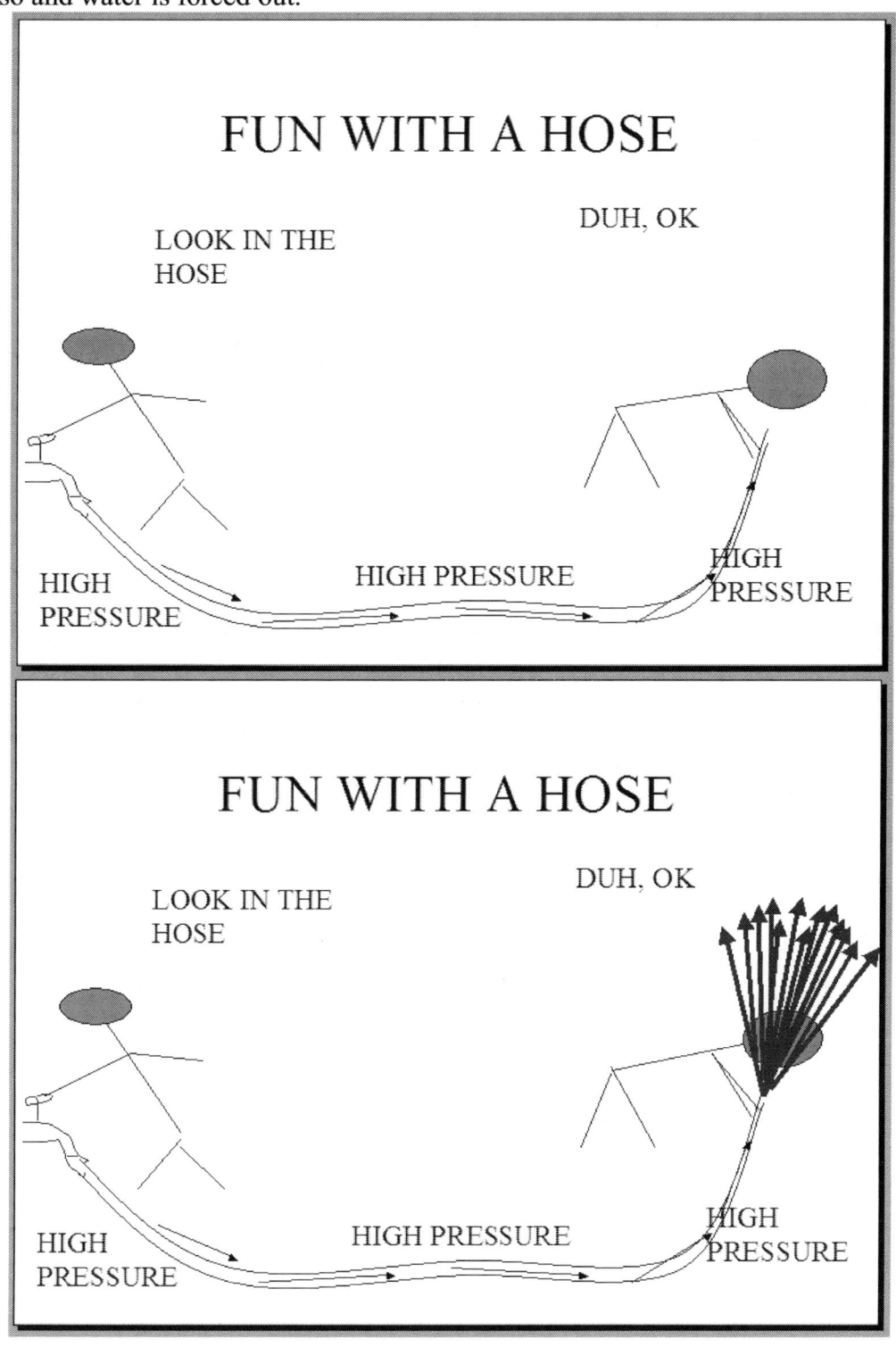

FUN WITH A HOSE

LOOK IN THE HOSE

DUH, OK

HIGH PRESSURE

HIGH PRESSURE

HIGH PRESSURE

FUN WITH A HOSE

LOOK IN THE HOSE

DUH, OK

HIGH PRESSURE

HIGH PRESSURE

HIGH PRESSURE

FUN WITH A HOSE

I FALL FOR THAT EVERY TIME

HEE HEE THAT IS PASCALS PRINCIPLE

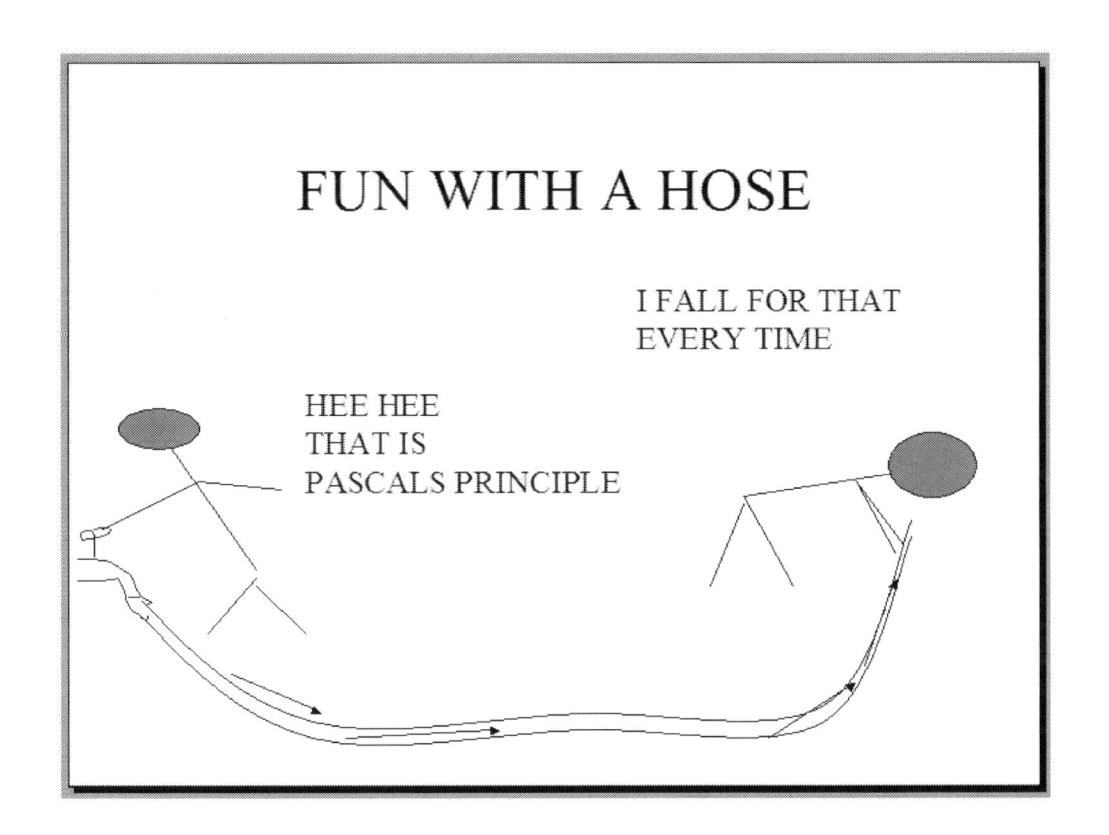

Hydraulics

The big application of Pascal's Principle is the invention of hydraulics. Hydraulics are liquid (<u>hydraulic fluids</u>) filled tubes used to transmit pressure. Imagine two syringes connected with a water filled tube. When you push one syringe in, the other syringe goes out. When one is pushed, the other one expands. This is hydraulics.

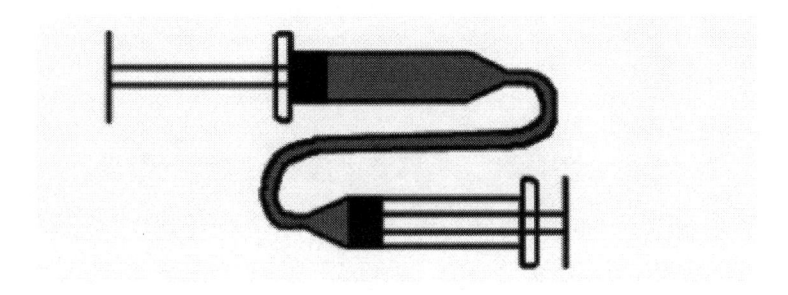

Now imagine if one syringe is larger than the other, the big one expands also, not as much, but it sill moves. This is called a <u>hydraulic press</u>.

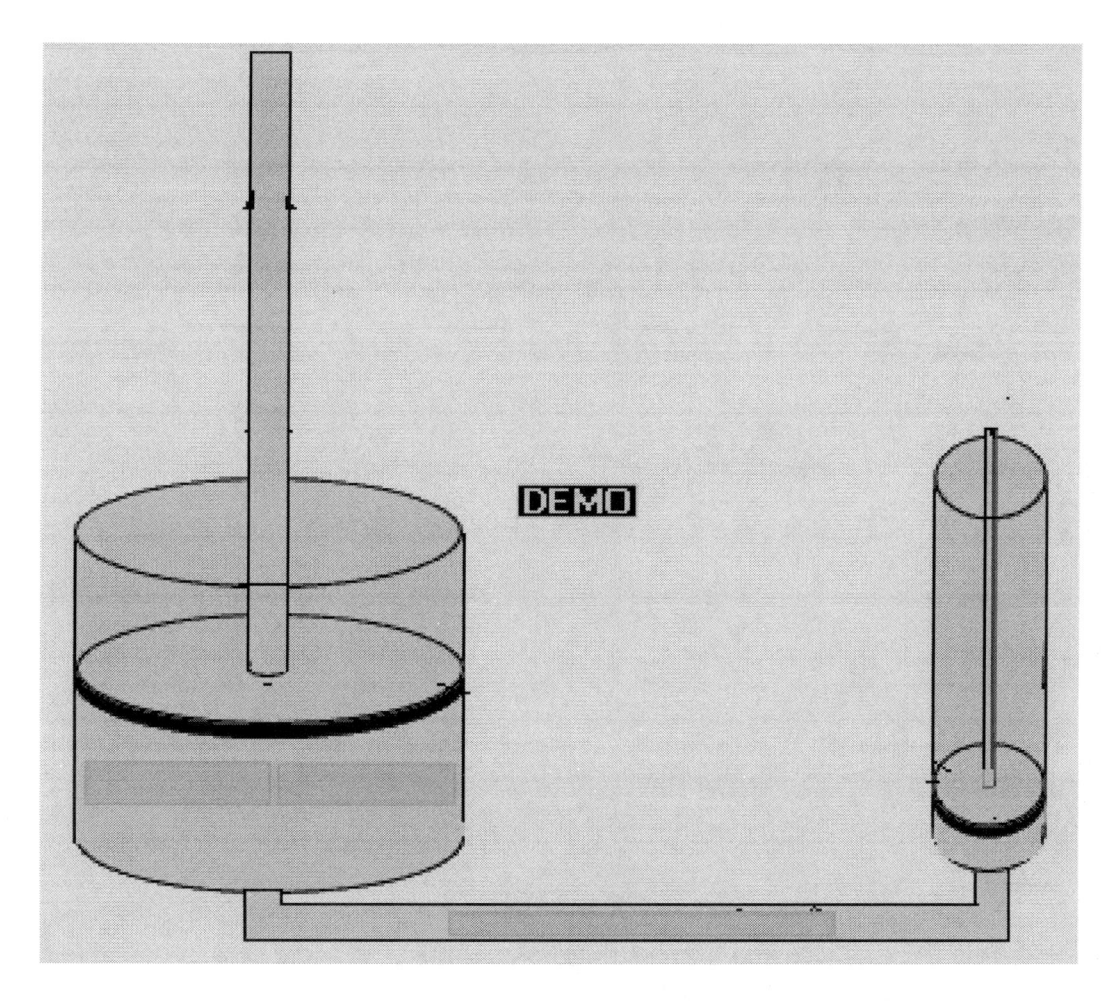

This is where the magic of Pascal's principle comes to play. The small syringe is very easy to push because it is skinny. Remember pressure is the force of an object per area (P=F/A). When the area gets smaller, the force to push it gets smaller too.

$P = F/A$

F_1 ↓ small-area piston

$P = F/A$

F_2 ↑ large-area piston

PRESSURE IS THE SAME EVERYWHERE

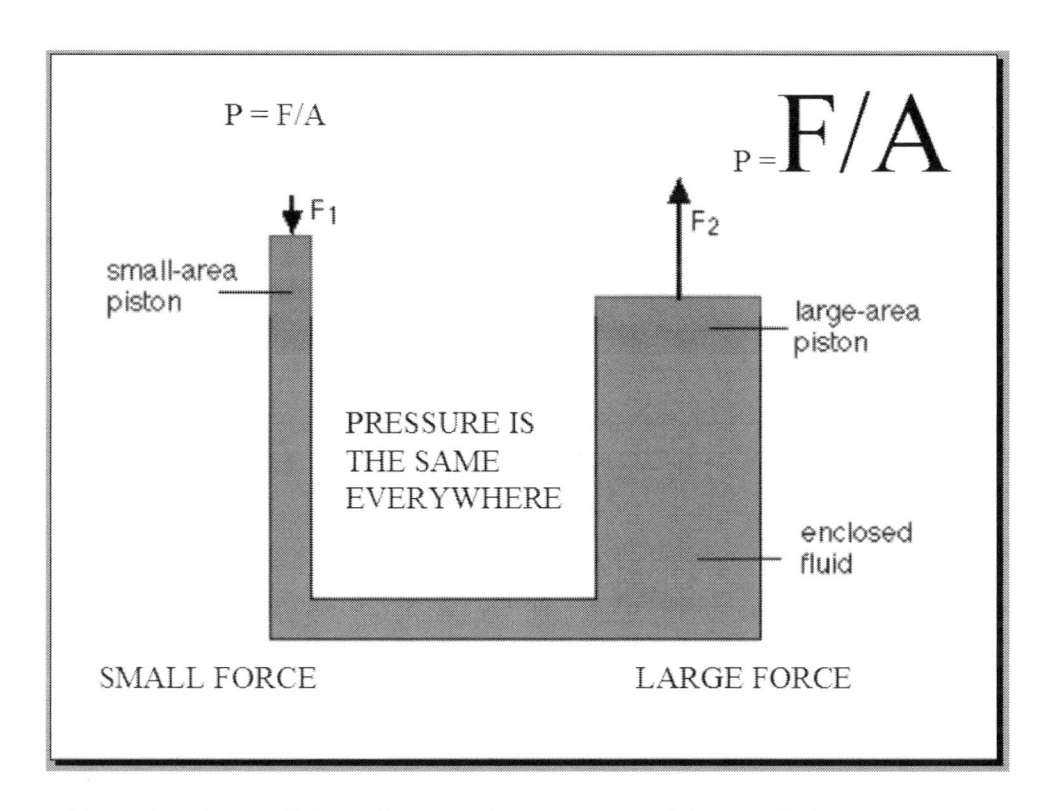

enclosed fluid

SMALL FORCELARGE FORCE

This makes it possible to lift very large forces with very little forces

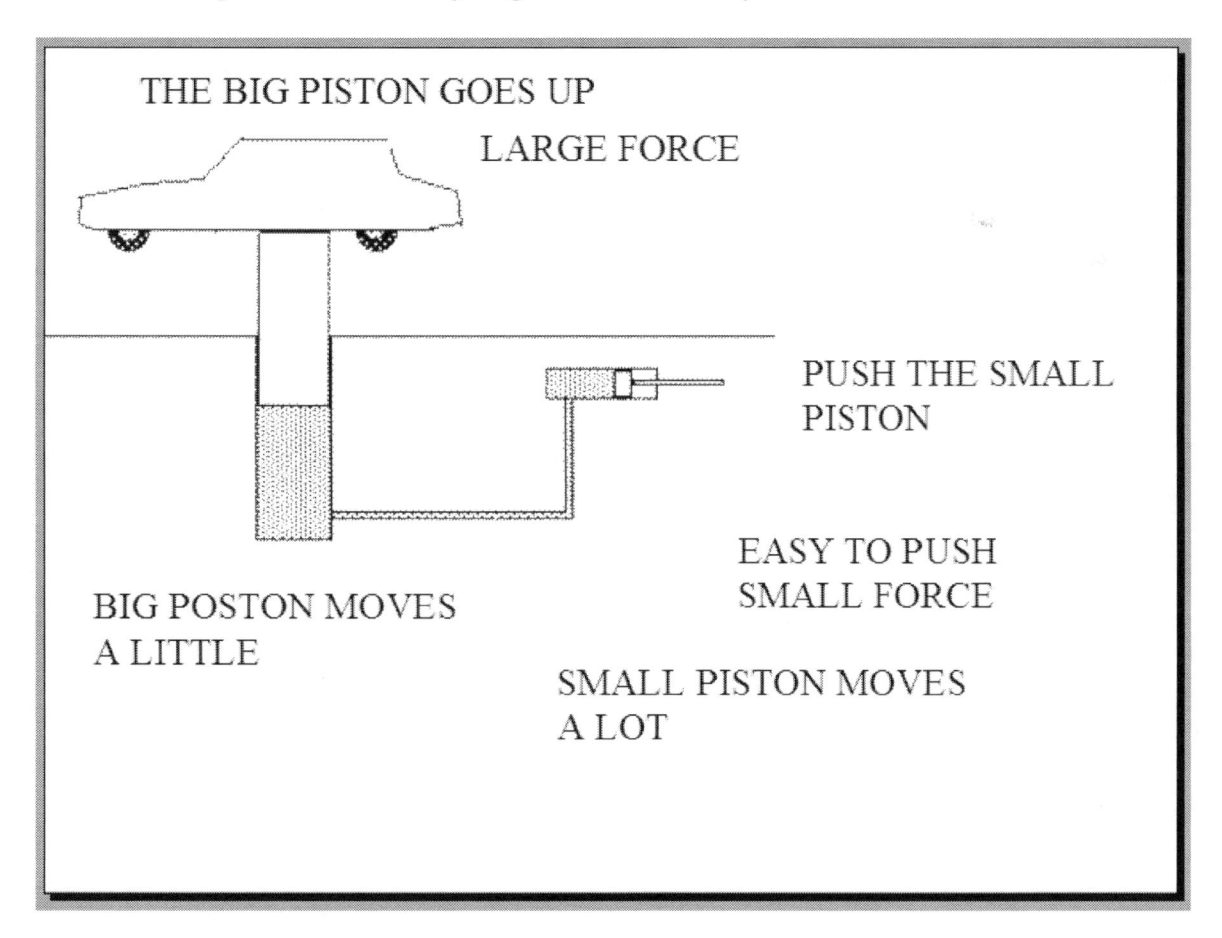

THE BIG PISTON GOES UP

LARGE FORCE

PUSH THE SMALL PISTON

EASY TO PUSH SMALL FORCE

BIG POSTON MOVES A LITTLE

SMALL PISTON MOVES A LOT

DEMO

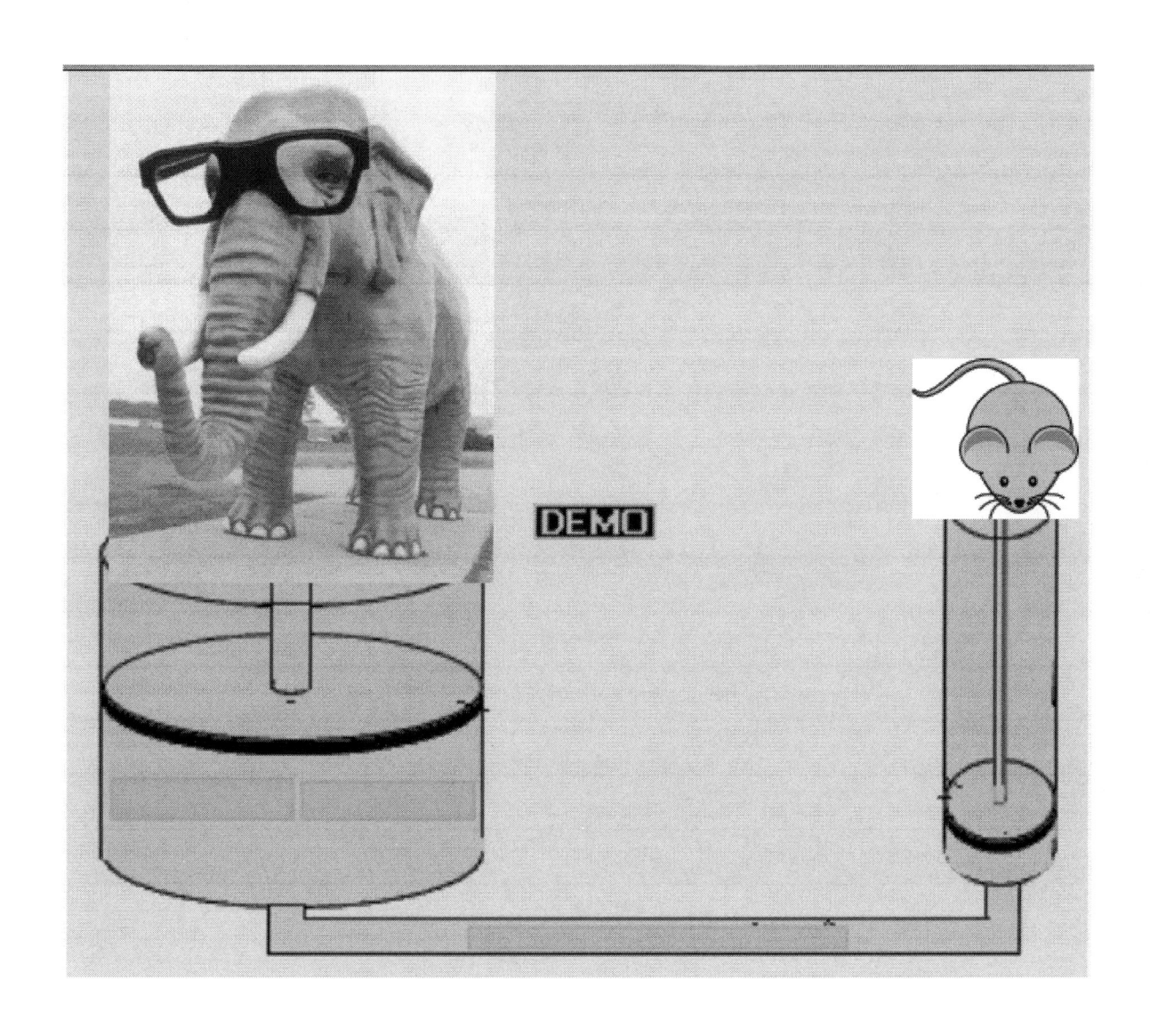

FULL
OF
LIQUID

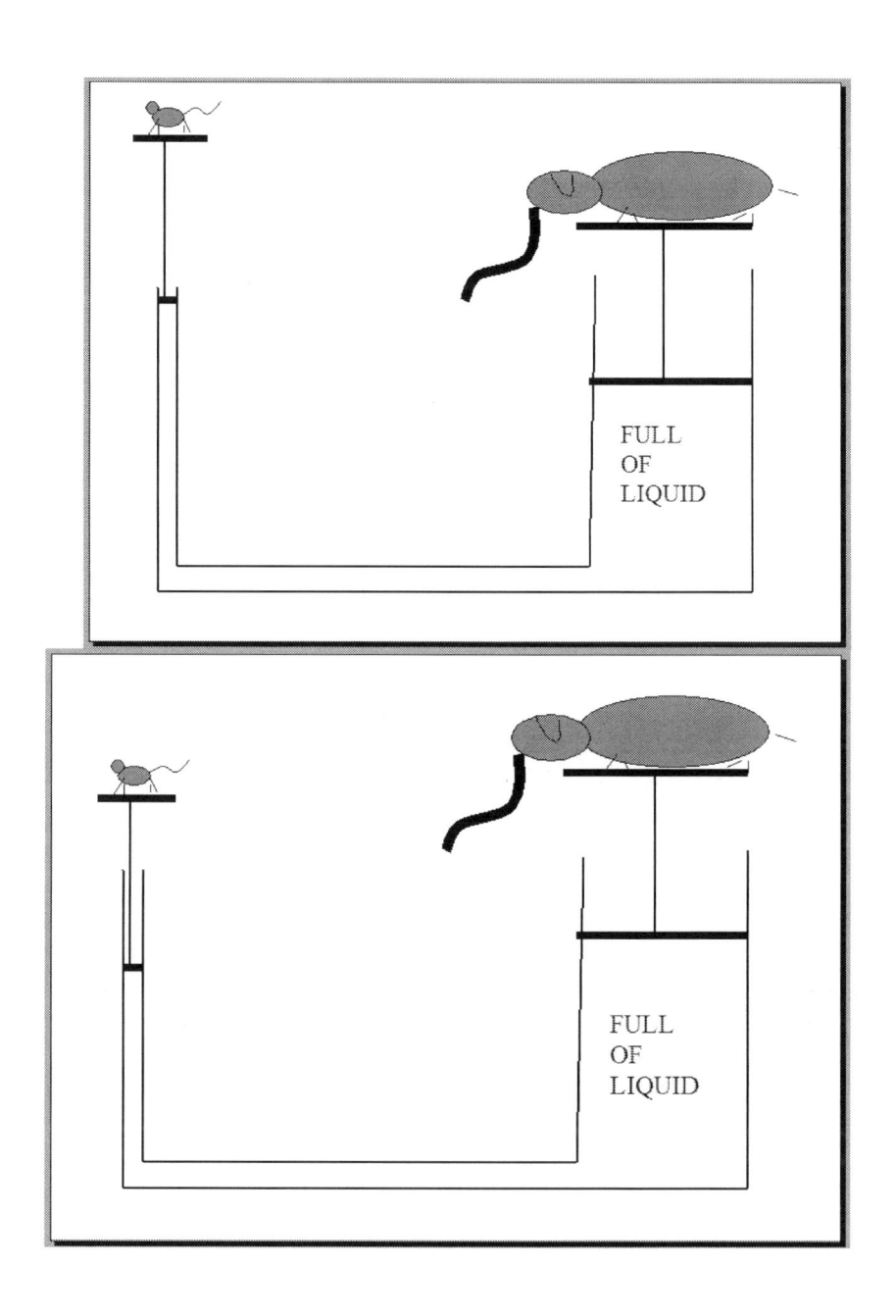

FULL
OF
LIQUID

299

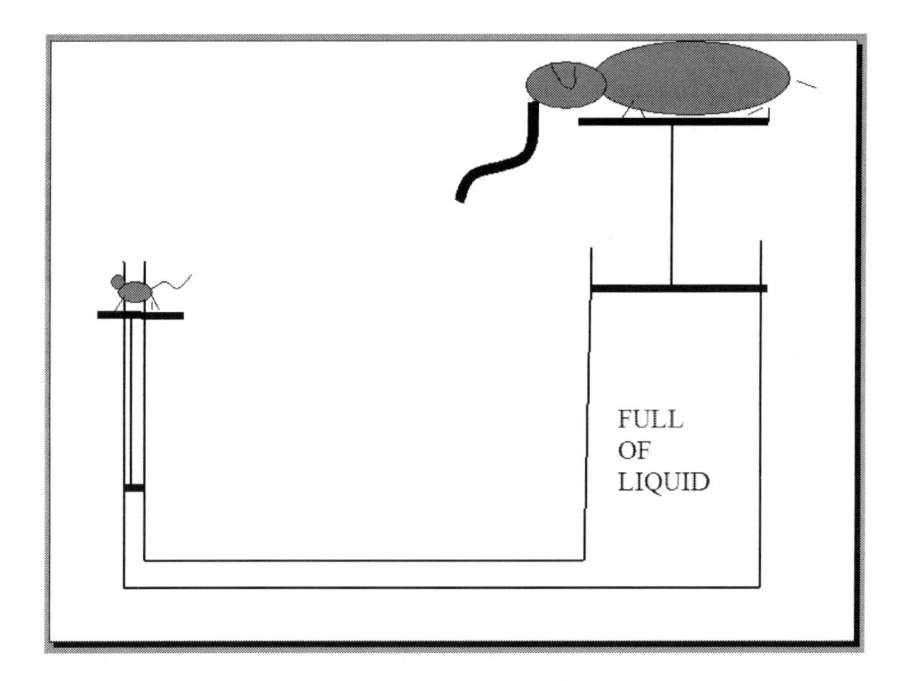

Hydraulics are not a kind of simple machine but they act like one. They increase force. Hydraulics are used in heavy machinery like bulldozers and cranes, to lift heavy weights, car jacks, and hundreds of other machines.

Julius Sumner Miller: Lesson 14 - Pascal's Principle - The Properties of Liquids: https://www.youtube.com/watch?v=FyCONdQLwBg

Review

Pascal's principle says the pressure in a closed container is the same everywhere in the container. Hydraulics are liquid filled tubes that transmit pressure and are used to increase force.

Flashcards at Quizlet: https://quizlet.com/176082853/chapter-40-pascals-principle-and-hydraulics-flash-cards/?new

A day in the life of Earwig Hickson III

A child prodigy. That is what I put on the form. It was one of those useless things you have to fill out in middle school. The question was "describe" yourself. I had heard of child prodigies but never met one. Under "goals in life", I put graduate from a University by age 15. Why not I thought, small dreams are no fun. Otis wanted to raise chickens. There were a lot of other questions about interests and talents; it was not long before I began making designs with the fill in dots. I had no clue that a rather expensive and sophisticated computer would analyze my answers and tell the world about me. Once a computer says something it is considered fact, and those facts fan out everywhere on the Internet and cannot be stopped. Computers rule the Earth.

There was to be a big meeting of all the smartest people in my town. The computer said I should be sent to the Big University 4000 miles away in California. The big meeting would decide my fate, the taxpayers would pay my tuition, so they had a say in my future.

I figured I had a chance; after all I was the most famous person in my town. Everyone had heard of me. I had been on the front page of the local paper over a dozen times. I was famous. If you Googled my name, Earwig, you came up with a giant list of amazing facts. Mostly about how I should be exterminated and was a yard pest. A bit disappointing, but famous none-the-less.

My father was very calm before the big meeting. He had a short speech written and guaranteed my acceptance in the Big University 4000 miles away. He was very confident in my abilities. He also added that it was my only chance to go to a university since he spent all my college money on something called Life, Liability and Homeowners insurance.

When the big day began the Mayor spoke first. He reminded everyone of how I made a giant quick sand pit that trapped and embarrassed his oldest son Oliver and how I had dug a whole through the Earth and his limousine fell in along with all his tax receipts. He reminded the fair citizens of the generators on the playground equipment and the wishing well incident. He said I did not deserve this honor. The Police chief went next and although he was happy that he was allowed to hire two new deputies since I had turned 12, he also reminded everyone of the Bigfoot, UFO sled, and giant snowman incidents, and the civil terror it created. He did not think I deserved any honors either. My science teacher went next, all he said was. "If you do anything stupid, and you will, use plastic". He reminded everyone of the giant hamster ball and how it squished all his strawberry plants. He did not think I should be rewarded either. The FBI was even there and they were against me too, claiming I was one experiment from the ten most wanted list. The CIA was even less helpful claiming they had to devote a whole satellite to track my location and how it took all their best agents to get the *build your own universe kit* out of my hands and locked away in Fort Knox. Otis was my character witness but when he was asked about a certain human centrifuge and a pool noodle boat, he ran away. My sister described a harmless car waxing in an unflattering way along with a sister control device. The neighbors all agreed that small children floating above houses was not natural and not something to be rewarded. The town was against me. Even my own mother described how her anniversary was ruined by a few gallons of liquid nitrogen, and her house from a Rube Goldberg Device. The only one on my side was my school principle who said he would gladly sign any papers necessary for my instant graduation of his school, C average or not.

It looked bad until my father started his speech. It was a very short speech, crafted in such a way to convert even the most anti-Earwig citizen to my side. It was genius. At that moment I realized it was my father who was the prodigy.

He put a large trashcan in front of the hostile crowd with the words "Send Earwig to the Big University donations." Then he said the following words.

"As I see things we can either donate money and send Earwig to the Big University **4000 miles away**, or he can live in **our town for the rest of his life**."

The money poured in.

Links

Water vascular system: https://en.wikipedia.org/wiki/Water_vascular_system

Tube feet: https://en.wikipedia.org/wiki/Tube_feet

Blaize Pascal: https://en.wikipedia.org/wiki/Blaise_Pascal

Pascal's Law: https://en.wikipedia.org/wiki/Pascal%27s_law

Hydraulic press: https://en.wikipedia.org/wiki/Hydraulic_press

Pascal's law - Animated and explained with 3d program: https://www.youtube.com/watch?v=qGQ4fojjwvQ

Application of Pascal's Law - Hydraulic Press/Machine: https://www.youtube.com/watch?v=5RCeLYbiZCk

Application of Pascal's Law: https://www.youtube.com/watch?v=E3hVuxX0lkk

Pascal's Law syringes: https://www.youtube.com/watch?v=kVw9apdu3Ng

Lesson 14 - Pascal's Principle - The Properties of Liquids - Demonstrations in Physics: https://www.youtube.com/watch?v=8ma4kW3xVT0

Julius Sumner Miller: Lesson 14 - Pascal's Principle - The Properties of Liquids:
https://www.youtube.com/watch?v=FyCONdQLwBg

Fun things to Google
Hydraulic cars
Lowrider car
Blaise Pascal
Car jack

This ends Volume Two of *The World's Greatest Physical Science Textbook for Middle School Students in The Known Universe and Beyond!*

The World's Greatest Physical Science Textbook for Middle School Students in the Known Universe and Beyond! Volume One can be found here.
E-book version
Paperback version

Farther adventures of an adult Earwig Hickson III can be found here.
Method is Everything: A sportsman's Reflections and Misadventures, A Memoir,* by *N.Y. Best Sellers
E-book
Paperback

Made in the USA
Lexington, KY
09 November 2017